BIBLIOGRAPHIC RECORD TARGET

\AN8938
UL FMT B RT a BL m T/C DT 07/15/88 R/DT 02/09/93 CC STAT mm E/L 1
035/1: : |a (RLIN)MIUG84-B50942
035/2: : |a (CaOTULAS)160099306
040: : |c RPB |d RPB |d MiU
050/1:0 : |a QA3 |b .A292
100:1 : |a Airy, George Biddell, |c Sir, |d 1801-1892.
245:10: |a Mathematical tracts on the lunar and planetary theories, |b the
figure of the earth, precession and nutation, the calculus of variations, and
the undulatory theory of optics.
250: : |a 3rd ed., |b corrected and improved.
260: : |a Cambridge, |b J. & J.J. Deighton; |a London, |b J.W. Parker, |c
1842.
300/1: : |a vii, [1], 390 p. |b 5 fold. plates. |c 23 cm.
650/1: 0: |a Celestial mechanics.
650/2: 0: |a Calculus of variations
650/3: 0: |a Geometrical optics.

MATHEMATICAL TRACTS.

MATHEMATICAL TRACTS

ON

THE LUNAR AND PLANETARY THEORIES,

THE FIGURE OF THE EARTH,

PRECESSION AND NUTATION,

THE CALCULUS OF VARIATIONS,

AND

THE UNDULATORY THEORY OF OPTICS.

DESIGNED FOR THE USE OF STUDENTS IN THE UNIVERSITY

BY GEORGE BIDDELL AIRY, M.A.

ASTRONOMER ROYAL,

FORMERLY FELLOW OF TRINITY COLLEGE, AND LATE PLUMIAN PROFESSOR OF ASTRONOMY AND EXPERIMENTAL PHILOSOPHY IN THE UNIVERSITY OF CAMBRIDGE.

THIRD EDITION, CORRECTED AND IMPROVED.

CAMBRIDGE:

PRINTED AT THE UNIVERSITY PRESS;

PUBLISHED BY J. & J. J. DEIGHTON, CAMBRIDGE;

AND

JOHN W. PARKER, LONDON.

M.DCCC.XLII.

PREFACE TO THE SECOND EDITION.

IN this Edition, a Tract on the Planetary Theory and one on the Undulatory Theory of Optics have been introduced. The former of these subjects, it is hoped, will (independently of its vast importance) be acceptable to the class of students for whom it is intended, for the general beauty of its results and the comparative facility of its investigations. Indeed it is certain that to the accomplished mathematician the Planetary Theory is much less laborious than the Lunar Theory; and it would at first sight seem that it ought to precede it. But the methods used in the Planetary Theory imply the possession of much greater mathematical knowledge, and of a more refined kind, than those of the Lunar Theory: the nature of the conclusions from the latter is also known to most students, from the circumstance that a general explanation of the whole and an accurate calculation of some is attempted in the *Principia*. For these reasons I have thought it desirable to keep the Lunar Theory first in order.

The Undulatory Theory of Optics is presented to the reader as having the same claims to his attention as the Theory of Gravitation: namely that it is certainly true, and that, by mathematical operations of general elegance, it leads to results of great interest. With regard to the evidence for this theory; if the simplicity of a hypothesis, which explains with accuracy a vast variety of phænomena of the most complicated kind, can be considered a proof of its correctness, I believe there is no physical theory so firmly established as the theory in question. This can be felt completely, perhaps, only by the person who has both observed the phænomena and made the calculations; as to my own pretensions to the former qualification, I shall merely state that I have repeated nearly every experiment alluded to in the following Tract. This character of certainty I conceive to belong only to what may be called the *geometrical* part of the theory: the hypothesis, namely, that light consists of undulations depending on transversal vibrations, and that these travel with certain velocities in different media according to the laws here explained. The *mechanical* part of the theory, as the suppositions relative to the constitution of the ether, the computation of the intensity of reflected and refracted rays, &c., though generally probable, I conceive to be far from certain.

The plan of this Tract has therefore been to include those phænomena only which admit of calculation. Many subjects are thus excluded (for instance, the absorption of light by coloured media) for which supplementary theories are still wanting. On the other hand, the investigations are applied only to phænomena which actually have been observed: as I have thought it useless to suppose imaginary combinations, where the real conditions of experiment offer so great variety.

The second investigation of the intensity of light reflected from a glass surface, and that of the nature of light reflected internally and totally from glass, were written as a conjectural restoration of Fresnel's investigations, when his paper was supposed to be lost. That paper has since been found and published: the only alteration which it appeared necessary to make is contained in the note attached to the latter.

The few alterations in the other Tracts have been made only with the view of rendering the investigations more simple, and (in one instance) of introducing greater uniformity in the notation.

Observatory, Cambridge,
Oct. 5, 1831.

PREFACE TO THE THIRD EDITION.

THIS Edition has been carefully revised: the wording of sentences has been frequently altered; and (in a few instances) entire paragraphs have been introduced as additions, or substituted for those of the former Edition. No alteration, however, is made in the general arrangement, or in the order of the numbers attached to the several Articles.

ROYAL OBSERVATORY, GREENWICH,
Aug. 14, 1841.

LUNAR

AND

PLANETARY THEORIES.

INTRODUCTION.

In our succeeding investigations, the following equation will several times occur, and it will therefore be convenient to premise its solution.

As the cases of it which will present themselves are distinguished by some peculiarities, we shall here consider each of them separately.

By the notation $\int_\theta \cos n\theta \,.\, \Theta$, $\int_\theta \cos n\theta \,.\, \cos \overline{m\theta + D}$, &c. we mean what are usually written

$$\int \cos n\theta \,.\, \Theta \,.\, d\theta, \quad \int \cos n\theta \,.\, \cos \overline{m\theta + D} \,.\, d\theta, \text{ \&c.};$$

they are the quantities whose differential coefficients, with respect to θ, are $\cos n\theta \,.\, \Theta$, $\cos n\theta \,.\, \cos \overline{m\theta + D}$, &c.

1. Prop. 1. To solve the equation

$$\frac{d^2 u}{d\theta^2} + n^2 u + \Theta = 0;$$

Θ being a function of θ and constants only.

Multiply the equation by $\cos n\theta$,*

$$\text{then } \cos n\theta \,.\, \frac{d^2 u}{d\theta^2} + n^2 \cos n\theta \,.\, u + \cos n\theta \,.\, \Theta = 0;$$

---

* $\sin n\theta$, $\sin \overline{n\theta + D}$, or $\cos \overline{n\theta + D}$, would have served as well for multiplier.

integrate the first term by parts;

$$\text{then } \int_\theta \cos n\theta \cdot \frac{d^2u}{d\theta^2} = \cos n\theta \cdot \frac{du}{d\theta} + n \int_\theta \sin n\theta \cdot \frac{du}{d\theta}$$

$$= \cos n\theta \cdot \frac{du}{d\theta} + n \sin n\theta \cdot u - n^2 \int_\theta \cos n\theta \cdot u.$$

Hence, the integral of the whole is

$$\cos n\theta \cdot \frac{du}{d\theta} + n \cdot \sin n\theta \cdot u + \int_\theta \cos n\theta \cdot \Theta = 0,$$

the arbitrary constant being included in the sign of integration. Divide this equation by $\cos^2 n\theta$:

$$\text{then } \frac{\frac{du}{d\theta}}{\cos n\theta} + \frac{n \sin n\theta \cdot u}{\cos^2 n\theta} + \frac{1}{\cos^2 n\theta} \int_\theta \cos n\theta \cdot \Theta = 0:$$

$$\text{integrating, } \frac{u}{\cos n\theta} + \int_\theta \frac{1}{\cos^2 n\theta} \int_\theta \cos n\theta \cdot \Theta = 0,$$

$$\text{or } u = -\cos n\theta \int_\theta \frac{1}{\cos^2 n\theta} \cdot \int_\theta \cos n\theta \cdot \Theta.$$

2. *Case* 1. Let $\Theta = 0$, or $\frac{d^2u}{d\theta^2} + n^2 u = 0$:

$$\text{then } \cos n\theta \cdot \Theta = 0; \int_\theta \cos n\theta \cdot \Theta = -C;$$

$$\therefore u = \cos n\theta \int_\theta \frac{C}{\cos^2 n\theta} = \cos n\theta \left(\frac{C}{n} \tan n\theta + C'\right)$$

$$= \frac{C}{n} \sin n\theta + C' \cos n\theta.$$

This may be put under the form $A \cos(n\theta - B)$,

$$\text{making } A \cos B = C', \ A \sin B = \frac{C}{n}.$$

Or under the form $E \sin(n\theta + F)$,

$$\text{making } E \cos F = \frac{C}{n}, \ E \sin F = C'.$$

3. *Case* 2. Let $\Theta = -a$;

then $\int_\theta \cos n\theta \,.\, \Theta = -\frac{a}{n}\sin n\theta - C$;

$$u = \cos n\theta \int_\theta \left(\frac{a}{n} \cdot \frac{\sin n\theta}{\cos^2 n\theta} + \frac{C}{\cos^2 n\theta}\right)$$

$$= \cos n\theta \,.\, \left(\frac{a}{n^2} \cdot \frac{1}{\cos n\theta} + \frac{C}{n}\tan n\theta + C'\right)$$

$$= \frac{a}{n^2} + \frac{C}{n}\sin n\theta + C'\cos n\theta,$$

$$\text{or } = \frac{a}{n^2} + A\cos(n\theta - B).$$

4. *Case* 3. Let $\Theta = b \,.\, \cos \overline{m\theta + D}$:

then $\int_\theta \cos n\theta . \Theta$, or $b\int_\theta \cos n\theta . \cos \overline{m\theta + D}$, (integrating by parts,)

$$= \frac{b}{n}\sin n\theta \,.\, \cos\overline{m\theta + D} + \frac{mb}{n}\int_\theta \sin n\theta \,.\, \sin\overline{m\theta + D}$$

$$= \frac{b}{n}\sin n\theta \,.\, \cos\overline{m\theta + D} - \frac{mb}{n^2}\cos n\theta \,.\, \sin\overline{m\theta + D}$$

$$+ \frac{m^2 b}{n^2}\int_\theta \cos n\theta \,.\, \cos\overline{m\theta + D}.$$

Hence $b\int_\theta \cos n\theta \,.\, \cos\overline{m\theta + D}$

$$= \frac{b}{m^2 - n^2}(m\cos n\theta \,.\, \sin\overline{m\theta + D} - n\sin n\theta \,.\, \cos\overline{m\theta + D}) - C.$$

Then

$$u = \cos n\theta \,.\, \frac{b}{m^2 - n^2}\int_\theta \left(-\frac{m\sin\overline{m\theta + D}}{\cos n\theta} + \frac{n\sin n\theta \,.\, \cos\overline{m\theta + D}}{\cos^2 n\theta}\right)$$

$$+ \cos n\theta \int_\theta \,.\, \frac{C}{\cos^2 n\theta}$$

$$= \frac{b}{m^2 - n^2} \,.\, \cos\overline{m\theta + D} + \frac{C}{n}\sin n\theta + C'\cos n\theta,$$

$$\text{or } = \frac{b}{m^2 - n^2} \,.\, \cos\overline{m\theta + D} + A\cos\overline{n\theta - B}.$$

5. *Case* 4. Let $\Theta = b \,.\, \cos \overline{n\theta + D}$.

Instead of multiplying by $\cos n\theta$, multiply by $\sin \overline{n\theta + D}$, (which will do equally well)

$$\therefore\ \sin \overline{n\theta + D} \,.\, \frac{d^2 u}{d\theta^2} + n^2 \,.\, \sin \overline{n\theta + D} \,.\, u + \frac{b}{2} \,.\, \sin \overline{2n\theta + 2D} = 0.$$

Integrating by parts, as in (1),

$$\sin \overline{n\theta + D} \,.\, \frac{du}{d\theta} - n \cos \overline{n\theta + D} \,.\, u - \frac{b}{4n} \,.\, \cos \overline{2n\theta + 2D} + C = 0,$$

$$\text{or } \sin \overline{n\theta + D} \,.\, \frac{du}{d\theta} - n \,.\, \cos \overline{n\theta + D} \,.\, u + \frac{b}{2n} \sin^2 \overline{n\theta + D} + C - \frac{b}{4n} = 0.$$

Dividing by $\sin^2 \overline{n\theta + D}$, and integrating,

$$\frac{u}{\sin \overline{n\theta + D}} + \frac{b}{2n} \theta - \left(\frac{C}{n} - \frac{b}{4n^2}\right) \cot \overline{n\theta + D} + C' = 0,$$

$$\text{or } u = -\frac{b}{2n} \theta \,.\, \sin \overline{n\theta + D} + \left(\frac{C}{n} - \frac{b}{4n^2}\right) \cos \overline{n\theta + D} - C' \sin \overline{n\theta + D},$$

which as before may be put under the form

$$u = -\frac{b}{2n} \,.\, \theta \,.\, \sin \overline{n\theta + D} + A \cos \overline{n\theta - B}.$$

6. *Remarks.* If Θ consisted of several terms, the expression for u would contain one term corresponding to each. The part which depends upon the abitrary constants, is entirely independent of Θ. The process above having shewn what is the form of the expression for u, we may sometimes solve the equation with greater ease, by assuming an expression with indeterminate coefficients. Thus, if we had the equation

$$\frac{d^2 u}{d\theta^2} + n^2 u + a + b \cos \overline{m\theta + B} + p \cos \overline{q\theta + Q} = 0,$$

we might assume

$$u = -\frac{a}{n^2} + A \cos (n\theta + C) + D \cos \overline{m\theta + B} + E \cos \overline{q\theta + Q},$$

and, substituting this series in the equation, determine the values of D and E.

7. When m does not differ much from n, it appears from (4), that the coefficient of $\cos m\theta + D$, in the expression for u, will be much greater than that in the original equation. This remark we shall find to be very important. The solution in the 4th case assumes a form different from any of the others: its peculiarity will materially affect our future operations.

MOTION OF TWO BODIES.

8. If the Sun were supposed to be at rest, the motion of a planet about it might be found by the formulæ for central forces. In the equation $\frac{d^2u}{d\theta^2} + u - \frac{P}{h^2u^2} = 0$, where h is a constant and $= \frac{1}{u^2} \cdot \frac{d\theta}{dt}$, (Whewell *on the Free Motion of Points*, Art. 24; Earnshaw's *Dynamics*, Art. 87; or Art. 32 below, if $T = 0$,) we must put for P the attraction of the Sun on the planet, and by solving the equation, we should find u in terms of θ, and the form of the orbit which the planet describes would then be known.

9. In the actual case of the Sun and a planet, these bodies move about their common center of gravity. But their relative motion will be the same as if we suppose the Sun to be at rest, provided we add to the accelerating forces which really act on the planet, another force equal and opposite to that which acts on the Sun. For if the same accelerating force be supposed to act on both, since the absolute motion which it communicates to both is the same, and in the same direction, their relative motion will be the same as if that force did not act: and if that force be equal and opposite to the force really acting on the Sun, the Sun will be at rest. Or, instead of this, if we add to the forces acting on the Sun, a force equal and opposite to that acting on the planet, the planet will be at rest, and the relative motions will be unaltered. We shall generally make the latter supposition.

10. Prop. 2. The orbit which the Sun appears to describe about a planet is a conic section.

Let M = mass of Sun (estimated by the accelerating force which its attraction produces at distance 1), M' = that of the planet: let their distance = r. The accelerating force on the Sun, according to the law of gravitation, $= \frac{M'}{r^2}$: that on the planet $= \frac{M}{r^2}$: if then we suppose this force applied in the opposite direction to the Sun, the whole accelerating force on the Sun, supposing the planet at rest, $= \frac{M+M'}{r^2} = (M+M')\,u^2$, if $u = \frac{1}{r}$. Substituting this for P in the equation above,

$$\frac{d^2u}{d\theta^2} + u - \frac{M+M'}{h^2} = 0,$$

the solution of which, by (3), is

$$u = \frac{M+M'}{h^2} + A\cos\overline{\theta - B},$$

$$\text{or } r = \frac{1}{\frac{M+M'}{h^2} + A\cos\overline{\theta - B}},$$

which is the general polar equation to the conic sections, the focus being the pole.

11. The conic section which a planet appears to describe about the Sun, or the Sun about a planet, is found to be an ellipse. Let a and e be the semi-axis-major and eccentricity, B the longitude of the perihelion, then (Hamilton's *Conic Sections*, Art. 115; *Analytical Geometry*, Art. 145),

$$r = \frac{1}{\frac{1}{a(1-e^2)} + \frac{e}{a(1-e^2)}\cos\overline{\theta - B}}.$$

Comparing this with the expression above,

$$\frac{M+M'}{h^2}=\frac{1}{a(1-e^2)},$$

$$\text{or } h=\sqrt{a(1-e^2)(M+M')}.$$

12. PROP. 3. To find the time of describing any part of the ellipse, or to express $nt+\epsilon-B$ (the mean anomaly) in terms of $\theta-B$, (the true anomaly).

By Whewell *On the Free Motion*, Art. 25, or Earnshaw's *Dynamics*, Art. 88, $\frac{dt}{d\theta}=\frac{r^2}{h}=$ (in the present instance), $\frac{a^{\frac{3}{2}}.(1-e^2)^{\frac{3}{2}}}{\sqrt{M+M'}}.\frac{1}{(1+e\cos\overline{\theta-B})^2}$. The most convenient form into which this can be expanded, is a series of cosines of multiple arcs, as

$$A+C\cos\overline{\theta-B}+D\cos 2.\overline{\theta-B}+E\cos 3.\overline{\theta-B}+\&c.$$

To effect this, we shall first expand $\frac{1}{1+e\cos\overline{\theta-B}}$ in a similar series.

13. If for $\cos\overline{\theta-B}$, we put $\frac{x+\frac{1}{x}}{2}$ (where $x=\epsilon^{\overline{\theta-B}\sqrt{-1}}$), we have

$$\frac{1}{1+e\cos\overline{\theta-B}}=\frac{1}{1+\frac{e}{2}\left(x+\frac{1}{x}\right)},$$

$$\text{which will }=\frac{1}{(\alpha+\beta x)\left(\alpha+\frac{\beta}{x}\right)},\text{ if } \alpha^2+\beta^2=1,\ \alpha\beta=\frac{e}{2}.$$

From these equations $\alpha=\frac{1}{2}(\sqrt{1+e}+\sqrt{1-e})$:

$$\beta=\frac{1}{2}(\sqrt{1+e}-\sqrt{1-e});$$

$$\therefore\ \alpha^2=\frac{1+\sqrt{1-e^2}}{2};\ \frac{\beta}{\alpha}=\frac{e}{1+\sqrt{1-e^2}}.$$

Put λ for $\frac{\beta}{a}$.

Then $\frac{1}{1+e\cos\overline{\theta-B}} = \frac{2}{1+\sqrt{1-e^2}} \cdot \frac{1}{(1+\lambda x)\left(1+\frac{\lambda}{x}\right)}$

$$= \frac{2}{1+\sqrt{1-e^2}} \cdot \frac{x}{(1+\lambda x)(x+\lambda)}$$

$$= \frac{2}{1+\sqrt{1-e^2}} \cdot \frac{1}{1-\lambda^2} \cdot \left(\frac{1}{1+\lambda x} - \frac{\lambda}{x+\lambda}\right)$$

$$= \frac{2}{1+\sqrt{1-e^2}} \cdot \frac{1}{1-\lambda^2} \cdot \left\{\frac{1}{1+\lambda x} - \frac{\frac{\lambda}{x}}{1+\frac{\lambda}{x}}\right\}.$$

Expanding these fractions, and observing that

$$x + \frac{1}{x} = 2\cos\overline{\theta - B}, \quad x^2 + \frac{1}{x^2} = 2\cos 2\,.\,\overline{\theta - B}, \quad \&c.,$$

$$\frac{1}{1+e\cos\overline{\theta-B}} = \frac{2}{1+\sqrt{1-e^2}} \cdot \frac{1}{1-\lambda^2} \times$$

$$(1 - 2\lambda\cos\overline{\theta-B} + 2\lambda^2.\cos 2\,.\,\overline{\theta-B} - 2\lambda^3.\cos 3\,.\,\overline{\theta-B} + \&c.).$$

But $1-\lambda^2 = \frac{2\sqrt{1-e^2}}{1+\sqrt{1-e^2}}$; hence

$$\frac{1}{1+e\cos\overline{\theta-B}} = \frac{1}{\sqrt{1-e^2}} \cdot (1 - 2\lambda\,.\,\cos\overline{\theta-B} + 2\lambda^2.\cos 2\,.\,\overline{\theta-B} - \&c.)$$

14. Now $\frac{1}{(1+e\cos\overline{\theta-B})^2} = \frac{d}{de}\left(\frac{e}{1+e\cos\overline{\theta-B}}\right)$ (see (88));

observing, then, that

$$\frac{d}{de}\left(\frac{e}{\sqrt{1-e^2}}\,.\,\lambda^p\right) = \frac{e^p\,(1+p\sqrt{1-e^2})}{(1+\sqrt{1-e^2})^p.\,(1-e^2)^{\frac{3}{2}}},$$

we have

$$\frac{1}{(1+e\cos\overline{\theta-B})^2}=\frac{1}{(1-e^2)^{\frac{3}{2}}}\cdot\left\{1-2.\frac{e(1+\sqrt{1-e^2})}{1+\sqrt{1-e^2}}\cos\overline{\theta-B}\right.$$

$$+2\,\frac{e^2(1+2\sqrt{1-e^2})}{(1+\sqrt{1-e^2})^2}\cos 2.\overline{\theta-B}$$

$$\left.-2\,\frac{e^3(1+3\sqrt{1-e^2})}{(1+\sqrt{1-e^2})^3}\cos 3.\overline{\theta-B}+\&c.\right\}.$$

15. Hence (12), $\dfrac{dt}{d\theta}=\dfrac{a^{\frac{3}{2}}}{\sqrt{M+M'}}\times$

$$\left\{1-2e\cos\overline{\theta-B}+\frac{2e^2(1+2\sqrt{1-e^2})}{(1+\sqrt{1-e^2})^2}\cos 2.\overline{\theta-B}-\&c.\right\}.$$

Integrating and correcting, so as to make $nt+\epsilon-B=0$, when $\theta-B=0$, ϵ being a constant depending on the time when the planet passed the perihelion, and putting

$$\frac{\sqrt{M+M'}}{a^{\frac{3}{2}}}=n,$$

$$nt+\epsilon-B=\overline{\theta-B}-2e\sin\overline{\theta-B}+\frac{e^2(1+2\sqrt{1-e^2})}{(1+\sqrt{1-e^2})^2}\sin 2.\overline{\theta-B}-\&c.$$

$$\pm\frac{2e^p.(1+p\sqrt{1-e^2})}{p(1+\sqrt{1-e^2})^p}\sin p.\overline{\theta-B}\mp\&c.$$

For a whole revolution, suppose θ increased by 2π: this change makes no alteration in the values of $\sin\overline{\theta-B}$, $\sin 2.\overline{\theta-B}$, &c.: then if T be the periodic time, t will be increased by T, and

$$nT=2\pi,\quad\text{or}\quad T=\frac{2\pi}{n}=\frac{2\pi.a^{\frac{3}{2}}}{\sqrt{M+M'}}.$$

16. The term *anomaly* is used generally to denote the angular distance of that body which is supposed to move, from its apse. The true anomaly, then, in the present case, is $\theta-B$.

But $nt + \epsilon - B$ is an angle which increases proportionally to the time, and is that which the body would have described at the end of the time t, if, commencing with the angle $\epsilon - B$, it had moved uniformly with such an angular velocity, that it would have performed a revolution in the same time in which it does perform it. It is therefore called the *mean* anomaly.

17. If we expand these coefficients as far as e^3,

$$nt + \epsilon = \theta - 2e \sin \overline{\theta - B} + \tfrac{3}{4} e^2 \sin 2 . \overline{\theta - B} - \frac{e^3}{3} \sin 3 . \overline{\theta - B}.$$

18. PROP. 4. To find $\overline{\theta - B}$ in terms of $nt + \epsilon - B$, or the true anomaly in terms of the mean. This must be done by Lagrange's theorem. If $y = z + x . \phi(y)$, then

$$y = z + \phi(z) . \frac{x}{1} + \frac{d}{dz} \left\{ \overline{\phi(z)}|^2 \right\} . \frac{x^2}{1.2} + \frac{d^2}{dz^2} \left\{ \overline{\phi(z)}|^3 \right\} . \frac{x^3}{1.2.3} + \&c.$$

Here, carrying the approximation no farther than e^3 and putting y for $\theta - B$, z for $nt + \epsilon - B$,

$$y = z + e\left(2 \sin y - \frac{3e}{4} \sin 2y + \frac{e^2}{3} \sin 3y\right);$$

$$\therefore \ \phi(z) = 2 \sin z - \frac{3e}{4} \sin 2z + \frac{e^2}{3} \sin 3z;$$

$$\therefore \ \overline{\phi(z)}|^2 = \left(2 \sin z - \frac{3e}{4} \sin 2z\right)^2$$

$$= 2 - 2 \cos 2z - \frac{3e}{2} \cos z + \frac{3e}{2} \cos 3z;$$

$$\therefore \ \frac{d}{dz} \left\{ \overline{\phi(z)}|^2 \right\} = 4 \sin 2z + \frac{3e}{2} \sin z - \frac{9e}{2} . \sin 3z,$$

$$\overline{\phi(z)}|^3 = 8 \sin^3 z = 6 \sin z - 2 \sin 3z;$$

$$\therefore \ \frac{d^2}{dz^2} \left\{ \overline{\phi(z)}|^3 \right\} = -6 \sin z + 18 \sin 3z;$$

$$\therefore\ y = z + \left(2 \sin z - \frac{3e}{4} \sin 2z + \frac{e^2}{3} \sin 3z\right) \frac{e}{1}$$

$$+ \left(4 \sin 2z + \frac{3e}{2} \sin z - \frac{9e}{2} . \sin 3z\right) \frac{e^2}{2}$$

$$+ (18 \sin 3z - 6 \sin z) \frac{e^3}{6}$$

$$= z + \left(2e - \frac{e^3}{4}\right) \sin z + \frac{5e^2}{4} \sin 2z + \frac{13}{12} e^3 . \sin 3z,$$

carrying the approximation as far as e^3.

$$\text{Or}\ \ \overline{\theta - B} = nt + \epsilon - B + \left(2e - \frac{e^3}{4}\right) \sin \overline{nt + \epsilon - B}$$

$$+ \frac{5e^2}{4} \sin \overline{2nt + 2\epsilon - 2B} + \frac{13}{12} e^3 \sin \overline{3nt + 3\epsilon - 3B} + \&c.$$

19. The mean anomaly, then, is that part of the true anomaly, which is independent of periodical terms, as sines or cosines. This is the usual signification of the word *mean*, in Astronomy.

The expression for θ is

$$nt + \epsilon + \left(2e - \frac{e^3}{4}\right) \sin \overline{nt + \epsilon - B} + \frac{5e^2}{4} \sin \overline{2nt + 2\epsilon - 2B} + \&c.$$

θ being called the longitude, or the *true* longitude, $nt + \epsilon$ (in conformity with the remark above) is called the *mean* longitude. The aggregate of the terms following $nt + \epsilon$ is called the *equation of the center*.

LUNAR THEORY.

PERTURBATION OF THE EARTH'S MOTION.

20. If the Earth, in its revolution round the Sun, were unaccompanied by any other body, it would accurately describe an ellipse. By the attraction of the Moon, the orbit will be altered: to assist us in the discovery of the orbit really described, the following theorem will be useful.

21. Prop. 5. The center of gravity of the Earth and Moon describes about the Sun, very nearly, an ellipse in one plane, and the area passed over by its radius vector is very nearly proportional to the time.

Let E and M, (fig. 1.) be the Earth and Moon, m' the Sun; G the center of gravity of the Earth and Moon; join $m'E$, $m'G$, $m'M$; and draw EH, MK, perpendicular to $m'G$; let

$$m'G = r', \quad EM = r, \quad m'E = y, \quad m'M = y', \quad \angle m'GM = \omega.$$

Now the accelerating force of m' on E in the direction Em' is $\frac{m'}{y^2}$; therefore the moving force in that direction $= \frac{m'.E}{y^2}$; therefore the moving force in direction parallel to Gm' is

$$\frac{m'.E}{y^2} \times \frac{m'H}{y}$$

$$= \frac{m'.E}{y^2} \cdot \frac{r' + GE . \cos\omega}{y}$$

$$= \frac{m'.E\,(r' + GE . \cos\omega)}{y^3}.$$

Similarly, the moving force on M in direction parallel to Gm'

$$= \frac{m'.M}{y'^2} \cdot \frac{m'K}{y'}$$

$$= \frac{m'.M\,(r' - GM . \cos\omega)}{y'^3};$$

therefore the accelerating force on the center of gravity in the direction Gm' is

$$\frac{m'}{M+E}\left\{\frac{E(r'+GE\cos\omega)}{y^3}+\frac{M(r'-GM\cos\omega)}{y'^3}\right\}.$$

And the moving force of m' on E in direction perpendicular to Gm' is $\frac{m'.E}{y^2}.\frac{EH}{y}$

$$=\frac{m'.E}{y^2}.\frac{GE\sin\omega}{y}$$

$$=\frac{m'.E}{y^3}GE.\sin\omega:$$

that on $M=-\frac{m'.M}{y'^2}.\frac{MK}{y'}$

$$=-\frac{m'.M}{y'^3}GM.\sin\omega;$$

therefore the accelerating force on the center of gravity in direction perpendicular to Gm' is

$$\frac{m'}{M+E}.\left(\frac{E.GE.\sin\omega}{y^3}-\frac{M.GM.\sin\omega}{y'^3}\right).$$

22. Now $\frac{1}{y^3}=\frac{1}{(r'^2+2r'.GE.\cos\omega+GE^2)^{\frac{3}{2}}}$

$$=\left(\text{putting for }\cos^2\omega\text{ its equal }\frac{1+\cos 2\omega}{2}\right)$$

$$\frac{1}{r'^3}\left\{1-3\frac{GE}{r'}\cos\omega+\frac{GE^2}{r'^2}\left(\frac{9}{4}+\frac{15}{4}\cos 2\omega\right)+\&c.\right\};$$

$$\therefore\ \frac{r'+GE.\cos\omega}{y^3}=\frac{1}{r'^2}\left\{1-2.\frac{GE}{r'}\cos\omega\right.$$

$$\left.+\frac{GE^2}{r'^2}\left(\frac{3}{4}+\frac{9}{4}\cos 2\omega\right)+\&c.\right\}.$$

Similarly, $\frac{r'-GM\cos\omega}{y'^3}=\frac{1}{r'^2}\left\{1+2.\frac{GM}{r'}\cos\omega\right.$

$$\left.+\frac{GM^2}{r'^2}\left(\frac{3}{4}+\frac{9}{4}\cos 2\omega\right)+\&c.\right\}.$$

But $GE=\frac{M}{E+M}r$; $\quad GM=\frac{E}{E+M}r$;

hence, accelerating force on center of gravity in direction Gm'

$$= \frac{m'}{r'^2}\left\{1 + \frac{EM^2 + ME^2}{(E+M)^3} \cdot \frac{r^2}{r'^2}\left(\frac{3}{4} + \frac{9}{4}\cos 2\omega\right) + \&c.\right\}$$

$$= \frac{m'}{r'^2}\left\{1 + \frac{ME}{(E+M)^2} \cdot \frac{r^2}{r'^2}\left(\frac{3}{4} + \frac{9}{4}\cos 2\omega\right) + \&c.\right\}.$$

Now this differs from $\frac{m'}{r'^2}$ only by a quantity which is multiplied by $\frac{r^2}{r'^2}$, and by $\frac{E}{E+M} \cdot \frac{M}{E+M}$, and which, in the lunar theory, $\left(\text{where } \frac{r}{r'} = \frac{1}{400} \text{ nearly, and } \frac{M}{E+M} = \frac{1}{80}\right)$ is quite insensible. In the same manner, we find the accelerating force perpendicular to Gm'

$$= -\frac{m'}{r'^2} \cdot \frac{3}{2} \cdot \frac{EM^2 + ME^2}{(E+M)^3} \cdot \frac{r^2}{r'^2} \sin 2\omega,$$

$$\text{or } \; -\frac{m'}{r'^2} \cdot \frac{3}{2} \cdot \frac{ME}{(M+E)^2} \cdot \frac{r^2}{r'^2} \cdot \sin 2\omega,$$

which, for the same reason, is too small to be perceptible in its effects. Hence, the only accelerating force acting on the center of gravity is $\frac{m'}{r'^2}$, and is in the direction of r'; and therefore the center of gravity moves, very nearly, in the same manner in which a body would move placed at G. Similarly it appears, that the force on the Sun, and the motion of the Sun, are the same, as if a mass $= E + M$ were collected at G; therefore the relative motion of the center of gravity about the Sun, is the same as that of a mass $= E + M$; that is, it will very nearly describe an ellipse in one plane, making the areas proportional to the times.

23. Cor. The Sun's apparent longitude, therefore, is not that found by the elliptic theory, for that is his longitude as seen from G; but must be obtained by adding to the longitude so found, the angle $Em'G$. Now $\sin Em'G$ or $Em'G$, (since it is a small angle, never exceeding $10''$) $= \frac{EG}{y} \sin EGm' = \frac{M}{E+M} \cdot \frac{r}{r'} \sin \omega$ very nearly: and, since

the orbits of the Earth and Moon are nearly circular, this angle varies as $\sin\omega$ very nearly. And if the Moon be above the plane of the ecliptic, the Earth will be below it, and the Sun will appear to have a latitude, which can be calculated from the latitude of the Moon.

PERTURBATION OF THE MOON'S MOTION.

24. If the Sun did not attract the Earth and Moon, or if it attracted them equally, their relative motions would not be disturbed, and the Moon would accurately describe an ellipse about the Earth. But the Sun attracts them unequally, and in different directions; so that not only is the force altered in the direction of the radius vector, but a force also acts perpendicular to it. And as the Moon's orbit is inclined to the ecliptic, the disturbing force draws the Moon from the plane in which she is moving, and thus the plane of her orbit is perpetually changing. There appears to be no better mode of estimating the disturbing force, than by resolving it into three parts, one in the direction of the projection of the radius vector on the ecliptic, another in the plane of the ecliptic, perpendicular to this projection, and a third perpendicular to the plane of the ecliptic.

25. Prop. 6. To find the resolved part of the Sun's disturbing force on the Moon, in the direction of the projection of the radius vector on the ecliptic.

Let E, M, m', (fig. 2.), be the Earth, Moon, and Sun: G the center of gravity of the Earth and Moon, which by Prop. 5., describes an ellipse in one plane about the Sun, or about which the Sun appears to describe an ellipse in one plane: draw MB, EA, perpendicular to the plane of the ecliptic; join $m'M$, $m'B$, BGA: let $m'G = r'$, $AB = \rho$, $EM = r$, $m'E = y$, $m'M = y'$, $\tan MGB = s$. Then AB is the projection of EM on the plane of the ecliptic. The force of m' upon M is $\frac{m'}{y'^2}$ in the direction Mm', which is equivalent to $\frac{m'}{y'^2} \cdot \frac{MG}{y'}$ in direction MG, and $\frac{m'}{y'^2} \cdot \frac{Gm'}{y'}$ in direction parallel to Gm'.

Resolving the force $\frac{m'.MG}{y'^3}$ into one parallel to MB, and one parallel to BG,

$$\text{the latter} = \frac{m'.MG}{y'^3} \times \cos MGB = \frac{m'.MG}{y'^3\sqrt{1+s^2}}:$$

and resolving the force $\frac{m'.Gm'}{y'^3}$ into one parallel to BG, and another perpendicular to BG, in the plane of the ecliptic,

$$\text{the former} = -\frac{m'.Gm'}{y'^3}.\cos m'GB.$$

Let θ be the longitude of M, seen from G; θ' the longitude of m': then $\angle m'GB = \theta - \theta'$*: and the part of $\frac{m'.Gm'}{y'^3}$ parallel to BG

$$= -\frac{m'.Gm'}{y'^3}\cos\overline{\theta-\theta'}.$$

Hence, the whole force on M in the direction BG, produced by the Sun's attraction, is

$$m'\left(\frac{MG}{y'^3\sqrt{1+s^2}} - \frac{Gm'}{y'^3}\cos\overline{\theta-\theta'}\right).$$

Similarly, the whole force on E, estimated in the same direction, is

$$m'\left(-\frac{EG}{y^3\sqrt{1+s^2}} - \frac{Gm'}{y^3}\cos\overline{\theta-\theta'}\right).$$

If, then, in the same manner as in (9), we suppose this force applied to M in the opposite direction, we have, for the whole disturbing force on M, in direction of the projection of the radius vector,

$$m'\left\{\frac{MG}{y'^3\sqrt{1+s^2}} + \frac{EG}{y^3\sqrt{1+s^2}} - Gm'\cos\overline{\theta-\theta'}\left(\frac{1}{y'^3}-\frac{1}{y^3}\right)\right\}.$$

---

* It is always to be understood that the orbital motions of the Earth and the Moon are in the direction opposite to that in which the hands of a watch revolve, and the angles are therefore estimated positive in that direction.

26. Prop. 7. To find the resolved part of the disturbing force, which is parallel to the plane of the ecliptic, and perpendicular to the projection of the radius vector.

The only force which acts in this direction on M is the resolved part of the force $\frac{m'.Gm'}{y'^3}$ acting parallel to Gm', which

$$= -\frac{m'.Gm'}{y'^3}.\sin\overline{\theta-\theta'},$$

(if we suppose the Moon to move so that the angle θ increases, and if we consider the force as positive when it tends to accelerate the Moon's motion.) And the only force on E, in the same direction, is

$$-\frac{m'.Gm'}{y^3}\sin\overline{\theta-\theta'}.$$

Supposing this applied to M in the opposite direction, the whole disturbing force, perpendicular to the projection of the radius vector

$$= -m'.Gm'.\sin\overline{\theta-\theta'}\left(\frac{1}{y'^3}-\frac{1}{y^3}\right).$$

27. Prop. 8. To find the resolved part of the disturbing force, which is perpendicular to the plane of the ecliptic.

The only force on M, perpendicular to the plane of the ecliptic, is the resolved part of the force in MG. It is, therefore,

$$= \frac{m'.MG}{y'^3}.\sin MGB$$

$$= \frac{m'.MG}{y'^3}.\frac{s}{\sqrt{1+s^2}}.$$

The force on E, in the same direction, is

$$-\frac{m'.EG}{y^3}.\frac{s}{\sqrt{1+s^2}}.$$

Applying this to M in the opposite direction, we have, for the whole disturbing force perpendicular to the plane of the ecliptic,

$$m' \frac{s}{\sqrt{1+s^2}}\left(\frac{MG}{y'^3}+\frac{EG}{y^3}\right).$$

28. PROP. 9. To find the whole force upon M in these directions: or to find P, T, and S.

Besides the disturbing forces, we must also find the forces resulting from the mutual attraction of E and M. The attraction of E upon $M = \frac{E}{r^2}$: that of M upon $E = \frac{M}{r^2}$, in the opposite direction: applying the latter to M with its direction changed, we have, for the whole force on M, $\frac{E+M}{r^2}$ acting in the direction ME. The resolved part of this, in the direction of the projection of the radius vector, is $\frac{E+M}{r^2} \times \cos MGB$

$$= \frac{E+M}{\rho^2 (1+s^2)^{\frac{3}{2}}};$$

the resolved part in the plane of the ecliptic, perpendicular to this projection, $= 0$:

the resolved part, perpendicular to the plane of the ecliptic,

$$= \frac{(E+M)s}{\rho^2 (1+s^2)^{\frac{3}{2}}}.$$

If, then, we put P, T, and S, for the whole forces on M, parallel to the projection of the radius, perpendicular to the projection of the radius, and perpendicular to the ecliptic, supposing E at rest, we have

$$P = \frac{E+M}{\rho^2 (1+s^2)^{\frac{3}{2}}}$$

$$+ m' \left\{\frac{1}{\sqrt{(1+s^2)}}\left(\frac{MG}{y'^3}+\frac{EG}{y^3}\right) - r'.\cos(\theta-\theta')\left(\frac{1}{y'^3}-\frac{1}{y^3}\right)\right\},$$

$$T = - m'.r'.\sin(\theta-\theta').\left(\frac{1}{y'^3}-\frac{1}{y^3}\right),$$

$$S = \frac{(E+M)s}{\rho^2(1+s^2)^{\frac{3}{2}}} + \frac{m's}{\sqrt{(1+s^2)}}\left(\frac{MG}{y'^3} + \frac{EG}{y^3}\right).$$

And $MG = \frac{E}{E+M}\cdot\rho\sqrt{(1+s^2)}$; $EG = \frac{M}{E+M}\rho\sqrt{(1+s^2)}$.

29. We have now the values of the forces upon M in three directions, considering E as fixed. We must, therefore, investigate the differential equations, for the motion of a body about a fixed center, acted on by forces in these directions.

30. PROP. 10. To find the differential equations for the motion of M about the fixed center E.

Let $E\Upsilon$, Fig. 3, be a straight line in the plane of the ecliptic, drawn from E towards the first point of Aries: Mb perpendicular to the ecliptic: then, ΥEb = Moon's longitude $= \theta$. Also, if x, y, and z be rectangular co-ordinates (z being perpendicular to the plane of the ecliptic, and x measured on the line drawn towards the first point of Aries),

$$x = \rho\cos\theta, \quad y = \rho\sin\theta, \quad z = \rho s.$$

And X, or the force in direction of x, $= -P\cos\theta - T\sin\theta$,

Y.................................. $= -P\sin\theta + T\cos\theta$,

Z.................................. $= -S$.

Hence the equations of motion

$$\frac{d^2x}{dt^2} = X, \quad \frac{d^2y}{dt^2} = Y, \quad \frac{d^2z}{dt^2} = Z,$$

are changed into the following,

$$\frac{d^2x}{dt^2} = -P\frac{x}{\rho} - T\frac{y}{\rho},$$

$$\frac{d^2y}{dt^2} = -P\frac{y}{\rho} + T\frac{x}{\rho},$$

$$\frac{d^2z}{dt^2} = -S.$$

Hence $2\frac{dx}{dt}\cdot\frac{d^2x}{dt^2}+2\frac{dy}{dt}\cdot\frac{d^2y}{dt^2}=-\frac{2P}{\rho}\left(x\frac{dx}{dt}+y\frac{dy}{dt}\right)$

$$+\frac{2T}{\rho}\left(x\frac{dy}{dt}-y\frac{dx}{dt}\right).$$

But $2.\frac{dx}{dt}\cdot\frac{d^2x}{dt^2}+2.\frac{dy}{dt}\cdot\frac{d^2y}{dt^2}=\frac{d}{dt}\left\{\left(\frac{dx}{dt}\right)^2+\left(\frac{dy}{dt}\right)^2\right\}$

$=\frac{d}{dt}\left\{\left(\frac{d\rho}{dt}\right)^2+\rho^2\left(\frac{d\theta}{dt}\right)^2\right\}$, since $\frac{dx}{dt}=\frac{d\rho}{dt}\cos\theta-\rho\sin\theta\frac{d\theta}{dt}$,

$\frac{dy}{dt}=\frac{d\rho}{dt}\sin\theta+\rho\cos\theta\frac{d\theta}{dt}$. And $x\frac{dx}{dt}+y\frac{dy}{dt}=\rho\frac{d\rho}{dt}$;

$$x\frac{dy}{dt}-y\frac{dx}{dt}=\rho^2\frac{d\theta}{dt};$$

$$\therefore\ \frac{d}{dt}\left\{\left(\frac{d\rho}{dt}\right)^2+\rho^2\left(\frac{d\theta}{dt}\right)^2\right\}=-2P.\frac{d\rho}{dt}+2T\rho.\frac{d\theta}{dt}\ \ldots\ldots(a).$$

Also $x\frac{d^2y}{dt^2}-y\frac{d^2x}{dt^2}=T\frac{x^2+y^2}{\rho}=T\rho.$

But $x\frac{d^2y}{dt^2}-y\frac{d^2x}{dt^2}=\frac{d}{dt}\left(x\frac{dy}{dt}-y\frac{dx}{dt}\right)=\frac{d}{dt}\left(\rho^2.\frac{d\theta}{dt}\right)$;

$$\therefore\ \frac{d}{dt}\left(\rho^2.\frac{d\theta}{dt}\right)=T\rho\ldots\ldots\ldots\ldots(b).$$

And $\frac{d^2(\rho s)}{dt^2}=-S\ldots\ldots\ldots\ldots(c)$.

31. These equations (a), (b), (c), appear to be the simplest equations to which the motion of a point can be reduced. In their present form, however, it is not possible to integrate them: we must obtain equations between ρ, θ, and s, independent of t. This is the object of the next proposition.

32. PROP. 11. To eliminate t from the differential equations.

Since, in their present form, t is the independent variable, we must take some other quantity for the independent variable. Let this be θ.

Now by equation (b), $\dfrac{d}{dt}\left(\rho^2 \cdot \dfrac{d\theta}{dt}\right) = T\rho$;

multiplying each side by $\rho^2 \dfrac{d\theta}{dt}$,

$$\rho^2 \frac{d\theta}{dt} \cdot \frac{d}{dt}\left(\rho^2 \frac{d\theta}{dt}\right) = T\rho^3 \frac{d\theta}{dt};$$

$$\therefore \frac{1}{2} \cdot \frac{d}{dt}\left(\rho^2 \cdot \frac{d\theta}{dt}\right)^2 = T\rho^3 \cdot \frac{d\theta}{dt};$$

$$\therefore \frac{1}{2} \cdot \frac{d}{d\theta}\left(\rho^2 \frac{d\theta}{dt}\right)^2 = T\rho^3;$$

$$\therefore \left(\rho^2 \frac{d\theta}{dt}\right)^2 = h^2 + 2\textstyle\int_\theta T\rho^3,$$

h^2 being a constant quantity;

hence $\rho^2 \dfrac{d\theta}{dt} = (h^2 + 2\int_\theta T\rho^3)^{\frac{1}{2}}$, and $\dfrac{dt}{d\theta} = \dfrac{\rho^2}{\sqrt{(h^2 + 2\int_\theta T\rho^3)}}$.

Now to transform equation (a), if we multiply both sides by $\dfrac{dt}{d\theta}$ it becomes

$$\frac{d}{d\theta}\left\{\left(\frac{d\rho}{dt}\right)^2 + \rho^2\left(\frac{d\theta}{dt}\right)^2\right\} = -2P \cdot \frac{d\rho}{d\theta} + 2T\rho,$$

$$\text{or } \frac{d}{d\theta}\left\{\left(\frac{d\rho}{d\theta}\right)^2 + \rho^2 \cdot \left(\frac{d\theta}{dt}\right)^2\right\} = -2P \cdot \frac{d\rho}{d\theta} + 2T\rho,$$

$$\text{or } \frac{d}{d\theta}\left[\left\{\frac{1}{\rho^4}\left(\frac{d\rho}{d\theta}\right)^2 + \frac{1}{\rho^2}\right\} \cdot (h^2 + 2\textstyle\int_\theta T\rho^3)\right] = -2P \cdot \dfrac{d\rho}{d\theta} + 2T\rho.$$

Let $\rho = \dfrac{1}{u}$; $\quad \dfrac{d\rho}{d\theta} = -\dfrac{1}{u^2} \cdot \dfrac{du}{d\theta}$; $\quad \dfrac{1}{\rho^2} \cdot \dfrac{d\rho}{d\theta} = -\dfrac{du}{d\theta}$;

$$\therefore \frac{d}{d\theta}\left[\left\{\left(\frac{du}{d\theta}\right)^2 + u^2\right\} \cdot \left(h^2 + 2\int_\theta \frac{T}{u^3}\right)\right] = \frac{2P}{u^2} \cdot \frac{du}{d\theta} + \frac{2T}{u},$$

or performing the differentiation,

$$2\frac{du}{d\theta}\left(\frac{d^2u}{d\theta^2}+u\right).\left(h^2+2\int_\theta\frac{T}{u^3}\right)+\left\{\left(\frac{du}{d\theta}\right)^2+u^2\right\}.\frac{2T}{u^3}$$

$$=\frac{2P}{u^2}.\frac{du}{d\theta}+\frac{2T}{u};$$

$$\therefore\ \left(\frac{d^2u}{d\theta^2}+u\right).\left(h^2+2\int_\theta\frac{T}{u^3}\right)+\frac{T}{u^3}.\frac{du}{d\theta}=\frac{P}{u^2};$$

$$\therefore\ \frac{d^2u}{d\theta^2}+u-\frac{\frac{P}{u^2}-\frac{T}{u^3}.\frac{du}{d\theta}}{h^2+2\int_\theta\frac{T}{u^3}}=0\ldots\ldots\ldots\ldots(d).$$

33. In the same manner, putting u for $\frac{1}{\rho}$ in equation (c),

$$\text{we have }\frac{d^2\left(\frac{s}{u}\right)}{dt^2}=-S.$$

$$\text{Now }\frac{d\left(\frac{s}{u}\right)}{dt}=\frac{d\left(\frac{s}{u}\right)}{d\theta}.\frac{d\theta}{dt}=\frac{1}{u^2}\left(u\frac{ds}{d\theta}-s\frac{du}{d\theta}\right).\frac{d\theta}{dt}$$

$$=\left(u\frac{ds}{d\theta}-s\frac{du}{d\theta}\right)\sqrt{h^2+2\int_\theta\frac{T}{u^3}}.$$

$$\text{And }\frac{d^2\left(\frac{s}{u}\right)}{dt^2}=\frac{d}{dt}\left\{\left(u\frac{ds}{d\theta}-s\frac{du}{d\theta}\right)\sqrt{h^2+2\int_\theta\frac{T}{u^3}}\right\}$$

$$=\frac{d\theta}{dt}.\frac{d}{d\theta}\left\{\left(u\frac{ds}{d\theta}-s\frac{du}{d\theta}\right)\sqrt{h^2+2\int_\theta\frac{T}{u^3}}\right\}$$

$$=u^2\sqrt{h^2+2\int_\theta\frac{T}{u^3}}.\left\{\left(u\frac{ds}{d\theta}-s\frac{du}{d\theta}\right).\frac{\frac{T}{u^3}}{\sqrt{h^2+2\int_\theta\frac{T}{u^3}}}\right.$$

$$+\left(u\frac{d^2s}{d\theta^2}-s\frac{d^2u}{d\theta^2}\right)\sqrt{h^2+2\int_\theta\frac{T}{u^3}}\Bigg\}$$

$$=\left(u\frac{ds}{d\theta}-s\frac{du}{d\theta}\right)\frac{T}{u}+\left(u\frac{d^2s}{d\theta^2}-s\frac{d^2u}{d\theta^2}\right)u^2\left(h^2+2\int_\theta\frac{T}{u^3}\right);$$

$$\therefore\ u\frac{d^2s}{d\theta^2}-s\frac{d^2u}{d\theta^2}+\frac{\frac{T}{u^3}\left(u\frac{ds}{d\theta}-s\frac{du}{d\theta}\right)+\frac{S}{u^2}}{h^2+2\int_\theta\frac{T}{u^3}}=0.$$

But multiplying equation (d) by s we find

$$s\frac{d^2u}{d\theta^2}+su+\frac{-\frac{Ps}{u^2}+\frac{Ts}{u^3}\cdot\frac{du}{d\theta}}{h^2+2\int_\theta\frac{T}{u^3}}=0.$$

Adding this to the last, and dividing the sum by u

$$\frac{d^2s}{d\theta^2}+s+\frac{\frac{S-Ps}{u^3}+\frac{T}{u^3}\cdot\frac{ds}{d\theta}}{h^2+2\int_\theta\frac{T}{u^3}}=0\ldots\ldots\ldots(e).$$

34. By the solution of the two equations (d) and (e), we must express u and s in terms of θ, which will give the form and position of the orbit. The time of describing any part will then be found by integrating the equation

$$\frac{dt}{d\theta}=\frac{\rho^2}{\sqrt{(h^2+2\int_\theta T\rho^3)}}=\frac{1}{u^2\sqrt{h^2+2\int_\theta\frac{T}{u^3}}}.$$

35. PROP. 12. To expand the expressions for

$$\frac{P}{u^2},\quad\frac{T}{u^3},\quad\text{and}\quad\frac{S-Ps}{u^3}.$$

The expressions for P and T were found in Prop. 9: and

$$S-Ps=m'.r'.s.\cos(\theta-\theta')\left(\frac{1}{y'^3}-\frac{1}{y^3}\right).$$

Now, $y^2 = (m'A)^2 + (AE)^2$ (fig. 2.)

$$= r'^2 + 2r'.GA.\cos(\theta - \theta') + GA^2 + GA^2.s^2.$$

Expanding as far as $\left(\frac{GA}{r'}\right)^2$, and rejecting $\left(\frac{GA}{r'}\right)^2.s^2$,

$$\frac{1}{y^3} = \frac{1}{r'^3}\left[1 - 3.\frac{GA}{r'}\cos(\theta - \theta') + \frac{GA^2}{r'^2}\left\{\frac{9}{4} + \frac{15}{4}\cos 2(\theta - \theta')\right\}*\right].$$

But $GA = \frac{M}{E+M}.\rho$; $\therefore \frac{1}{y^3}$

$$= \frac{1}{r'^3}\left\{1 - 3\frac{M}{E+M}.\frac{\rho}{r'}.\cos\overline{\theta-\theta'} + \frac{M^2}{(E+M)^2}.\frac{\rho^2}{r'^2}\left(\frac{9}{4} + \frac{15}{4}\cos 2.\overline{\theta-\theta'}\right)\right\}.$$

Similarly, $\frac{1}{y'^3}$

$$= \frac{1}{r'^3}\left\{1 + 3\frac{E}{E+M}.\frac{\rho}{r'}\cos(\theta-\theta') + \frac{E^2}{(E+M)^2}.\frac{\rho^2}{r'^2}\left(\frac{9}{4} + \frac{15}{4}\cos 2(\theta-\theta')\right)\right\};$$

$$\therefore \frac{1}{y'^3} - \frac{1}{y^3}$$

$$= \frac{1}{r'^3}\left\{3\frac{\rho}{r'}\cos(\theta-\theta') + \frac{E-M}{E+M}.\frac{\rho^2}{r'^2}\left(\frac{9}{4} + \frac{15}{4}\cos 2(\theta-\theta')\right)\right\}.$$

Hence, $\cos(\theta-\theta')\left(\frac{1}{y'^3} - \frac{1}{y^3}\right)$

$$= \frac{1}{r'^3}\left\{\frac{3}{2}.\frac{\rho}{r'}(1+\cos 2.\overline{\theta-\theta'}) + \frac{E-M}{E+M}.\frac{\rho^2}{r'^2}.\left(\frac{33}{8}\cos\overline{\theta-\theta'} + \frac{15}{8}\cos 3.\overline{\theta-\theta'}\right)\right\},$$

$$\sin(\theta-\theta')\left(\frac{1}{y'^3} - \frac{1}{y^3}\right)$$

$$= \frac{1}{r'^3}\left\{\frac{3}{2}.\frac{\rho}{r'}\sin 2.(\theta-\theta') + \frac{E-M}{E+M}.\frac{\rho^2}{r'^2}.\left(\frac{3}{8}\sin(\theta-\theta') + \frac{15}{8}\sin 3.(\theta-\theta')\right)\right\}.$$

---

* The reader who wishes to proceed with the approximation to the Lunar motions no farther than it is *complëted* in this Tract may omit the last term in these expansions.

And since $EG = AG\sqrt{(1+s^2)} = \dfrac{M}{E+M}\rho\sqrt{(1+s^2)}$,

$$\text{and } MG = \frac{E}{E+M}\rho\sqrt{(1+s^2)},$$

we find

$$\frac{1}{\sqrt{(1+s^2)}}\left(\frac{MG}{y'^3}+\frac{EG}{y^3}\right) = \frac{\rho}{r'^3}\left\{1+3\frac{E-M}{E+M}\cdot\frac{\rho}{r'}\cos(\theta-\theta')\right\},$$

neglecting powers of ρ above the square.

Substituting these values, putting $E+M=\mu$, and expanding $\dfrac{1}{(1+s^2)^{\frac{3}{2}}}$ as far as s^4,

$$P = \frac{\mu}{\rho^2}\left(1-\frac{3}{2}s^2+\frac{15}{8}s^4\right)$$

$$-m'\left\{\frac{\rho}{r'^3}\left(\frac{1}{2}+\frac{3}{2}\cos 2.(\theta-\theta')\right)+\frac{E-M}{\mu}.\frac{\rho^2}{r'^4}\left(\frac{9}{8}\cos(\theta-\theta')+\frac{15}{8}\cos 3.(\theta-\theta')\right)\right\},$$

$$T = -m'\left\{\frac{3}{2}.\frac{\rho}{r'^3}.\sin 2.(\theta-\theta')+\frac{E-M}{\mu}.\frac{\rho^2}{r'^4}\left(\frac{3}{8}\sin(\theta-\theta')+\frac{15}{8}\sin 3.(\theta-\theta')\right)\right\},$$

$$S - Ps = m's.\frac{\rho}{r'^3}\left\{\frac{3}{2}+\frac{3}{2}\cos 2.(\theta-\theta')\right\},$$

(omitting the products of s by ρ^2).

Now, putting $\dfrac{1}{u}$ for ρ, and $\dfrac{1}{u'}$ for r', we find, at length,

$$\frac{P}{u^2} = \mu\left(1-\frac{3}{2}s^2+\frac{15}{8}s^4\right)$$

$$-m'\left\{\frac{u'^3}{u^3}\left(\frac{1}{2}+\frac{3}{2}\cos 2.\overline{\theta-\theta'}\right)+\frac{E-M}{\mu}.\frac{u'^4}{u^4}\left(\frac{9}{8}\cos\overline{\theta-\theta'}+\frac{15}{8}\cos 3.\overline{\theta-\theta'}\right)\right\}\ldots(f),$$

$$\frac{T}{u^3} = -m'\left\{\frac{3}{2}.\frac{u'^3}{u^4}\sin 2.\overline{\theta-\theta'}+\frac{E-M}{\mu}.\frac{u'^4}{u^5}\left(\frac{3}{8}\sin\overline{\theta-\theta'}+\frac{15}{8}\sin 3.\overline{\theta-\theta'}\right)\right\}\ldots(g),$$

$$\frac{S-Ps}{u^3} = m's.\frac{u'^3}{u^4}.\left\{\frac{3}{2}+\frac{3}{2}\cos 2(\theta-\theta')\right\}\ldots\ldots\ldots\ldots(h).$$

36. It appears, then, that upon substituting these values in the equation (d), it will be reduced to this form, $\frac{d^2u}{d\theta^2} + u + \Pi = 0$, Π being a complicated function of u, s, and θ. No method of directly solving such an equation is known: but we have seen in Prop. 1, that it could be solved, if Π were a function of θ only. This suggests the method of solving by successive substitution. Find a value of u in terms of θ, which is nearly the true one: substitute this value for u, in the terms of small magnitude; Π will then be a function of θ only, and the equation may be solved, and a more approximate value of u found. Substitute this for u in Π, and again solve the equation, and a value will be found still nearer the truth. Proceed in the same manner to find the value of s.

37. But, in order to carry on this process with facility, it is necessary to establish some rule with regard to the comparative value of small quantities, so that, fixing upon some quantity as a standard, our first approximation may include its first power, and the first powers of quantities nearly as great; our second approximation may comprehend its square, and the squares of the others, and the products of any two, &c. Thus, let e be the eccentricity of the lunar orbit: e' that of the solar orbit; k the tangent of the mean inclination of the lunar orbit to the ecliptic: m the ratio of the Sun's mean motion to the Moon's mean motion*. Here $e = \frac{1}{20}$ nearly; $e' = \frac{1}{60}$; $k = \frac{1}{12}$; $m = \frac{1}{13}$: taking e, then, as our standard, e', k, and m, are small quantities, not differing much in magnitude from e, and are therefore said to be small quantities of the first order. But, $\frac{\rho}{r'}$ or $\frac{u'}{u}$ is little more than $\frac{1}{400}$, and therefore admits better of being compared with e^2 than with e: it is on that account considered to be a small quantity of the second order: m^2e, $\frac{\rho}{r'}e'$, &c. would be called of the third order; &c.

* By the term *mean motion* is meant, the velocity with which the mean longitude increases. The mean motion varies therefore inversely as the periodic time.

38. It is of importance to determine what is the order of the disturbing force on the Moon, compared with the force which is independent of the Sun's action. Upon examining the expressions for P, T, and S, it will be seen that the mutual attraction of the Earth and Moon is expressed by $\frac{\mu}{\rho^2}$, while the disturbing force is given by a multiple of $\frac{m'\rho}{r'^3}$. We must therefore find the order of $\frac{m'\rho}{r'^3}$, as compared with $\frac{\mu}{\rho^2}$, or of $\frac{m'}{r'^3}$, compared with $\frac{\mu}{\rho^3}$. Now, by (15), if T be the time of a revolution of the Moon about the Earth, in a circular orbit, (the disturbing force and the ellipticity being very small) $T = \frac{2\pi \cdot \rho^{\frac{3}{2}}}{\mu^{\frac{1}{2}}}$ nearly: and, if T' be the time in which the system of the Earth and Moon (supposed very small in comparison with the Sun) revolves round the Sun, $T' = \frac{2\pi \cdot r'^{\frac{3}{2}}}{m'^{\frac{1}{2}}}$ nearly:

$$\text{hence } \frac{m'}{r'^3} : \frac{\mu}{\rho^3} :: \frac{1}{T'^2} : \frac{1}{T^2} :: m^2 : 1:$$

and the disturbing force is of the second order.

39. Prop. 13. To integrate the differential equations, first approximation.

We propose here to include small terms of the first order. Since the disturbing force is of the second order, we shall not take any terms arising from it. Thus, we have

$$\frac{P}{u^2} = \mu; \qquad \frac{T}{u^3} = 0; \qquad \frac{S - Ps}{u^3} = 0:$$

and the equations (d) and (e) become

$$\frac{d^2u}{d\theta^2} + u - \frac{\mu}{h^2} = 0, \qquad \frac{d^2s}{d\theta^2} + s = 0.$$

The solution of the first is

$$u = \frac{\mu}{h^2}(1 + e\cos\overline{\theta - \alpha}) = a(1 + e\cos\overline{\theta - \alpha}),$$

putting $\frac{\mu}{h^2} = a$; that of the second,

$$s = k \cdot \sin(\theta - \gamma).$$

The first shews that the Moon's orbit is an ellipse; the second, that the tan latitude $= k$. sin longitude from node, and therefore, that she apparently moves in a great circle. For, if we conceive the earth to be in the center of a sphere, and if upon the sphere we describe a spherical right-angled triangle, in which one of the sides including the right angle (and representing the Moon's longitude on the ecliptic as measured from a certain point) is $\theta - \gamma$, and the other side including the right angle (and representing the Moon's latitude) is ϕ, where $\tan\phi = k \cdot \sin\overline{\theta - \gamma}$; and if we draw a great circle for the hypothenuse of the triangle, then the angle ψ opposite to the side ϕ is determined by this equation

$$\tan\psi = \frac{\tan\phi}{\sin\overline{\theta - \gamma}}.$$

The value found for s or $\tan\phi$ reduces this equation to the form

$$\tan\psi = k.$$

Therefore, from whatever point of the Moon's path we draw a great circle to a certain point on the ecliptic, the angle at which it meets it is invariable: consequently the same great circle must pass through every point of the Moon's path, or the Moon moves apparently in a great circle.

40. Prop. 14. To integrate the differential equations, second approximation.

As terms of the second order are to be included, we shall here have the first terms of the disturbing force.

$$\frac{P}{u^2} \text{ therefore} = \mu\left(1 - \frac{3}{2}s^2\right) - \frac{m'u'^3}{u^3}\left\{\frac{1}{2} + \frac{3}{2}\cos 2 \cdot (\theta - \theta')\right\},$$

$$\frac{T}{u^3} = -\frac{3}{2} \cdot \frac{m'u'^3}{u^4} \sin 2 \cdot (\theta - \theta') \; : \; \frac{S - Ps}{u^3} = 0.$$

We have just found for u the expression $a\left\{1 + e\cos(\theta - \alpha)\right\}$; but it is evident that, in the substitution of this value in terms of the second order, the terms containing e will be of the third order. We shall therefore for u put a: and for u' shall put a' (a' in the Sun's orbit corresponding to a in the Moon's). Also, for θ' we shall put the value which it would have, if the motions of the Sun and Moon were both uniform, that is, $m\theta + \beta$, β being the Sun's mean longitude, when $\theta = 0$. Thus,

$$\frac{P}{u^2} = \mu\left(1 - \frac{3k^2}{4} + \frac{3k^2}{4}\cos 2\,.\,\overline{\theta - \gamma}\right)$$

$$- \frac{m'a'^3}{a^3}\left\{\frac{1}{2} + \frac{3}{2}\cos\overline{(2-2m)\theta - 2\beta}\right\};$$

$$\frac{T}{u^3} = -\frac{3}{2}\cdot\frac{m'a'^3}{a^4}\,.\sin\overline{(2-2m)\theta - 2\beta};$$

$$\frac{S - Ps}{u^3} = 0.$$

41. In this and succeeding approximations, it will be found most convenient to put the differential equations (d) and (e) into the following form;

$$\frac{d^2u}{d\theta^2} + u + \left(\frac{d^2u}{d\theta^2} + u\right)2\,.\int_\theta \frac{T}{h^2u^3} - \frac{P}{h^2u^2} + \frac{T}{h^2u^3}\cdot\frac{du}{d\theta} = 0\ldots\ldots(k),$$

$$\frac{d^2s}{d\theta^2} + s + \left(\frac{d^2s}{d\theta^2} + s\right)2\,.\int_\theta \frac{T}{h^2u^3} + \frac{S - Ps}{h^2u^3} + \frac{T}{h^2u^3}\cdot\frac{ds}{d\theta} = 0\ldots\ldots(l).$$

Now, with the assumed value of u, $\dfrac{d^2u}{d\theta^2} + u = a$,

$$\int_\theta \frac{T}{h^2a^3} = \frac{3}{2(2-2m)}\cdot\frac{m'a'^3}{h^2a^4}\cos\overline{(2-2m)\theta - 2\beta}$$

$$= \frac{3}{4}\cdot\frac{m'a'^3}{h^2a^4}\cos\overline{(2-2m)\theta - 2\beta},$$

(using $\dfrac{3}{2}$ instead of $\dfrac{3}{2-2m}$, because the preservation of m in the denominator, on expanding the fraction, would introduce a term of the third order)

$$\therefore \left(\frac{d^2u}{d\theta^2} + u\right) 2\int_\theta \frac{T}{h^2 a^3} = \frac{3}{2} \cdot \frac{m' a'^3}{h^2 a^3} \cos \overline{(2 - 2m)\,\theta - 2\beta}.$$

$$\text{Then } \frac{-P}{h^2u^2} = -a\left\{1 - \frac{3k^2}{4} + \frac{3k^2}{4}\cos 2\,.\,(\theta - \gamma)\right\}$$

$$+ \frac{m' a'^3}{h^2 a^3}\left\{\frac{1}{2} + \frac{3}{2}\cos \overline{(2 - 2m)\,\theta - 2\beta}\right\}.$$

And $\frac{T}{h^2 u^2} \cdot \frac{du}{d\theta}$, since $\frac{du}{d\theta}$ involves e, is of the third order.

Hence, the equation becomes

$$0 = \frac{d^2u}{d\theta^2} + u - a + \frac{3k^2 a}{4} + \frac{m' a'^3}{2h^2 a^3} - \frac{3k^2 a}{4}\cos 2\,.\,(\theta - \gamma)$$

$$+ 3\,\frac{m' a'^3}{h^2 a^3}\cos \overline{(2 - 2m)\,\theta - 2\beta}.$$

The integral of this equation gives, by (4), $u =$

$$a - \frac{3k^2 a}{4} - \frac{m' a'^3}{2h^2 a^3} + ae\,.\cos(\theta - \alpha) - \frac{3k^2 a}{4\,(2^2 - 1)}\cos 2\,.\,(\theta - \gamma)$$

$$+ \frac{3m' a'^3}{h^2 a^3\,(\overline{2 - 2m}|^2 - 1)}\cos \overline{(2 - 2m)\,\theta - 2\beta}.$$

Or, taking the last term to the second order only, $u =$

$$a - \frac{3k^2 a}{4} - \frac{m' a'^3}{2h^2 a^3} + ae\cos(\theta - \alpha) - \frac{k^2 a}{4}\cos 2\,.\,(\theta - \gamma)$$

$$+ \frac{m' a'^3}{h^2 a^3}\cos \overline{(2 - 2m)\,\theta - 2\beta}.$$

To take away $\frac{m'}{h^2}$, we observe that a' and a are nearly the reciprocals of r' and ρ, which occur in (38): and, since

$$\frac{m'}{r'^3} = m^2 \cdot \frac{\mu}{\rho^3},$$

we have $m' a'^3 = m^2 \,.\, \mu \,.\, a^3$; $\quad \therefore \frac{m' a'^3}{h^2 a^3} = m^2 \cdot \frac{\mu}{h^2} = m^2 a.$

Thus, the value of u becomes

$$a\left\{1-\frac{3k^2}{4}-\frac{m^2}{2}+e\cos\overline{\theta-\alpha}-\frac{k^2}{4}.\cos 2.\overline{\theta-\gamma}+m^2\cos\overline{(2-2m)\theta-2\beta}\right\}.$$

The value of s is still $k.\sin(\theta-\gamma)$, all the terms of its equation being of the third order.

42. PROP. 15. To integrate the differential equations, third approximation.

Since the disturbing force is now to be taken to the third order, the value of $\frac{m'u'^3}{h^2u^3}$, or $\frac{m'a'^3}{h^2a^3}\cdot\frac{u'^3}{a'^3}\cdot\frac{a^3}{u^3}$, which occurs in $\frac{P}{h^2u^2}$, must be taken to the third order; and as $\frac{m'a'^3}{h^2a^3}$ is of the second order, $\frac{u'^3}{a'^3}$ and $\frac{a^3}{u^3}$ must each be expanded to the first power of e' and e. Putting for u the value found in Prop. 13, namely, $a(1+e\cos\overline{\theta-\alpha})$,

$$\frac{a^3}{u^3}=1-3e\cos(\theta-\alpha).$$

We will stop a moment to consider the effect of this term.

43. In consequence of the introduction of this term, our equation will have the following form;

$$0=\frac{d^2u}{d\theta^2}+u+\&c.+Ae\cos(\theta-\alpha)+\&c.$$

Its integral therefore, by (5), will contain, in the expression for u, one term of the form $-\frac{Ae}{2}\theta.\sin(\theta-\alpha)$. The peculiarity of this term is, that while all the variable terms which have yet occurred, being sines or cosines, were periodical, and never exceeded a certain value, this term contains a factor θ, which admits of increase without limit; and the value of the term, instead of being confined within certain limits, may be of any magnitude. Our assumed expression then for u, viz.,

$$a(1+e\cos\overline{\theta-\alpha}),$$

was not an approximate one, since terms will be added to it, whose value may exceed its own; and, as the operations of the last proposition were carried on upon the supposition that the assumed value of u was near the truth, the results of these operations fall to the ground.

44. But a slight alteration in the form of our assumption will extricate us from this difficulty. If we assume

$$u = a(1 + e\cos\overline{c\theta - \alpha}),$$

and suppose c to differ very little from 1,

$$\text{then } \frac{d^2u}{d\theta^2} + u = a(1 + \overline{1 - c^2}.e.\cos\overline{c\theta - \alpha}),$$

which, as far as quantities of the first order, $= a$ (since $1 - c^2$ is very small); and therefore the equation used in Prop. 13., is satisfied as well as it was before, and this expression for u may therefore, with propriety, be used for substitution in the equation including terms of the second order. And the equation of this proposition, viz.

$$0 = \frac{d^2u}{d\theta^2} + u + \&c. + Ae\cos(c\theta - \alpha) + \&c.$$

$$\text{or } 0 = a(1 + \overline{1 - c^2}.e\cos\overline{c\theta - \alpha}) + \&c. + Ae\cos(c\theta - \alpha) + \&c.$$

gives, by the comparison of coefficients of the same cosine,

$$1 - c^2 = -\frac{A}{a},$$

and there is now no arc in any part of the expression for u.

*44. It is not however to be imagined, that the peculiar form of the assumption $u = a(1 + e\cos\overline{c\theta - \alpha})$ is in any degree left to our choice. If, instead of assuming any specific form, we make $u = a(1 + w)$, and substitute this expression in the differential equation, we shall have

$$\frac{a^3}{u^3} = 1 - 3w,$$

and the equation will therefore have the form

$$0 = a\frac{d^2w}{d\theta^2} + aw + \&c. + Aw + \&c.$$

$$\text{or } 0 = \frac{d^2w}{d\theta^2} + \left(1 + \frac{A}{a}\right)w + \&c.$$

which, solved as in Article (2) or (6), gives

$$w = e.\cos\left(\theta\sqrt{1+\frac{A}{a}} - \alpha\right) + \&c.$$

$$\text{or } u = a + ae\cos\overline{c\theta - \alpha} + \&c.$$

$$\text{where } c^2 = 1 + \frac{A}{a}, \text{ or } 1 - c^2 = -\frac{A}{a},$$

the same equation as that found in (44).

45. Suppose, now, we substituted $a(1+e\cos\overline{c\theta-\alpha})$ for u, in the expression for $\frac{P}{h^2u^2}$ (f) Art. 35. The second term depending on the disturbing force is $-\frac{3}{2}.\frac{m'u'^3}{h^2u^3}.\cos 2.(\theta-\theta')$. Now u', being the reciprocal of the radius vector in the elliptical orbit which the Sun appears to describe about the Earth, will be expressed by $a'(1+e'\cos\overline{\theta'-\zeta})$, ζ being the longitude of the Sun's perigee. And as $\theta' = m\theta + \beta$, nearly, by (40),

$$u' = a'(1 + e'.\cos\overline{m\theta + \beta - \zeta}), \text{ nearly};$$

$$\therefore \frac{u'^3}{u^3} = \frac{a'^3\{1 + e'\cos\overline{m\theta+\beta-\zeta}\}^3}{a^3(1+e\cos\overline{c\theta-\alpha})^3}$$

$$= \frac{a'^3}{a^3}(1 + 3e'\cos\overline{m\theta+\beta-\zeta} - 3e\cos\overline{c\theta-\alpha}),$$

neglecting e^2, &c. in the expansion. Hence, the term in question

$$= -\frac{3}{2}.\frac{m'a'^3}{h^2a^3}.\left(1+3e'\cos\overline{m\theta+\beta-\zeta}-3e\cos\overline{c\theta-\alpha}\right).\cos 2.(\theta-\theta').$$

Now $\cos 2.(\theta-\theta') = \cos\overline{(2-2m)\theta - 2\beta}$, nearly; and, consequently, the product of $\cos(c\theta-\alpha)$, and $\cos 2.(\theta-\theta')$ will contain $\cos\overline{(2-2m+c)\theta-2\beta-\alpha}$, and $\cos\overline{(2-2m-c)\theta-2\beta+\alpha}$. But we have seen in (4), that upon solving the equation

$$0=\frac{d^2u}{d\theta^2}+u+\&c.+A\cos(b\theta+B),$$

there will be in the expression for u, a term

$$\frac{A}{b^2-1}\cos(b\theta+B).$$

If then b differs little from 1, there will be a large term in the value of u. Now $2-2m-c$ is in this case; for c very nearly $=1$; $\therefore\ 2-2m-c=1-2m$, nearly; $\therefore\ (2-2m-c)^2-1=-4m$, nearly. And, since this term in the differential equation is of the third order, it will rise in the value of u to the second order. Our integration therefore to the second order is not correct, and we must repeat it, examining all the terms of the third order, and not rejecting those in which the coefficient of θ is nearly $=1$.

46. We may also remark, that the first term of $\frac{P}{h^2u^2}$ which results from the disturbing force, $-\frac{1}{2}\cdot\frac{m'u'^3}{h^2u^3}$, will contain $\cos(\theta'-\zeta)$ or $\cos(m\theta+\beta-\zeta)$, nearly, multiplied by a quantity of the third order. Since m is not nearly $=1$, the resulting term in the expression for u will also be of the third order. But when, after determining u, we proceed to integrate the expression

$$\frac{dt}{d\theta}=\frac{1}{hu^2\left(1+2\int_\theta\frac{T}{h^2u^3}\right)^{\frac{1}{2}}},$$

upon expanding this fraction, there will be one term $C.\cos(m\theta+\beta-\zeta)$, C being a quantity of the third order, the integral of which will give, in the expression for t, a term $\frac{C}{m}\sin(m\theta+\beta-\zeta)$, the coefficient of which is of the second order. Now the principal object, in the lunar theory, is to find θ in terms of t; for which purpose, t must be found in terms of θ. It will be proper, then, to include in our equation all those terms of the third order, in which the coefficient of θ is small, as well as those in which it is nearly $= 1$.

47. Upon examining the equation for s, (l), (Art. 41.), and observing that the principal term of $\frac{S - Ps}{u^3}$ is a multiple of s, (Art. 35.), it will be seen, that the same remarks apply to it, as to that which determines u. Instead of taking $s = k \sin(\theta - \gamma)$, we must take $s = k \,.\, \sin(g\theta - \gamma)$: and must preserve, among the terms of the third order, all those in which the coefficient of θ is nearly $= 1$. We shall now integrate our equations accurately to the second order.

48. Prop. 16. To find the value of $\theta - \theta'$, $\sin 2 \,.\, (\theta - \theta')$, and $\cos 2 \,.\, (\theta - \theta')$ to the first order.

Since these, in every place in which they occur, are multiplied by a quantity of the second order, we do not want to find their values to a higher order than the first. And to find θ' in terms of θ to this order, we must find t in terms of θ, and θ' in terms of t.

$$\text{Now } \frac{dt}{d\theta} = \frac{1}{hu^2 \left(1 + 2 \int_\theta \frac{T}{h^2 u^3}\right)^{\frac{1}{2}}};$$

$$\text{which to the first order} = \frac{1}{hu^2}$$

$$= \frac{1}{ha^2} \frac{1}{\{1 + e \cos(c\theta - \alpha)\}^2}$$

$$= \frac{1}{ha^2} \{1 - 2e \cos(c\theta - \alpha)\}:$$

integrating, and supposing $t = 0$, when the Moon's mean longitude $= 0$,

$$t = \frac{1}{ha^2} (\theta - \frac{2e}{c} \sin \overline{c\theta - \alpha})$$

$$= \frac{1}{ha^2} (\theta - 2e \sin \overline{c\theta - \alpha}).$$

Also by (19), as the Sun's mean motion, since t was $= 0$, is nt, and therefore his mean longitude $= nt + \beta$ (β being his mean longitude, when $t = 0$,) and his mean anomaly, consequently, $= nt + \beta - \zeta$, we have $\theta' =$ Sun's true longitude

$$= nt + \beta + 2e' . \sin(nt + \beta - \zeta).$$

$$\text{Now } nt + \beta = \frac{n}{ha^2}(\theta - 2e \sin \overline{c\theta - \alpha}) + \beta:$$

or, since the coefficient of θ must be m,

$$nt + \beta = m\theta + \beta,$$

(neglecting $- 2me . \sin \overline{c\theta - \alpha}$, which is of the second order;)

$$\therefore \sin(nt + \beta - \zeta) = \sin(m\theta + \beta - \zeta);$$

$$\theta' = (m\theta + \beta) + 2e' . \sin(m\theta + \beta - \zeta),$$

$$\text{and } (\theta - \theta') = \overline{(1 - m)\theta - \beta} - 2e' . \sin(m\theta + \beta - \zeta):$$

$$2 . (\theta - \theta') = \overline{(2 - 2m)\theta - 2\beta} - 4e' . \sin(m\theta + \beta - \zeta),$$

which we shall call $\overline{(2 - 2m)\theta - 2\beta - p}$.

49. We have now to find to the first order

$$\sin\{\overline{(2 - 2m)\theta - 2\beta - p}\}, \text{ and } \cos\{\overline{(2 - 2m)\theta - 2\beta - p}\},$$

p being of the first order.

$$\text{Now } \sin\{\overline{(2 - 2m)\theta - 2\beta - p}\}$$

$$= \sin\overline{(2 - 2m)\theta - 2\beta} . \cos p - \cos\overline{(2 - 2m)\theta - 2\beta} . \sin p.$$

But $\cos p$, or $1 - \frac{p^2}{2} +$ &c. differs from 1 only by a quantity of the second order; and $\sin p$, or $p - \frac{p^3}{6} +$ &c. differs from p only by a quantity of the third order;

$$\therefore \sin\{\overline{(2 - 2m)\theta - 2\beta - p}\}$$

$$= \sin\overline{(2 - 2m)\theta - 2\beta} - p . \cos\overline{(2 - 2m)\theta - 2\beta}:$$

which, putting for p its value $4e'.\sin(m\theta+\beta-\zeta)$, gives

$$\sin 2.(\theta-\theta') = \sin\overline{(2-2m)\theta-2\beta}$$

$$-4e'.\sin(m\theta+\beta-\zeta).\cos\overline{(2-2m)\theta-2\beta}$$

$$=\sin\overline{(2-2m)\theta-2\beta}-2e'\sin\overline{(2-m)\theta-\beta-\zeta}$$

$$+2e'.\sin\overline{(2-3m)\theta-3\beta+\zeta}.$$

Similarly,

$$\cos 2.(\theta-\theta') = \cos\overline{(2-2m)\theta-2\beta}-2e'.\cos\overline{(2-m)\theta-\beta-\zeta}$$

$$+2e'.\cos\overline{(2-3m)\theta-3\beta+\zeta}.$$

50. Prop. 17. To find the value of $\frac{P}{h^2u^2}$ to the third order.

The first part is $\frac{\mu}{h^2}(1-\frac{3}{2}s^2)$,

$$\text{or } a\left(1-\frac{3k^2}{4}+\frac{3k^2}{4}\cos\overline{2g\theta-2\gamma}\right).$$

The second part is $-\frac{m'u'^3}{2h^2u^3}$,

$$\text{or } -\frac{m'a'^3}{2h^2a^3}.(1+e'\cos\overline{\theta'-\zeta})^3.(1+e\cos\overline{c\theta-\alpha})^{-3}$$

$$=-\frac{m'a'^3}{2h^2a^3}.(1+3e'\cos\overline{m\theta+\beta-\zeta}-3e\cos\overline{c\theta-\alpha})$$

$$=\text{as in (41.)}, -\frac{m^2a}{2}.(1+3e'\cos\overline{m\theta+\beta-\zeta}-3e\cos\overline{c\theta-\alpha}).$$

Since the coefficient of θ in the first arc is small, and in the second is nearly $=1$, all the terms must be preserved.

The third part is

$$-\frac{3}{2}.\frac{m'u'^3}{h^2u^3}.\cos 2.(\theta-\theta'),$$

$$\text{or } -\frac{3}{2}m^2a.\{1+3e'\cos\overline{m\theta+\beta-\zeta}-3e\cos\overline{c\theta-\alpha}\}$$

$$\times\{\cos\overline{(2-2m)\theta-2\beta}-2e'\cos\overline{(2-m)\theta-\beta-\zeta}$$

$$+2e'.\cos\overline{(2-3m)\theta-3\beta+\zeta}\}.$$

Multiplying together these series, and setting down all the terms of the third order, we have

$$-\frac{3}{2}m^2a\left\{\begin{aligned}&\cos\overline{(2-2m)\theta-2\beta}+\frac{3e'}{2}\cos\overline{(2-m)\theta-\beta-\zeta}\\&+\frac{3e'}{2}\cos\overline{(2-3m)\theta-3\beta+\zeta}\\&-\frac{3e}{2}\cos\overline{(2-2m+c)\theta-2\beta-\alpha}\\&-\frac{3e}{2}\cos\overline{(2-2m-c)\theta-2\beta+\alpha}\\&-2e'\cos\overline{(2-m)\theta-\beta-\zeta}\\&+2e'\cos\overline{(2-3m)\theta-3\beta+\zeta}.\end{aligned}\right.$$

Of this the only part which must be preserved, is

$$-\frac{3}{2}m^2a\{\cos\overline{(2-2m)\theta-2\beta}-\frac{3e}{2}\cos\overline{(2-2m-c)\theta-2\beta+\alpha}\},$$

because the coefficients of θ in the other terms (all which are of the third order), namely $2-m$, $2-3m$, $2-2m+c$, are neither small nor nearly equal to 1.

The remaining terms of $\frac{P}{h^2u^2}$ are of the fourth order; hence, our value of $\frac{P}{h^2u^2}$ is

$$a\,(1-\frac{3k^2}{4}+\frac{3k^2}{4}\cos\overline{2g\theta-2\gamma})$$

$$-\frac{m^2a}{2}\{1+3e'\cos\overline{m\theta+\beta-\zeta}-3e\cos\overline{c\theta-\alpha}+3\cos\overline{(2-2m)\theta-2\beta}$$

$$-\frac{9e}{2}\cos\overline{(2-2m-c)\theta-2\beta+\alpha}\}.$$

51. PROP. 18. To find the values of

$$\frac{T}{h^2u^3},\quad \int_\theta \frac{T}{h^2u^3},\quad \left(\frac{d^2u}{d\theta^2}+u\right) 2\int_\theta \frac{T}{h^2u^3},\quad \text{and}\quad \frac{T}{h^2u^3}\cdot\frac{du}{d\theta},$$

to the third order.

The only term of $\frac{T}{h^2u^3}$ to be taken here, is

$$-\frac{3}{2}\cdot\frac{m'u'^3}{h^2u^4}\cdot\sin 2\,.\,(\theta-\theta'),$$

(the others being of the fourth order), or, as in (50),

$$-\frac{3}{2}m^2\left\{1+3e'\cos\overline{m\theta+\beta-\zeta}-4e\cos\overline{c\theta-\alpha}\right\}\times$$

$$\left\{\sin\overline{(2-2m)\theta-2\beta}-2e'.\sin\overline{(2-m)\theta-\beta-\zeta}\right.$$
$$\left.+2e'.\sin\overline{(2-3m)\theta-3\beta+\zeta}\right\},$$

which

$$=-\frac{3}{2}m^2\left\{\begin{aligned}&\sin\overline{(2-2m)\theta-2\beta}+\frac{3e'}{2}\sin\overline{(2-m)\theta-\beta-\zeta}\\&+\frac{3e'}{2}\sin\overline{(2-3m)\theta-3\beta+\zeta}\\&-2e\sin\overline{(2-2m+c)\theta-2\beta-\alpha}\\&-2e\sin\overline{(2-2m-c)\theta-2\beta+\alpha}\\&-2e'\sin\overline{(2-m)\theta-\beta-\zeta}\\&+2e'\sin\overline{(2-3m)\theta-3\beta+\zeta}.\end{aligned}\right.$$

52. Now, since $u=a(1+e\cos\overline{c\theta-\alpha})$,

$$\frac{du}{d\theta}=-cae\sin(c\theta-\alpha)=-ae\sin(c\theta-\alpha),\ \text{very nearly};$$

As this is a quantity of the first order, we shall use only the first term of $\frac{T}{h^2u^3}$ to multiply with it (because the product of ae by the other terms would be of the fourth order);

$$\therefore \frac{T}{h^2u^3} \cdot \frac{du}{d\theta} = \tfrac{3}{4} m^2 a e \left\{\cos \overline{(2-2m-c)\theta - 2\beta + a}\right.$$

$$\left. - \cos \overline{(2-2m+c)\theta - 2\beta - a}\right\}.$$

The only term to be preserved, is

$$\tfrac{3}{4} am^2e \,.\, \cos \overline{(2-2m-c)\theta - 2\beta + a}.$$

53. Also $\int_\theta \frac{T}{h^2u^3}$

$$= \frac{3}{2} m^2 \left\{ \begin{aligned} &\frac{1}{2-2m} \cos \overline{(2-2m)\theta - 2\beta} \\ &- \frac{e'}{4-2m} \cos \overline{(2-m)\theta - \beta - \zeta} \\ &+ \frac{7e'}{4-6m} \cos \overline{(2-3m)\theta - 3\beta + \zeta} \\ &- \frac{2e}{2-2m+c} \cos \overline{(2-2m+c)\theta - 2\beta - a} \\ &- \frac{2e}{2-2m-c} \cos \overline{(2-2m-c)\theta - 2\beta + a}, \end{aligned} \right.$$

where the first term is of the second order, and all the others are of the third order.

And $\frac{d^2u}{d\theta^2} + u$, to the first order, is a;

$\therefore 2\left(\frac{d^2u}{d\theta^2} + u\right)\int_\theta \frac{T}{h^2u^3}$, retaining only that term of the third order which will be increased on integration, and putting $\frac{1}{2}$ for $\frac{1}{2-2m}$ in the coefficient of the term of the second order, and $2e$ for $\frac{2e}{2-2m-c}$ (45) in the last term, becomes

$$3m^2a\left\{\tfrac{1}{2}\cos \overline{(2-2m)\theta - 2\beta} - 2e\cos \overline{(2-2m-c)\theta - 2\beta + a}\right\}.$$

54. Prop. 19. To form the differential equation for u.

Collecting the terms, and substituting them in the equation of (41), we have

$$\text{First,}\quad \frac{d^2u}{d\theta^2}+u$$

$$=\frac{d^2u}{d\theta^2}+u.$$

$$\text{Second,}\quad -\left(\frac{d^2u}{d\theta^2}+u\right)2\int_\theta \frac{T}{h^2u^3}\,.$$

$$=3m^2a\left\{\tfrac{1}{2}\cos\overline{(2-2m)\theta-2\beta}\right.$$

$$\left.-2e\cos\overline{(2-2m-c)\theta-2\beta+a}\right\}\quad \text{(Prop. 18.)}$$

$$\text{Third,}\quad -\frac{P}{h^2u^2}$$

$$=-a\left(1-\frac{3k^2}{4}+\frac{3k^2}{4}\cos\overline{2g\theta-2\gamma}\right)+\frac{m^2a}{2}$$

$$+\frac{m^2a}{2}\left(3e'.\cos\overline{m\theta+\beta-\zeta}-3e\cos\overline{c\theta-a}\right)$$

$$+\frac{m^2a}{2}\left\{3\cos\overline{(2-2m)\theta-2\beta}-\frac{9e}{2}\cos\overline{(2-2m-c)\theta-2\beta+a}\right\}$$

(Prop. 17.)

$$\text{Fourth,}\quad +\frac{T}{h^2u^3}\cdot\frac{du}{d\theta}$$

$$=+\frac{3}{4}m^2ae.\cos\overline{(2-2m-c)\theta-2\beta+a}.\quad \text{(Prop. 18.)}$$

Taking the sum, by (41),

$$0=\frac{d^2u}{d\theta^2}+u-a\left(1-\frac{3k^2}{4}-\frac{m^2}{2}\right)$$

$$-\frac{3m^2ae}{2}.\cos(c\theta-a)-\frac{3k^2a}{4}\cos(2g\theta-2\gamma)$$

$$+3m^2a.\cos\overline{(2-2m)\theta-2\beta}-\frac{15}{2}.m^2ae\cos\overline{(2-2m-c)\theta-2\beta+a}$$

$$+\frac{3}{2}m^2ae'.\cos(m\theta+\beta-\zeta).$$

55. Prop. 20. To integrate accurately to the second order the differential equation for u.

$$\text{Assume } u = a\left\{1 - \frac{3k^2}{4} - \frac{m^2}{2} + e \,.\cos(c\theta - \alpha)\right.$$

$$+ A\cos(2g\theta - 2\gamma) + B\cos\overline{(2-2m)\theta - 2\beta}$$

$$\left. + C\cos\overline{(2-2m-c)\theta - 2\beta + \alpha} + D\cos(m\theta + \beta - \zeta)\right\},$$

according to the direction in (6).

Substituting this value in the differential equation, and making $=0$ the coefficient of each cosine,

$$ae(1-c^2) - \frac{3m^2 ae}{2} = 0;$$

$$\therefore\ c^2 = 1 - \frac{3m^2}{2},\quad c = 1 - \frac{3m^2}{4},\ \text{nearly}.$$

$$aA(1-4g^2) - \frac{3k^2 a}{4} = 0;$$

$$\therefore\ A = \frac{3k^2}{4(1-4g^2)} = -\frac{k^2}{4},\ \text{nearly, since } g \text{ nearly} = 1.$$

$$aB(1 - \overline{2-2m}|^2) + 3m^2 a = 0;$$

$$\therefore\ B = \frac{3m^2}{(2-2m)^2 - 1} = m^2,\ \text{nearly}.$$

$$aC(1 - \overline{2-2m-c}|^2) - \frac{15}{2}m^2 ae = 0;$$

$$\therefore\ C = \frac{15}{2}\cdot\frac{m^2 e}{1-(2-2m-c)^2} = \frac{15}{2}\cdot\frac{m^2 e}{1-(1-2m)^2},\ \text{nearly},$$

$$(\text{since } c \text{ nearly} = 1) = \frac{15}{8}me,\ \text{nearly}.$$

$$aD(1-m^2) + \frac{3}{2}m^2 ae' = 0;$$

$$\therefore\ D = -\frac{3}{2}\cdot\frac{m^2 e'}{1-m^2} = -\frac{3}{2}m^2 e',\ \text{nearly}.$$

And u therefore $= a\left\{1 - \frac{3k^2}{4} - \frac{m^2}{2} + e\cos(c\theta - \alpha)\right.$

$$- \frac{k^2}{4}\cos(2g\theta - 2\gamma) + m^2\cos\overline{(2-2m)\theta - 2\beta}$$

$$\left. + \frac{15}{8} me\cos\overline{(2-2m-c)\theta - 2\beta + \alpha} - \frac{3}{2}m^2e'.\cos(m\theta + \beta - \zeta)\right\}.$$

56. Prop. 21. To form the differential equation for s.

First, s, as we have observed, will be approximately represented by $k\sin(g\theta - \gamma)$, where g differs little from 1; the difference, which is caused entirely by the disturbing force, being of the second order. Hence,

$$\frac{d^2 s}{d\theta^2} + s, \text{ or } k(1 - g^2)\sin(g\theta - \gamma),$$

will be small, of the third order: consequently, the term

$$\left(\frac{d^2 s}{d\theta^2} + s\right) 2\int_\theta \frac{T}{h^2 u^3},$$

in equation (l) of Art. 41, will be of the fifth order, and is not to be considered.

Second, $\dfrac{S - Ps}{h^2 u^3}$, or $m's\dfrac{u'^3}{u^4}\left(\dfrac{3}{2} + \dfrac{3}{2}\cos 2.\overline{\theta - \theta'}\right)$

$$= \frac{m'a'^3}{h^2 a^4}\cdot\frac{(1 + e'\cos\overline{m\theta + \beta - \zeta})^3}{(1 + e\cos\overline{c\theta - \alpha})^4}.k\sin\overline{g\theta - \gamma}.\left(\frac{3}{2} + \frac{3}{2}\cos 2.\overline{\theta - \theta'}\right).$$

Now $\dfrac{m'a'^3}{h^2a^4}k$ or m^2k, form a product of the third order: hence, in the quantities which multiply them, all small terms are to be rejected;

$$\therefore \frac{S - Ps}{h^2 u^3} = m^2k.\sin(g\theta - \gamma)\left\{\frac{3}{2} + \frac{3}{2}\cos\overline{(2-2m)\theta - 2\beta}\right\}$$

$$= m^2k\left\{\frac{3}{2}\sin(g\theta - \gamma) + \frac{3}{4}\sin\overline{(2 - 2m + g)\theta - 2\beta - \gamma}\right.$$

$$\left. - \frac{3}{4}\sin\overline{(2 - 2m - g)\theta - 2\beta + \gamma}\right\}.$$

The only terms to be preserved are

$$m^2k\left\{\frac{3}{2}\sin(g\theta-\gamma)-\frac{3}{4}\sin\overline{(2-2m-g)\theta-2\beta+\gamma}\right\};$$

because g is very nearly $=1$, and $2-2m-g=1-2m$, nearly, and therefore differs little from 1: and therefore, by (45) and (47), both these terms, on integration, will be important.

Third, $\frac{ds}{d\theta}=kg\,.\cos(g\theta-\gamma)=k\,.\cos(g\theta-\gamma)$ nearly,

which is of the first order; taking therefore the first term only of the expression for $\frac{T}{h^2u^3}$, that is, (51)

$$-\frac{3}{2}m^2.\sin\overline{(2-2m)\theta-2\beta},$$

we have

$$\frac{T}{h^2u^3}\cdot\frac{ds}{d\theta}=-\frac{3}{2}m^2k\,.\cos(g\theta-\gamma)\,.\sin\overline{(2-2m)\theta-2\beta}$$

$$=-\frac{3}{4}m^2k\left\{\sin\overline{(2-2m+g)\theta-2\beta-\gamma}+\sin\overline{(2-2m-g)\theta-2\beta+\gamma}\right\}.$$

The only term to be preserved is

$$-\frac{3}{4}m^2k\,.\sin\overline{(2-2m-g)\theta-2\beta+\gamma}.$$

Collecting these parts, the equation of (41) becomes

$$0=\frac{d^2s}{d\theta^2}+s+\frac{3}{2}m^2k\,.\sin(g\theta-\gamma)-\frac{3}{2}m^2k\sin\overline{(2-2m-g)\theta-2\beta+\gamma}.$$

57. PROP. 22. To integrate the differential equation for s.

Assume $s=k\left\{\sin\overline{g\theta-\gamma}+A\sin\overline{(2-2m-g)\theta-2\beta+\gamma}\right\}$, and substitute in the equation above: then, making $=0$ the coefficient of each sine,

$$k(1-g^2)+\frac{3}{2}m^2k=0;\qquad\therefore\ g^2=1+\frac{3}{2}m^2;$$

$$kA(1 - \overline{2 - 2m - g}|^2) - \frac{3}{2} m^2 k = 0;$$

$$\therefore \; A = \frac{3}{2} \cdot \frac{m^2}{1 - (2 - 2m - g)^2}$$

$$= \frac{3}{2} \cdot \frac{m^2}{1 - (1 - 2m)^2} \text{ nearly } = \frac{3}{2} \cdot \frac{m^2}{4m} \text{ nearly } = \frac{3m}{8};$$

$$\therefore \; s = k \left\{ \sin \overline{g\theta - \gamma} + \frac{3m}{8} \cdot \sin \overline{(2 - 2m - g)\theta - 2\beta + \gamma} \right\},$$

to the second order.

58. PROP. 23. To find t in terms of θ to the second order.

We must expand $\frac{dt}{d\theta}$ or $\dfrac{1}{hu^2 \left(1 + 2 \int_\theta \frac{T}{h^2 u^3}\right)^{\frac{1}{2}}}$ to the second order, preserving those terms of the third order in which the coefficient of θ is small. Putting then for u the value found in Prop. 20, and for $\int_\theta \frac{T}{h^2 u^3}$ the first term of its value in Prop. 18,

$$\frac{dt}{d\theta} = \frac{1}{ha^2} \left\{ 1 + \frac{3k^2}{2} + m^2 - 2e \cos (c\theta - a) \right.$$

$$+ \frac{3e^2}{2} + \frac{3e^2}{2} \cos (2c\theta - 2a) + \frac{k^2}{2} \cos (2g\theta - 2\gamma)$$

$$- 2m^2 \cos \overline{(2 - 2m)\theta - 2\beta} - \frac{15}{4} me \cos \overline{(2 - 2m - c)\theta - 2\beta + a}$$

$$\left. + 3m^2 e' \cos (m\theta + \beta - \zeta) - \frac{3}{4} m^2 \cos \overline{(2 - 2m)\theta - 2\beta} \right\}.$$

Integrating this, and taking the coefficients to the second order only,

$$t = \frac{1}{ha^2} \left\{ \left(1 + \frac{3k^2}{2} + m^2 + \frac{3e^2}{2}\right)\theta - 2e \sin (c\theta - a) \right.$$

$$+ \frac{3e^2}{4} \sin (2c\theta - 2a) + \frac{k^2}{4} \sin (2g\theta - 2\gamma)$$

$$-\frac{11}{8}m^2\sin\overline{(2-2m)\theta-2\beta}$$

$$-\frac{15}{4}me.\sin\overline{(2-2m-c)\theta-2\beta+\alpha}+3me'\sin(m\theta+\beta-\zeta)\Big\},$$

$$\text{Let } \frac{ha^2}{1+\frac{3k^2}{2}+m^2+\frac{3e^2}{2}}=p: \text{ then,}$$

$$pt=\theta-2e\sin(c\theta-\alpha)+\frac{3e^2}{4}\sin(2c\theta-2\alpha)$$

$$+\frac{k^2}{4}\sin(2g\theta-2\gamma)-\frac{11}{8}m^2.\sin\overline{(2-2m)\theta-2\beta}$$

$$-\frac{15}{4}me.\sin\overline{(2-2m-c)\theta-2\beta+\alpha}+3me'.\sin(m\theta+\beta-\zeta),$$

omitting the divisor $1+\frac{3k^2}{2}+m^2+\frac{3e^2}{2}$ in all the terms after the first, as their value is not sensibly altered by it.

59. Prop. 24. To find θ in terms of t to the second order.

This must be done by Lagrange's theorem. Applying it, we find

$$\theta=pt+2e.\sin(cpt-\alpha)+\frac{5e^2}{4}\sin(2cpt-2\alpha)$$

$$-\frac{k^2}{4}\sin(2gpt-2\gamma)+\frac{11}{8}m^2.\sin\overline{(2-2m)pt-2\beta}$$

$$+\frac{15}{4}me.\sin\overline{(2-2m-c)pt-2\beta+\alpha}-3me'.\sin(mpt+\beta-\zeta).$$

60. Prop. 25. To find an expression for the Moon's parallax.

The Moon's parallax varies inversely as her distance, that is, it $\propto\frac{1}{r}$,

$$\text{or } \propto\frac{1}{\rho(1+s^2)^{\frac{1}{2}}},\quad \text{or } \propto\frac{u}{(1+s^2)^{\frac{1}{2}}},\quad \text{or } \propto u\left(1-\frac{s^2}{2}\right),$$

$$\text{or} \propto u\left\{1 - \frac{k^2}{4} + \frac{k^2}{4}\cos(2g\theta - 2\gamma)\right\}.$$

Taking for u the expression found in Prop. 20, parallax

$$\propto a\left\{1 - k^2 - \frac{m^2}{2} + e\cos(c\theta - \alpha) + m^2\cos\overline{(2-2m)\theta - 2\beta}\right.$$

$$\left. + \frac{15}{8}me \,.\, \cos\overline{(2 - 2m - c)\theta - 2\beta + \alpha}\right\}:$$

omitting the term $-\frac{3}{2}m^2e' \,.\, \cos(m\theta + \beta - \zeta)$ which is of the third order; or, if P be the mean parallax, that is, that part of the expression independent of cosines, the parallax

$$= P\left\{1 + e\cos(c\theta - \alpha) + m^2 \,.\, \cos\overline{(2-2m)\theta - 2\beta}\right.$$

$$\left. + \frac{15}{8}me \,.\, \cos\overline{(2 - 2m - c)\theta - 2\beta + \alpha}\right\}.$$

61. Prop. 26. To explain the effect of the different terms in these expressions.

The first and greatest inequality of parallax is

$$e\cos \overline{c\left(\theta - \frac{\alpha}{c}\right)}.$$

This, though similar to the inequality which would exist in an elliptic orbit, is not exactly the same, but it would be the same if it depended on the angle $\theta - \frac{\alpha}{c}$ instead of $c\left(\theta - \frac{\alpha}{c}\right)$. Let then EA, fig. 4, be the Moon's least distance: EM any other distance: $\angle AEM = \theta - \frac{\alpha}{c}$: let AmB be an ellipse, whose latus rectum is $\frac{1}{a}$, and eccentricity e: take $\angle AEm = c \times \angle AEM$ (c being < 1): then, EM will $= Em$. For

$$\frac{1}{Em} = a(1 + e\cos AEm) = a(1 + e\cos\overline{c\theta - \alpha}) = \frac{1}{EM}.$$

If now an ellipse aMb be described similar and equal to AmB, whose major axis ab is inclined to AB at an angle equal to mEM, or $(1-c)\left(\theta-\frac{\alpha}{c}\right)$, M will evidently be found in its circumference, and the arc aM will be $=Am$. Hence, the motion of M may be represented by supposing it to move in an ellipse, and supposing that ellipse to revolve in the same direction, with an angular velocity which is to the whole angular velocity of M as $1-c : 1$. The perigee of the Moon's orbit therefore is not fixed, but (while we neglect the other terms of the parallax) moves almost uniformly in the direction of the Moon's motion. (*Newton*, Lib. I, Prop. 66, Cor. 7.)

62. For the explanation of the next term,

$$m^2 . \cos \overline{(2-2m)\theta - 2\beta},$$

we observe that $\overline{(1-m)\theta-\beta}$ (Prop. 15.) is nearly the difference of longitude of the Sun and Moon, and consequently,

$$\cos \overline{(2-2m)\theta - 2\beta}$$

is greatest, when the Moon is in syzygies, and least (or has its greatest negative value), when she is in quadratures; and between these situations, it has all the intermediate values. The Moon's parallax therefore, setting aside the elliptic inequality, is greatest in syzygies, and least in quadratures, and therefore her distance is least in syzygies, and greatest in quadratures. (*Newton*, Prop. 66, Cor. 5.)

63. The next term,

$$\frac{15}{8} me \cos \overline{(2-2m-c)\theta - 2\beta + \alpha},$$

shews its influence by the alteration of the eccentricity of the Moon's orbit. To prove this, let us suppose that when $\theta=0$, the Moon, and the axis of the Moon's orbit, were in syzygies: that is, suppose $\alpha=0$ (which makes the anomaly or $c\theta-\alpha=0$ when $\theta=0$); and suppose $\beta=0$, or 180°, (which makes the Sun's longitude, or $m\theta+\beta$, $=0$ or 180°, when $\theta=0$): then, (considering only the elliptic inequality, and the present one) parallax

$$= P\left\{1 + e\cos c\theta + \frac{15}{8} me \cos \overline{(2-2m-c)\theta}\right\}:$$

and, since c and $2-2m-c$ each nearly $=1$, the Moon's parallax, for one revolution, will be nearly represented by the expression

$$P\left(1+\overline{e+\frac{15}{8}me}\,.\cos\theta\right),$$

which is the same as in an orbit, whose eccentricity

$$=e\left(1+\frac{15}{8}m\right).$$

Again, suppose that, when $\theta=0$, the Moon is at her apse, and the Sun's longitude $=-90^0$, or $\beta=-90^0$: then, when the Moon's longitude $=\theta$, her parallax

$$=P\left\{1+e\cos c\theta+\frac{15}{8}me\cos\overline{(2-2m-c)\theta+180^0}\right\}:$$

which, for a single revolution, is nearly represented by

$$P\left\{1+e\cos\theta+\frac{15}{8}me\cos(\theta+180^0)\right\},$$

$$\text{or}\quad P\left(1+\overline{e-\frac{15}{8}me}\,.\cos\theta\right):$$

and here the apparent eccentricity is $e\left(1-\frac{15}{8}m\right)$. The eccentricity therefore is increased by this term when the axis of the Moon's orbit is in syzygies, and diminished when it is in quadratures (*Newton*, Prop. 66, Cor. 9.): its effect in intermediate situations we shall consider presently.

64. Corresponding to these inequalities of parallax, there are, in the expression for the Moon's longitude, the terms

$$2e\,.\sin(cpt-a)+\frac{5e^2}{4}\sin(2cpt-2a),$$

$$\frac{11}{8}m^2\sin\overline{(2-2m)pt-2\beta},$$

$$\text{and}\quad\frac{15}{4}me\,.\sin\overline{(2-2m-c)pt-2\beta+a},\quad\text{(Prop. 24.)}.$$

The first two depend on the eccentricity e and the mean distance from the perigee $cpt - \alpha$: their sum constitutes the *elliptic inequality* in longitude. The next term, which is called the *variation*, is proportional to $\sin 2 . (pt - mpt + \beta)$, or sin 2 (Moon's mean longitude − Sun's mean longitude). It is therefore = 0 when the Moon is in conjunction; it is greatest and positive when the difference of longitudes $= 45^0$; it is again = 0 when the difference $= 90^0$, or when the Moon is at quadrature: it has its greatest negative value when that difference $= 135^0$; and is again = 0 when the difference of longitudes $= 180^0$, or when the Moon is in opposition. The Moon's true place therefore is before the mean place from syzygy to quadrature, and behind it from quadrature to syzygy. (*Newton*, Lib. III. Prop. 29.) The last term, depending on $\sin (2 - 2m - c) pt - 2\beta + \alpha$, is called the *evection:* it appears to increase the elliptic inequality when the axis of the Moon's orbit is in syzygies, and to diminish it when that axis is in quadratures: the reasoning of the last article applies to it in every respect.

There are, besides, (Prop. 24.) these terms,

$$-\frac{k^2}{4} (\sin 2gpt - 2\gamma), \text{ and } -3me' \sin (mpt + \beta - \zeta).$$

The former of these depends upon the Moon's distance from the mean place of her node, and is nearly the difference between her longitude, measured on her orbit, and her longitude, measured on the ecliptic: it is called the *reduction.* The latter depends on the Sun's mean anomaly: it appears that, while the Sun (apparently) goes from perigee to apogee, the Moon's true place is behind her mean place: while the Sun goes from apogee to perigee, the Moon's true place is before her mean place. (*Newton*, Prop. 66. Cor. 6.) This is called the *annual equation.* The alteration in the parallax, from this cause, is very small, being of the third order (see the last term in the expression for u, Art. 55.).

65. In respect of magnitude, the evection is far the most important of the inequalities which are produced by the disturbing force of the Sun. And its effect on the

position and eccentricity of the Moon's orbit is so remarkable, that we shall here consider it a little more generally than in Art. 63.

66. PROP. 27. To determine the change in the position of the axis, and in the eccentricity of the Moon's orbit, produced by the evection.

The elliptic inequality and evection, in the expression for u, are together represented by

$$e\left\{\cos(c\theta - \alpha) + \frac{15}{8} m \cos \overline{(2 - 2m - c)\theta - 2\beta + \alpha}\right\},$$

where α = longitude of perigee if there were no evection, β = longitude of the Sun when $\theta = 0$. During a part of one revolution, we may, without great error, suppose the perigee and the Sun to be stationary: then, for $c\theta - \alpha$, we must put $\theta - \alpha$; and for $\overline{(2-2m)\theta - 2\beta - (c\theta - \alpha)}$, or twice the distance of the Sun and Moon – the Moon's anomaly, we must put

$$(2\theta - 2\beta) - (\theta - \alpha) = (\theta - \alpha) + (2\alpha - 2\beta).$$

And the united inequalities

$$= e\left\{\cos(\theta - \alpha) + \frac{15}{8} m \,.\, \cos \overline{(\theta - \alpha) + 2(\alpha - \beta)}\right\}$$

$$= e\left\{1 + \frac{15}{8} m \,.\, \cos 2(\alpha - \beta)\right\} . \cos(\theta - \alpha)$$

$$- e \,.\, \frac{15}{8} m \,.\, \sin 2(\alpha - \beta) \,.\, \sin(\theta - \alpha).$$

This may be put under the form $E \cos(\theta - \alpha + \delta)$, if

$$E \cos\delta = e\left\{1 + \frac{15}{8} m \,.\, \cos 2(\alpha - \beta)\right\},$$

$$E \sin\delta = e\,\frac{15}{8} m \,.\, \sin 2(\alpha - \beta).$$

From these equations

$$\delta \text{ or } \tan\delta = \frac{\frac{15}{8} m \sin 2(\alpha - \beta)}{1 + \frac{15}{8} m \cos 2(\alpha - \beta)}$$

$$= \frac{15}{8} m \sin 2(\alpha - \beta) \text{ nearly},$$

$$E = e\sqrt{\left\{1 + \frac{15}{8} m \cos 2(\alpha - \beta)\right\}^2 + \left\{\frac{15}{8} m \sin 2(\alpha - \beta)\right\}^2}$$

$$= e\left\{1 + \frac{15}{8} m \cos 2(\alpha - \beta)\right\} \text{ nearly}.$$

And the united inequalities are represented by

$$e\left\{1 + \frac{15}{8} m . \cos 2(\alpha - \beta)\right\} \cos\left\{\theta - \alpha + \frac{15}{8} m . \sin 2(\alpha - \beta)\right\}.$$

This is the same as the expression for the elliptic inequality in an orbit whose eccentricity

$$= e\left\{1 + \frac{15}{8} m . \cos 2(\alpha - \beta)\right\},$$

and in which the longitude of the perigee

$$= \alpha - \frac{15}{8} m . \sin 2(\alpha - \beta).$$

Hence, to find the Moon's place, when we have found the longitude of the perigee on the supposition of its uniform progression, we must subtract from that longitude

$$\frac{15}{8} m . \sin 2 (\text{long. perigee} - \text{long. Sun}),$$

and apply the equation due to an elliptic orbit, whose eccentricity

$$= e\left\{1 + \frac{15}{8} m . \cos 2 (\text{long. perigee} - \text{long. Sun})\right\}.$$

(*Newton*, Lib. III. *Scholium to the Lunar Theory.*)

67. Prop. 28. To explain the effect of the terms in the expression for s.

The first of these is $k.\sin g\left(\theta-\frac{\gamma}{g}\right)$. If this depended on $\theta-\frac{\gamma}{g}$, it would shew that the Moon moved in a plane (39). But, as it depends on $g\theta-\gamma$, or $\theta-(\gamma+\overline{g-1}\,\theta)$, the Moon's motion in latitude may be represented by supposing her to move in a plane, the tangent of whose inclination to the ecliptic is k, and supposing the intersection of this plane with the ecliptic to move with a retrograde motion which is to the whole motion of the Moon as $g-1 : 1$, and which therefore is nearly uniform. This is exactly analagous to the motion of the perigee in (61), with the single difference, that c being <1, and $g>1$, the motion in one case is direct, and in the other, retrograde.

68. The second term

$$k\,.\,\frac{3m}{8}\,.\,\sin\overline{(2-2m-g)\theta-2\beta+\gamma},$$

has precisely the same relation to the first, which the evection has to the elliptic inequality; and the alteration which it produces in the place of the node and the inclination of the orbit, may be found in the same manner. Thus, γ is the longitude of the node, if the second term did not exist, and β the longitude of the Sun, when $\theta=0$. Now, during the description of a portion only of the orbit, we may, without material error, suppose in our expressions, that the Sun and node are stationary: then, for $g\theta-\gamma$, the Moon's distance from the node, we must put $\theta-\gamma$; for $\overline{(2-2m)\theta-2\beta}$, double the excess of the Moon's longitude above the Sun's, we must put $2\theta-2\beta$; and for $\overline{(2-2m-g)\theta-2\beta+\gamma}$, we must put

$$(2\theta-2\beta-\theta+\gamma),\ \text{ or } (\theta-\gamma)+2\,.\,(\gamma-\beta).$$

Hence, $s=k\left\{\sin(\theta-\gamma)+\frac{3m}{8}\sin\overline{\theta-\gamma+2(\gamma-\beta)}\right\}$

$$=k\left\{1+\frac{3m}{8}\cos 2\,.\,\overline{\gamma-\beta}\right\}.\sin\overline{\theta-\gamma}+k\frac{3m}{8}\sin 2\,.\,\overline{\gamma-\beta}\,.\cos\overline{\theta-\gamma}.$$

This may be put under the form $K \sin(\theta - \gamma + \kappa)$,

$$\text{if } K \sin \kappa = k \frac{3m}{8} \sin 2(\gamma - \beta),$$

$$K \cos \kappa = k \left\{1 + \frac{3m}{8} \cos 2 . (\gamma - \beta)\right\}.$$

These equations give

$$\kappa \text{ or } \tan \kappa = \frac{3m}{8} \sin 2 . (\gamma - \beta),$$

$$K = k \left\{1 + \frac{3m}{8} \cos 2 . (\gamma - \beta)\right\},$$

$$\text{and } s = k \left\{1 + \frac{3m}{8} \cos 2 . (\gamma - \beta)\right\} . \sin\left\{\theta - \gamma + \frac{3m}{8} \sin 2 . (\gamma - \beta)\right\}.$$

This is the same expression that we should have had, if the longitude of the node were

$$\gamma - \frac{3m}{8} \sin 2 . (\gamma - \beta),$$

and the tangent of the inclination of the orbit

$$k \left\{1 + \frac{3m}{8} \cos 2 . (\gamma - \beta)\right\}.$$

When therefore the longitude of the node is found on the supposition of its uniform retrogradation, we must subtract from it

$$\frac{3m}{8} . \sin 2 (\text{long. node} - \text{long. Sun});$$

and, taking that for the true longitude of the node, we must suppose the tangent of inclination

$$= k \left\{1 + \frac{3m}{8} \cos 2 (\text{long. node} - \text{long. Sun})\right\}.$$

(*Newton*, Lib. III. Prop. 33. and 35.)

69. From this expression it is evident that the inclination of the orbit is greatest when $\gamma - \beta = 0$ or 180^0, that is, when the line of nodes is in syzygies; and least, when $\gamma - \beta = 90^0$ or 270^0, that is, when the line of nodes is in quadratures. (*Newton*, Prop. 66. Cor. 10.)

70. In the same manner in which we have approximated to the values of u and θ to the second order, we might go on to the third and higher orders. For the third order, it would be necessary to examine the terms of the equation to the fourth order, and thus the last terms, in the expressions for $\frac{P}{h^2u^2}$, &c., in Art. 35, would be employed. As the method of conducting all these approximations must be the same, we shall here mention the several steps.

(1) From the last approximate value, find t in terms of θ.

(2) Since θ' is, by the elliptic theory, found in terms of t, it can be expressed in terms of θ, and $\theta - \theta'$, $2(\theta - \theta')$, &c. can be expressed. u' also, which is known in terms of θ', can be expressed in terms of θ.

(3) Find expressions for $\sin 2(\theta - \theta')$, $\cos 2(\theta - \theta')$, &c. to as many orders as may be necessary.

(4) Substitute these values, and the last approximate value of u, in the expressions for $\frac{P}{h^2u^2}$, &c.

(5) When these are substituted in the equation, integrate it, as for the second order.

(6) Proceed in the same way in every respect, for the determination of s.

71. In carrying the approximation to higher orders, it frequently happens, that a term will rise, by integration, two orders*. This renders the operations very troublesome, and particular methods are sometimes necessary: but we cannot stop to explain them here.

72. We shall here mention some of the most interesting results of the next approximation. (1) The last terms in the expressions for $\frac{P}{h^2u^2}$, &c. introduce into the equation for u

* An instance of this will be seen in the Tract on the Figure of the Earth, Art. 71.

multiples of $m^2 . \frac{a'}{a} . \frac{E-M}{\mu} . \cos \overline{(1-m)\theta - \beta}$, of the fourth order; which, upon integration, rise to the third order (the coefficient of θ differing little from 1); and the corresponding inequality in longitude is of the third order. The comparison of this observed inequality with its computed value, gives us the means of determining $\frac{a'}{a}$ $\left(\frac{M}{E}\right.$ being known pretty exactly, and therefore, $\frac{E-M}{\mu}$ or $\frac{E-M}{E+M}$ being known$\left.\right)$, that is, the ratio of the Sun's parallax to the Moon's. The latter of these is very well known from observation: hence, the former can be found. Its quantity thus determined agrees very exactly with that determined by transits of Venus.

(2) The value of c, found by the second approximation, is $1 - \frac{3m^2}{4}$, which gives, for the progression of the perigee in a revolution of the Moon, $\frac{3m^2}{4} . 2\pi = 1^0 . 30'$. This is about half its true quantity, which is $3^0 . 2' . 22''$. By continuing the approximation, we find that c is expressed by a slowly converging series, of which the first terms are

$$1 - \frac{3}{4} m^2 - \frac{225}{32} m^3,$$

and the progression of the perigee in one revolution is

$$2\pi . \left(\frac{3}{4} m^2 + \frac{225}{32} m^3 + \&c.\right).$$

On calculating a sufficient number of terms, the theoretical value of the progression of the perigee is found to agree most accurately with the observed value.

The series for g, on the contrary, converges fast; it is

$$1 + \frac{3}{4} m^2 - \frac{9}{32} m^3 + \&c.$$

and the regression of the node in one revolution is

$$2\pi . \left(\frac{3}{4} m^2 - \frac{9}{32} m^3 + \&c.\right),$$

the computed value of which also agrees well with the observed value. The ratio of the motion of the perigee to the motion of the node, which is expressed by the fraction

$$\frac{\frac{3}{4}m^2 + \frac{225}{32}m^3 + \&c.}{\frac{3}{4}m^2 - \frac{9}{32}m^3 + \&c.}, \text{ or } \frac{1 + \frac{75}{8}m + \&c.}{1 - \frac{3}{8}m + \&c.},$$

is much greater for the Moon, where $m = \frac{1}{13}$, than for one of Jupiter's satellites,* where m is extremely small. This is alluded to by *Newton*, Lib. III. Prop. 23.

73. Upon continuing the approximations, it appears that p, the coefficient of t in the Moon's mean longitude, depends upon e', and consequently, an alteration in e' produces an alteration in the Moon's mean motion. Now e', the eccentricity of the Sun's or Earth's orbit, is slowly diminishing from the attraction of the other planets, and this causes an increase in the Moon's mean motion. It is remarkable, that the indirect effect on the Moon is much greater than the direct effect on the Earth.

74. The coefficients of inequalities of a high order cannot easily be calculated from theory. Even Laplace supposed it necessary, after finding the forms of the inequalities from theory, to discover the coefficients† from observation. Proceeding on this principle, he suggested that there must be, in the expression for the time in terms of the Moon's longitude, a term of the form

$$\sin \overline{(3 - 2g - c)\,\theta + \alpha + 2\gamma - 3\zeta}.$$

This could result only from the multiplication of sines or cosines of these arcs; $\overline{(3 - 3m)\,\theta - 3\beta}$, $(3m\theta + 3\beta - 3\zeta)$,

* That is, supposing the motions of the perigee and node of Jupiter's satellites to be caused by the disturbing force of the Sun. In reality, the principal part of these motions is occasioned by the oblate form of Jupiter.

† In Damoiseau's tables, and in Plana's and Lubbock's treatises, the coefficients are calculated from theory.

$(2g\theta - 2\gamma)$, $(c\theta - \alpha)$. The first would have for coefficient some multiple of $m^2 . \frac{a'}{a}$, which would therefore be of the fourth order: the second, some multiple of e'^3, which is of the third order: the third a multiple of k^2, of the second order: and the fourth a multiple of e, of the first order. Hence, the coefficient of

$$\sin \overline{(3 - 2g - c)\theta + \alpha + 2\gamma - 3\zeta}$$

in $\frac{T}{h^2u^3}$ would be of the tenth order. But, in the expression for

$$t = \int_\theta \frac{1}{hu^2\left(1 + 2\int_\theta \frac{T}{h^2u^3}\right)^{\frac{1}{2}}},$$

this would be twice integrated, and its coefficient would therefore be divided by $(3 - 2g - c)^2$. Now, by continued approximations, it is found that $c = 0{,}991548$, $g = 1{,}004022$; therefore $3 - 2g - c = 0{,}000407$, and the divisor is $(0{,}000407)^2$. It was supposed that the term would be so much increased by the exceeding smallness of this divisor, that it would become numerically sensible in the expression for t in terms of θ, and therefore that (on expressing θ in terms of t by means of Lagrange's theorem) there would be a sensible term of corresponding form in the expression for the true longitude in terms of the time. Observations appeared to shew that the coefficient was 15″: and this term, on the authority of Laplace, was adopted in Bürg's Lunar Tables. Further investigations however, by Plana and others, appear to shew that it is theoretically insensible.

75. In the preceding investigations, we have supposed the Sun's perigee, or the Earth's perihelion, to be stationary. It has in reality a slow progressive motion, which will be represented by putting in all the expressions $c'\theta' - \zeta$ for $\theta' - \zeta$, where $c' = 0{,}999990779$.

PLANETARY THEORY.

76. In the same manner in which the Su
(apparently) round the Earth disturbs the
round the Earth, each of the planets revolving round the Sun disturbs every other planet revolving round the Sun. But though the principle of perturbation is the same in both cases, the circumstances upon which the nature and magnitude of its effect depend are so different as to require investigations very different in form.

The disturbing force on the Moon is a resolved part of the Sun's attraction, that is, of the attraction of the largest body in the solar system. The disturbing force on a planet is the attraction of another planet, and the largest of these is very inconsiderable when compared with the Sun. The perturbations of the Moon are consequently very much greater than those of any planet: and this circumstance alone makes an important difference in the process of calculation. For the disturbing force depends itself upon the situation of the disturbed body. Now the disturbance of a planet is so small that for all ordinary purposes we may, in calculating the disturbing force, use the place in which the planet would have been if not disturbed, instead of that where it really is, without producing any sensible error. But we should in the lunar theory obtain most inaccurate results, if we expressed the forces in terms of any thing but the *true* co-ordinates of the Moon. The motion of the perihelion of a planet is scarcely sensible in ten years: the Moon's perigee has in that time gone through every sign of the zodiac. In this respect then the planetary theory is simpler than the lunar theory.

On the other hand we must observe that the expansions of P, T, S, or the equivalent expressions, are much shorter in the lunar than in the planetary theory. For these expansions proceed by powers of $\frac{\rho}{r}$ (35): now $\frac{\rho}{r}$ in the lunar theory is about $\frac{1}{400}$, but in the planetary theory it sometimes approaches to $\frac{3}{4}$. And while two terms are sufficient in the former case, thirteen or more are necessary in the latter. Here then the planetary theory is much more complicated than the lunar theory.

In following for this investigation a process somewhat different from that of the former, we shall introduce the reader to some of the most remarkable methods of modern analysts. To avoid complexity, we shall suppose that all the planets move in the same plane.

77. Prop. 29. To express the disturbing force of one planet upon another, by partial differential coefficients (taken with respect to the co-ordinates of the disturbed planet) of a function of their co-ordinates.

Let x and y be the co-ordinates of the disturbed planet (which we shall call m,) the Sun being the origin, and x being measured towards the first point of Aries: x' and y' those of the disturbing planet: m' the mass of the latter (expressed by the acceleration which it would produce at distance 1 in the unit of time). The difference of the co-ordinates of the two planets is $x'-x$ and $y'-y$ in the directions of x and y: and their distance is

$$\sqrt{(x'-x)^2+(y'-y)^2}.$$

The accelerating force of m' upon m is

$$\frac{m'}{(x'-x)^2+(y'-y)^2}$$

in the direction of the line joining them: the resolved parts of which are

$$\frac{m'(x'-x)}{\{(x'-x)^2+(y'-y)^2\}^{\frac{3}{2}}} \text{ and } \frac{m'(y'-y)}{\{(x'-x)^2+(y'-y)^2\}^{\frac{3}{2}}}$$

in the directions of x and y. These however are not the forces which *disturb* the planet. To obtain these, we must, as in (9) and (25), find the forces which m' exerts upon the Sun, and apply them with sign changed to the disturbed planet: the expressions will then represent the forces which disturb the *relative* motions of the Sun and planet, and which are in fact the objects of this investigation. The forces exerted by the planet on the Sun are; in the direction of x, $\dfrac{m'x'}{(x'^2+y'^2)^{\frac{3}{2}}}$, and in the direction of y, $\dfrac{m'y'}{(x'^2+y'^2)^{\frac{3}{2}}}$. Hence the disturbing forces on m are,

$$\text{in } x,\quad \frac{m'(x'-x)}{\{(x'-x)^2+(y'-y)^2\}^{\frac{3}{2}}}-\frac{m'x'}{(x'^2+y'^2)^{\frac{3}{2}}},$$

$$\text{in } y,\quad \frac{m'(y'-y)}{\{(x'-x)^2+(y'-y)^2\}^{\frac{3}{2}}}-\frac{m'y'}{(x'^2+y'^2)^{\frac{3}{2}}}.$$

Now let $R=-\dfrac{m'}{\sqrt{\{(x'-x)^2+(y'-y)^2\}}}+\dfrac{m'(x'x+y'y)}{(x'^2+y'^2)^{\frac{3}{2}}}$,

$$\text{then } \frac{dR}{dx}=\frac{-m'(x'-x)}{\{(x'-x)^2+(y'-y)^2\}^{\frac{3}{2}}}+\frac{m'x'}{(x'^2+y'^2)^{\frac{3}{2}}},$$

$$\frac{dR}{dy}=\frac{-m'(y'-y)}{\{(x'-x)^2+(y'-y)^2\}^{\frac{3}{2}}}+\frac{m'y'}{(x'^2+y'^2)^{\frac{3}{2}}},$$

Consequently the disturbing force in $x=-\dfrac{dR}{dx}$

.... in $y=-\dfrac{dR}{dy}$,

in which expressions the differential coefficients are to be taken as if x and y were independent of each other, and x' and y' did not vary when x and y varied.

78. If we put μ for the sum of the masses of the Sun and the disturbed planet, the force arising from their mutual attraction, as in (9), is $-\dfrac{\mu x}{(x^2+y^2)^{\frac{3}{2}}}$ in the direction of x, and $-\dfrac{\mu y}{(x^2+y^2)^{\frac{3}{2}}}$ in the direction of y. Hence the whole force acting on m is,

$$\text{in } x, \quad -\frac{\mu x}{(x^2+y^2)^{\frac{3}{2}}} - \frac{dR}{dx} = X,$$

$$\text{in } y, \quad -\frac{\mu y}{(x^2+y^2)^{\frac{3}{2}}} - \frac{dR}{dy} = Y.$$

79. Prop. 30. To transform the equations for rectangular co-ordinates into equations for polar co-ordinates.

The differential equations of motion are,

$$\frac{d^2 x}{dt^2} = X, \quad \frac{d^2 y}{dt^2} = Y.$$

Multiplying the former by $2\frac{dx}{dt}$, the latter by $2\frac{dy}{dt}$, and adding them, and putting for X and Y their values,

$$0 = \frac{d}{dt}\left\{\left(\frac{dx}{dt}\right)^2 + \left(\frac{dy}{dt}\right)^2\right\} + \frac{2\mu}{(x^2+y^2)^{\frac{3}{2}}}\left(x\frac{dx}{dt} + y\frac{dy}{dt}\right)$$
$$+ 2\left(\frac{dR}{dx}\cdot\frac{dx}{dt} + \frac{dR}{dy}\cdot\frac{dy}{dt}\right).$$

Again, subtracting the product of the first by y from that of the second by x,

$$0 = \frac{d}{dt}\left(x\frac{dy}{dt} - y\frac{dx}{dt}\right) + x\frac{dR}{dy} - y\frac{dR}{dx}.$$

Let $r_{,}$ be the true radius vector of m: $\theta_{,}$ its true longitude; r' and θ' those of m'.

Then $x = r_{,}\cos\theta_{,}$: $y = r_{,}\sin\theta_{,}$: $x' = r'\cos\theta'$: $y' = r'\sin\theta'$: and these equations become,

$$0 = \frac{d}{dt}\left\{\left(\frac{dr_{,}}{dt}\right)^2 + r_{,}^2\left(\frac{d\theta_{,}}{dt}\right)^2\right\} + \frac{2\mu}{r_{,}^2}\cdot\frac{dr_{,}}{dt} + 2\left(\frac{dR}{dx}\cdot\frac{dx}{dt} + \frac{dR}{dy}\cdot\frac{dy}{dt}\right),$$

$$0 = \frac{d}{dt}\left(r_{,}^2\frac{d\theta_{,}}{dt}\right) + x\frac{dR}{dy} - y\frac{dR}{dx}.$$

80. Now it must be remarked that the only way in which R can depend on the position of m is by depending on the two co-ordinates which determine the place of m.

If then the place of m be defined by rectangular co-ordinates, the complete differential coefficient of R with respect to t, as depending on the position of m only, is

$$\frac{dR}{dx}\cdot\frac{dx}{dt}+\frac{dR}{dy}\cdot\frac{dy}{dt}.$$

If it were defined by polar co-ordinates, the complete differential coefficient would be

$$\frac{dR}{dr_{,}}\cdot\frac{dr_{,}}{dt}+\frac{dR}{d\theta_{,}}\cdot\frac{d\theta_{,}}{dt}.$$

Either of these equivalent quantities may be expressed by $\frac{d(R)}{dt}$, where it must be distinctly remarked that the differentiation is to be performed with respect to t, only upon all those parts of R (in whatever form it may be expressed) which are functions of t, in consequence of their depending on the position of m. Those terms which depend only on the position of m' must not be subjected to this differentiation. We shall hereafter find a shorter method of expressing this restriction.

Substituting in the first equation,

$$0=\frac{d}{dt}\left\{\left(\frac{dr_{,}}{dt}\right)^2+r_{,}^2\left(\frac{d\theta_{,}}{dt}\right)^2\right\}+\frac{2\mu}{r_{,}^2}\cdot\frac{dr_{,}}{dt}+2\frac{d(R)}{dt}.$$

Integrating, $C=\left(\frac{dr_{,}}{dt}\right)^2+r_{,}^2\left(\frac{d\theta_{,}}{dt}\right)^2-\frac{2\mu}{r_{,}}+2\int_t\frac{d(R)}{dt}.$

81. Here it must be observed that $\int_t\frac{d(R)}{dt}$ is not the same thing as R. For R is an artificial quantity, introduced as a matter of convenience, under this condition only, that its differential coefficient with respect to t, not as contained in all its terms, but only as contained in those terms depending on the position of m, shall stand in the place of a larger expression. But there is not and cannot be any such limitation in the integration: The equation must therefore remain in the form in which we have put it.

82. For changing the other equation we must remark that, as R is now to be considered a function of $r_{,}$ and $\theta_{,}$, both of which depend on x and y, we must put

$$\text{for } \frac{dR}{dx}, \qquad \frac{dR}{dr_{,}} \cdot \frac{dr_{,}}{dx} + \frac{dR}{d\theta_{,}} \cdot \frac{d\theta_{,}}{dx};$$

$$\text{and for } \frac{dR}{dy}, \qquad \frac{dR}{dr_{,}} \cdot \frac{dr_{,}}{dy} + \frac{dR}{d\theta_{,}} \cdot \frac{d\theta_{,}}{dy};$$

where in obtaining $\frac{dr_{,}}{dx}$, $\frac{d\theta_{,}}{dx}$, &c. we must be careful to express $r_{,}$ in terms of x and y only without $\theta_{,}$: and $\theta_{,}$ in terms of x and y only, without $r_{,}$.

$$\text{Thus } r_{,}^2 = x^2 + y^2. \qquad 2r_{,}\frac{dr_{,}}{dx} = 2x;$$

$$\text{or } \frac{dr_{,}}{dx} = \frac{x}{r} = \cos\theta_{,}. \qquad \text{Similarly } \frac{dr_{,}}{dy} = \sin\theta_{,}.$$

$$\text{And } \tan\theta_{,} = \frac{y}{x}; \text{ whence } \frac{1}{\cos^2\theta_{,}} \cdot \frac{d\theta_{,}}{dx} = -\frac{y}{x^2} = -\frac{1}{r_{,}} \cdot \frac{\sin\theta_{,}}{\cos^2\theta_{,}},$$

$$\text{or } \frac{d\theta_{,}}{dx} = -\frac{\sin\theta_{,}}{r_{,}}.$$

$$\text{Similarly } \frac{1}{\cos^2\theta_{,}} \cdot \frac{d\theta_{,}}{dy} = \frac{1}{x} = \frac{1}{r_{,}\cos\theta_{,}}:$$

$$\text{whence } \frac{d\theta_{,}}{dy} = \frac{\cos\theta_{,}}{r_{,}}.$$

$$\text{Consequently } \frac{dR}{dx} = \frac{dR}{dr_{,}}\cos\theta_{,} - \frac{dR}{d\theta_{,}} \cdot \frac{\sin\theta_{,}}{r_{,}},$$

$$\frac{dR}{dy} = \frac{dR}{dr_{,}}\sin\theta_{,} + \frac{dR}{d\theta_{,}} \cdot \frac{\cos\theta_{,}}{r_{,}},$$

and the second equation becomes,

$$0 = \frac{d}{dt}\left(r_{,}^2 \cdot \frac{d\theta_{,}}{dt}\right) + \frac{dR}{d\theta_{,}},$$

where it must be recollected that $\frac{dR}{d\theta_{,}}$ means the differential coefficient with respect to $\theta_{,}$ only so far as $\theta_{,}$ is *explicitly* contained in R, and not considering $r_{,}$ as depending on $\theta_{,}$.

83. These two equations (analogous to those employed in the elliptic theory and in the lunar theory) are sufficient for solution of the problem, but the following is convenient. Since $r_{,}^{2} = x^2 + y^2$, we find by differentiating twice,

$$\frac{d^2 (r_{,}^{2})}{dt^2} = 2\left\{\left(\frac{dx}{dt}\right)^2 + \left(\frac{dy}{dt}\right)^2\right\}$$

$$+ 2\left(x\frac{d^2 x}{dt^2} + y\frac{d^2 y}{dt^2}\right).$$

But $\left(\frac{dx}{dt}\right)^2 + \left(\frac{dy}{dt}\right)^2$ or $\left(\frac{dr_{,}}{dt}\right)^2 + r_{,}^{2}\left(\frac{d\theta_{,}}{dt}\right)^2$

has been found

$$= C + \frac{2\mu}{r_{,}} - 2\int_t \frac{d(R)}{dt}:$$

and $x\frac{d^2 x}{dt^2} + y\frac{d^2 y}{dt^2} = xX + yY$

$$= -\frac{\mu}{r_{,}} - \frac{dR}{dx} r_{,} . \cos\theta_{,} - \frac{dR}{dy} . r_{,} . \sin\theta_{,}$$

$$= -\frac{\mu}{r_{,}} - r_{,} . \frac{dR}{dr_{,}}.$$

Consequently

$$\frac{d^2 (r_{,}^{2})}{dt^2} = 2C + \frac{2\mu}{r_{,}} - 4\int_t \frac{d(R)}{dt} - 2r_{,} . \frac{dR}{dr_{,}}.$$

84. The value of R, as will be found on substituting for x, y, x', and y', their values, is

$$-\frac{m'}{\sqrt{\{r'^2 - 2r'r_{,}\cos(\theta' - \theta_{,}) + r_{,}^{2}\}}} + \frac{m' r_{,} \cos(\theta' - \theta_{,})}{r'^2}.$$

85. There are two methods of deriving a result from these equations. One consists in obtaining immediately the alterations in the radius vector and the longitude produced by the disturbing force: this is the method commonly used for the periodical terms of short period. The other consists in obtaining the variation of the elements of the orbit: it is best adapted to the discovery of secular terms, whether periodical or permanent. We shall at present proceed with the former.

PERTURBATIONS OF RADIUS VECTOR AND LONGITUDE.

86. Prop. 31. To investigate very approximately the equation for the perturbation of the radius vector.

Let r be the length of the radius vector, and θ the longitude, in the orbit which m would have described if undisturbed: δr the perturbation of the radius vector: then $r + \delta r = r_{,}$. Substituting this in the last equation,

$$\frac{d^2 (r + \delta r)^2}{dt^2} = 2C + \frac{2\mu}{r + \delta r} - 4 \int_t \frac{d(R)}{dt} - 2 \,.\, (r + \delta r) \,.\, \frac{dR}{dr_{,}}.$$

Now δr is supposed to be so small that we may neglect its square. Also $\frac{dR}{dr_{,}}$ is itself a quantity depending on the disturbing force, and therefore we may neglect the product of δr and $\frac{dR}{dr_{,}}$. Again, as we have stated (76) the effect of the disturbing force in altering the body's place is so small that in calculating R, (the function whose differential coefficients express its value), we may take the undisturbed place of the body instead of the true one: that is, we may now suppose

$$R = \frac{m' r \,.\, \cos (\theta' - \theta)}{r'^2} - \frac{m'}{\sqrt{\{r'^2 - 2r'r \cos (\theta' - \theta) + r^2\}}},$$

and for $\frac{dR}{dr_{,}}$ may put $\frac{dR}{dr}$.

The equation now becomes

$$\frac{d^2(r^2 + 2r\delta r)}{dt^2} = 2C + 2\mu\left(\frac{1}{r} - \frac{\delta r}{r^2}\right) - 4\int_t . \frac{d(R)}{dt} - 2r . \frac{dR}{dr}.$$

If there had been no disturbance, the equation would have been

$$\frac{d^2 . r^2}{d t^2} = 2C + \frac{2\mu}{r}$$

where r has the same value as in the last equation.

Taking the remaining parts which relate to the disturbing force and the disturbance,

$$2\frac{d^2(r\delta r)}{dt^2} = -2\mu\frac{\delta r}{r^2} - 4\int_t . \frac{d(R)}{dt} - 2r\frac{dR}{dr},$$

$$\text{or} \quad \frac{d^2(r\delta r)}{dt^2} + \frac{\mu}{r^3}r\delta r + 2\int_t . \frac{d(R)}{dt} + r\frac{dR}{dr} = 0.$$

87. PROP. 32. To solve the equation for the perturbation of the radius vector.

Let* a be the semi-major axis of the approximate orbit of m: e its eccentricity: ϖ the longitude of the perihelion: ϵ the mean longitude of m when $t = 0$: $n = \left(\frac{\mu}{a^3}\right)^{\frac{1}{2}}$.

---

* The equation may be completely integrated in the following manner. In an undisturbed orbit,

$$\frac{d^2x}{dt^2} + \frac{\mu}{r^3}x = 0, \qquad \frac{d^2y}{dt^2} + \frac{\mu}{r^3}y = 0.$$

The form of these equations is nearly similar to that for $r\delta r$. Assume therefore $r\delta r = Mx + Ny$: and as there are two quantities M and N to be determined, assume

$$\frac{d(r\delta r)}{dt} = M\frac{dx}{dt} + N\frac{dy}{dt}:$$

$$\text{and let } 2\int_t \frac{d(R)}{dt} + r\frac{dR}{dr} = S.$$

Then from our last assumption, $\left(\text{since the complete expression for } \frac{d(r\delta r)}{dt} \text{ is } M\frac{dx}{dt} + x\frac{dM}{dt} + N\frac{dy}{dt} + y\frac{dN}{dt}\right)$,

$$x\frac{dM}{dt} + y\frac{dN}{dt} = 0:$$

Then, by (16) and (19), $nt+\epsilon$ is the mean longitude of m at the time t: $nt+\epsilon-\varpi$ is its mean anomaly; also $\theta-\varpi$ is the true anomaly; and the formula of (19) becomes

$$\theta = nt + \epsilon + (2e - \frac{e^3}{4} + \&c.)\sin(nt+\epsilon-\varpi)$$

$$+ \left(\frac{5e^2}{4} + \&c.\right)\sin(2nt + 2\epsilon - 2\varpi) + \&c.$$

$$\text{Also } r = \frac{a(1-e^2)}{1+e\,.\cos(\theta-\varpi)}.$$

Putting in $\theta-\varpi$ the value found above for θ, and taking its cosine, we get at length

$$r = a\left\{1 + \frac{e^2}{2} + \&c. - (e + \&c.)\cos(nt+\epsilon-\varpi)\right.$$

$$\left. - \left(\frac{e^2}{2} + \&c.\right)\cos(2nt+2\epsilon-2\varpi) + \&c.\right\}$$

which we shall call $a+v$. Similar expressions hold for the place of m'.

$$\text{Also } \frac{\mu}{r^3} = \frac{n^2a^3}{r^3} = n^2 + n^2\left\{3e\cos(nt+\epsilon-\varpi)\right.$$

$$\left. + \frac{3e^2}{2} + \frac{9e^2}{2}\cos(2nt+2\epsilon-2\varpi) + \&c.\right\}.$$

---

also by again differentiating $\frac{d(r\delta r)}{dt}$, and substituting in the equation $\frac{d^2(r\delta r)}{dt^2} + \frac{\mu}{r^3}r\delta r + S = 0$,

$$\frac{dx}{dt}\cdot\frac{dM}{dt} + \frac{dy}{dt}\cdot\frac{dN}{dt} + S = 0.$$

From these (as $x\frac{dy}{dt} - y\frac{dx}{dt} = r^2\frac{d\theta}{dt} = h$), $\quad \frac{dM}{dt} = \frac{1}{h}yS, \quad \frac{dN}{dt} = -\frac{1}{h}xS;$

and as $x = r\cos\theta$, $\quad y = r\sin\theta$,

we get $\delta r = \frac{1}{h}(\cos\theta\int_t r\sin\theta\,.\,S - \sin\theta\int_t r\cos\theta\,.\,S)$,

where the elliptical values are to be substituted for r and θ. It will easily be seen that we might have put

$$r\delta r = M(x\cos\varpi + y\sin\varpi) + N(y\cos\varpi - x\sin\varpi),$$

and this would have given

$$\delta r = \frac{1}{h}(\cos\overline{\theta-\varpi}\int_t r\sin\overline{\theta-\varpi}\,.\,S - \sin\overline{\theta-\varpi}\int_t r\cos\overline{\theta-\varpi}\,.\,S),$$

which in actual computation would be rather more convenient. The method in the text is the most practicable.

88. Before proceeding with the solution of our equation, we shall make one important remark. R, so far as it depends on the place of m, is a function of r and θ. Now r, when given in terms of no variable but t, is expressed entirely by constants and by cosines of multiples of $(nt + \epsilon - \varpi)$: θ is $nt + \epsilon +$ a series depending on constants and on sines of multiples of $(nt + \epsilon - \varpi)$. Consequently, whether we suppose R completely expanded or not, if it be expressed in terms of t, wherever we find nt we shall find ϵ*. Suppose then

$$R = A + B\cos(nt + \epsilon - E) + C\cos(2nt + 2\epsilon - F) + \&c.,$$

where A, B, C, &c. are any functions whatever of the co-ordinates of m': then taking $\frac{d(R)}{dt}$ with the restriction imposed on it in (80) and (81).

$$\frac{d(R)}{dt} = -n \,.\, B\sin(nt + \epsilon - E)$$

$$-2nC \,.\, \sin(2nt + 2\epsilon - F) - \&c.$$

$$\text{But } \frac{dR}{d\epsilon} = -B \,.\, \sin(nt + \epsilon - E)$$

$$-2C \,.\, \sin(2nt + 2\epsilon - F) - \&c.$$

$$\text{therefore } \frac{d(R)}{dt} = n\frac{dR}{d\epsilon},$$

where it is supposed† that, in order to take $\frac{dR}{d\epsilon}$, both r and θ have been expressed in terms of $nt + \epsilon$; which is in fact the easiest method for the planets. We have now got rid of the restriction with which we were before incumbered (80).

If we had developed A, B, C, &c. we should have had for each a series of such terms as $L\cos q(n't + \epsilon' - E')$ which when

* This is also true, if the inclinations of the orbits be taken into account.

† From this and other instances, the reader will observe the importance (in giving the partial differential coefficient of a function which admits of being expanded in terms of different quantities) of stating distinctly the supposition made relative to the expansion.

multiplied by one of the terms depending on $nt + \epsilon - E$ would have produced such terms as

$$P.\cos \overline{p(nt + \epsilon - E) \pm q(n't + \epsilon' - E')}.$$

The equation $\frac{d(R)}{dt} = n\frac{dR}{d\epsilon}$ would still have been found true, as the term $n't$ depends on t only because the place of m' varies with t, and therefore for our restricted value of $\frac{d(R)}{dt}$ it must not be differentiated.

Should the reader find any difficulty in conceiving a differential coefficient taken with regard to one of the quantities commonly considered constant, we beg him to observe that the variation of value, used in the definition at least of a differential coefficient, is hypothetical only and not necessarily true. We state that *if* we alter the value of one quantity, and *if* we find the corresponding alteration of the function, and *if* we divide the latter alteration by the former, and *if* we find the limit to which the quotient approaches on diminishing the former indefinitely, this limit is the differential coefficient. The whole of this operation can be performed equally well, whether the quantity be in the nature of things variable, or we be compelled to conceive a variation where none really exists.

89. To return to our equation: we may now put it under the form

$$0 = \frac{d^2(r\delta r)}{dt^2} + n^2.r\delta r + 2n\int_t \frac{dR}{d\epsilon} + r\frac{dR}{dr}$$

$$+n^2.r\delta r\left\{3e\cos(nt+\epsilon-\varpi) + \frac{3e^2}{2} + \frac{9e^2}{2}\cos(2nt+2\epsilon-2\varpi) + \&c.\right\}.$$

This is most easily solved by approximation: neglecting first the terms which are multiplied by e and its powers, which reduces the equation to

$$0 = \frac{d^2(r\delta r)}{dt^2} + n^2.r\delta r + 2n\int_t \frac{dR}{d\epsilon} + r\frac{dR}{dr},$$

then substituting the value so found in the terms multiplied by e and solving again, &c. The process is exactly similar to that for determining u in the Lunar Theory: the equations being in both cases of the form of that treated in (1) and (4).

90. Suppose for instance in the expansion of

$$2n\int_t \frac{dR}{d\epsilon} + r\frac{dR}{dr}$$

one term is $P\cos(pnt - qn't + Q)$. This will represent every term, p and q being whole numbers, (as we shall shew when we treat of the expansion of R, &c.). The equation, deprived of terms depending on e, is

$$\frac{d^2(r\delta r)}{dt^2} + n^2 . r\delta r + P\cos(pnt - qn't + Q) = 0.$$

The general solution, by (4), is

$$r\delta r = A\cos(nt - B) + \frac{P}{(pn - qn')^2 - n^2}\cos(pnt - qn't + Q).$$

The first or arbitrary term of this is to be rejected. For, considering that first term only,

$$\delta r = \frac{A}{r}\cos(nt - B) = \frac{A}{a}\cos(nt - B) + \text{smaller terms},$$

and consequently r or $a(1 - e\cos\overline{nt + \epsilon - \varpi} + \&c.)$ is reduced

$$\text{to } a\left\{1 - e\cos\overline{nt + \epsilon - \varpi} + \frac{A}{a^2}\cos\overline{nt - B} + \&c.\right\}.$$

The variable part under the bracket, which is the sum of the principal term of the eccentricity and of the arbitrary part, is

$$\left(-e + \frac{A}{a^2}\cos\overline{\epsilon - \varpi + B}\right)\cos(nt + \epsilon - \varpi)$$

$$+ \frac{A}{a^2}\sin(\epsilon - \varpi + B) . \sin(nt + \epsilon - \varpi)$$

$$= -\sqrt{\left\{e^2 - \frac{2Ae}{a^2}\cos(\epsilon - \varpi + B) + \frac{A^2}{a^4}\right\}}\cos(nt + \epsilon - \varpi + \Pi),$$

$$\text{where } \tan \Pi = \frac{A \cdot \sin(\epsilon - \varpi + B)}{a^2 e - A \cdot \cos(\epsilon - \varpi + B)}.$$

That is, the eccentricity and the place of the perihelion are altered, in consequence of this term, by *constant* quantities. As the alteration may be comprehended in the values of e and ϖ, this term may be neglected without any loss of results. The approximate value of $r\delta r$ is therefore

$$\frac{P}{(pn - qn')^2 - n^2} \cos(pnt - qn't + Q).$$

Now the term multiplying e is $3n^2 e \,.\, r\delta r \,.\, \cos(\overline{nt + \epsilon - \varpi})$: or, substituting the value just found for $r\delta r$,

$$\frac{3}{2} \cdot \frac{n^2 e P}{(pn - qn')^2 - n^2} \{\cos \overline{(p+1)nt - qn't + Q + \epsilon - \varpi}$$

$$+ \cos \overline{(p-1)nt - qn't + Q - \epsilon + \varpi}\}.$$

Adding these two terms to the equation of the last paragraph, and integrating it, we find

$$r\delta r = \frac{P}{(pn - qn')^2 - n^2} \{\cos \overline{pnt - qn't + Q}$$

$$+ \frac{3}{2} \cdot \frac{n^2 e}{(\overline{p+1} \,.\, n - qn')^2 - n^2} \cos \overline{(p+1)nt - qn't + Q + \epsilon - \varpi}$$

$$+ \frac{3}{2} \cdot \frac{n^2 e}{(\overline{p-1} \,.\, n - qn')^2 - n^2} \cos \overline{(p-1)nt - qn't + Q - \epsilon + \varpi}\}.$$

In the same manner we might proceed to a more approximate value. The value of $r\delta r$ being found, that of δr is found by multiplying it by

$$\frac{1}{a}(1 + e \cos \overline{nt + \epsilon - \varpi} + e^2 \,.\, \cos \overline{2nt + 2\epsilon - 2\varpi} + \&\text{c.})$$

$$\text{the value of } \frac{1}{r}.$$

91. One important remark must not be omitted. In the development of

$$2n \int_t \frac{dR}{d\epsilon} + r \frac{dR}{dr}$$

there will be one term $P.\cos(nt+Q)$, introduced by the first term of v. The equation for $r\delta r$, considering this term only, will be

$$0 = \frac{d^2(r\delta r)}{dt^2} + n^2.r\delta r + P.\cos(nt+Q).$$

The solution of the equation in this form, as in (5) and (43), would have terms depending on an arc, which would admit of unlimited increase, and therefore our original supposition of elliptic motion would not be even an approximation. To avoid this difficulty, we must, as in (44) and (44*), suppose the first term of v to depend on $\cos c(nt+\epsilon-\varpi)$, which will change the term in the equation to $P\cos(cnt+Q)$. This amounts to supposing, as in (61), that the perihelion has a constant motion. We shall not in this place proceed further with the investigation of its motion, as it will be better found by the method of *variation of elements*, to be explained hereafter.

There will also be one constant term in $r\delta r$: this shews that, the mean motion being given, the axis major is not the same as if there were no disturbing force.

92. PROP. 33. To examine the terms of R which are increased most in integrating the equation for $r\delta r$.

It is evident that the terms which are most increased by the integration are those in which

$$(pn-qn')^2-n^2, \text{ or } \overline{(p+1)n-qn'}.\overline{(p-1)n-qn'},$$

is small; that is, those in which one of the quantities

$$\overline{(p+1)n-qn'}, \quad \overline{(p-1)n-qn'},$$

is small: where p and q are whole numbers, and may be positive or negative. Consequently, if we can find two numbers $p+1$ and q, or $p-1$ and q, which make

$$(p+1)n-qn', \text{ or } (p-1)n-qn'$$

very small, we may expect that the term of δr depending on $\cos(pnt-qn't+Q)$ will be large. The train of terms introduced with it will all be affected by the same divisor, and therefore, though multiplied by e, e^2, &c. may be sensible.

This consideration saves much trouble, in the selection of the terms of R which are to be retained for integration: for, among the terms multiplied by e and e' or their powers, there are few that become sensible, except those which have a very small divisor. For instance, the periodic times of Jupiter and Saturn are very nearly in the ratio of 2 to 5. Consequently, supposing Jupiter to be disturbed by Saturn, as n is the angle described in the unit of time by Jupiter, and n' that described by Saturn,

$n : n' :: 5 : 2$ nearly, and therefore $2n - 5n'$ is very small.

If $2 = p + 1$, and $5 = q$, the term $\cos(pnt - qn't + Q)$ is $\cos(nt - 5n't + Q)$.

If $2 = p - 1$, and $5 = q$, the term $\cos(pnt - qn't + Q)$ is $\cos(3nt - 5n't + Q)$.

The terms of these two forms must be carefully sought out, as in the integration they will be divided by $2n - 5n'$. The latter however (for a reason which we shall explain hereafter) is much more important than the former.

The coefficient of the cosine of $(pnt - qn't + Q)$ is only *once* divided by the small quantity

$$(p + 1) . n - qn', \quad \text{or} \quad (p - 1) . n - qn'.$$

If $pn - qn'$ be small, the terms divided by

$$(\overline{p + 1} . n - qn')^2 - n^2 \quad \text{and} \quad (\overline{p - 1} . n - qn')^2 - n^2$$

will be much increased. The same remarks apply in all respects to them.

93. Prop. 34. To investigate the equation for the perturbation in longitude.

If we subtract the equation of (80) from double the equation of (83), we get

$$0 = 3\left(\frac{dr_,}{dt}\right)^2 + 4r_, \frac{d^2r_,}{dt^2}$$

$$- r_,^2\left(\frac{d\theta_,}{dt}\right)^2 - \frac{2\mu}{r_,} - 3C + 6\int_t \frac{d(R)}{dt} + 4r_, \frac{dR}{dr_,}.$$

Putting $r+\delta r$ for $r_{\prime}$, $\theta+\delta\theta$ for $\theta_{\prime}$, and $n\frac{dR}{d\epsilon}$ for $\frac{d(R)}{dt}$, and neglecting small quantities as in (86), we find (setting down those terms only which are the effect of the disturbing force),

$$0=\left(6\frac{dr}{dt}\cdot\frac{d.\delta r}{dt}+2\delta r.\frac{d^2r}{dt^2}+4r.\frac{d^2\delta r}{dt^2}\right)$$

$$-2r^2\frac{d\theta}{dt}\cdot\frac{d.\delta\theta}{dt}+2\frac{\delta r}{r}\left\{r\frac{d^2r}{dt^2}-r^2\left(\frac{d\theta}{dt}\right)^2+\frac{\mu}{r}\right\}$$

$$+6n\int_t\frac{dR}{d\epsilon}+4r\frac{dR}{dr}.$$

Now, neglecting the disturbing force,

$$\frac{\mu}{r}=-x\frac{d^2x}{dt^2}-y\frac{d^2y}{dt^2}=-\frac{d}{dt}\left(x\frac{dx}{dt}+y\frac{dy}{dt}\right)+\left(\frac{dx}{dt}\right)^2+\left(\frac{dy}{dt}\right)^2$$

$$=-\frac{d}{dt}\left(r\frac{dr}{dt}\right)+\left(\frac{dr}{dt}\right)^2+r^2\left(\frac{d\theta}{dt}\right)^2$$

$$=-r\frac{d^2r}{dt^2}+r^2\left(\frac{d\theta}{dt}\right)^2:$$

consequently, the multiplier of $\frac{2\delta r}{r}=0$.

Also $r^2\frac{d\theta}{dt}=h$ {by integration of the equation (*b*) in (30), or that in (82), omitting the term depending on disturbing force}

$$=\sqrt{a(1-e^2)\mu}\,(11)=a^2n\sqrt{(1-e^2)}.$$

Thus our equation becomes

$$0=\frac{d}{dt}\left(4r\frac{d.\delta r}{dt}+2\frac{dr}{dt}\delta r\right)-2a^2n\sqrt{(1-e^2)}.\frac{d.\delta\theta}{dt}$$

$$+6n\int_t\frac{dR}{d\epsilon}+4r\frac{dR}{dr}.$$

Integrating, and observing that the arbitrary constant must $=0$ (as if it had any value it might be united with the elliptical value of θ)

$$0 = 2r\frac{d.\delta r}{dt} + \frac{dr}{dt}\delta r - a^2 n\sqrt{(1-e^2)}.\delta\theta$$

$$+ 3n\int_t\int_t\frac{dR}{d\epsilon} + 2\int_t r\frac{dR}{dr}.$$

$$\text{Or } \delta\theta = \frac{1}{a^2 n\sqrt{(1-e^2)}}\left\{\frac{d(2r\delta r)}{dt} - \frac{1}{r}.\frac{dr}{dt}r\delta r\right\}$$

$$+ \frac{3}{a^2\sqrt{(1-e^2)}}\int_t\int_t\frac{dR}{d\epsilon} + \frac{2}{a^2 n\sqrt{(1-e^2)}}\int_t r\frac{dR}{dr}.$$

94. Prop. 35. To examine the terms of R which are increased most in integrating the equation for $\delta\theta$.

For the complete calculation of $\delta\theta$, the value of $r\delta r$ found in (89) and (90) is necessary. The terms which have been increased in the integration for $r\delta r$ will not have their value materially altered in the integration for $\delta\theta$. For, let $\cos(pnt - qn't + Q)$ be one of these; then (92) $pn - qn'$ does not differ much from $\pm n$: and therefore the first term in the value of $\delta\theta$ will not be altered by the differentiation.

95. But the terms which are most important in $\delta\theta$ are those which depend upon terms in R of the form

$$P.\cos\overline{p(nt+\epsilon) - q(n't+\epsilon') + Q},$$

where $pn - qn'$ is very small. For, with respect to this term,

$$\frac{dR}{d\epsilon} = -pP.\sin\overline{p(nt+\epsilon) - q(n't+\epsilon') + Q};$$

$$\text{whence } \frac{3}{a^2\sqrt{(1-e^2)}}\int_t\int_t\frac{dR}{d\epsilon}$$

$$= \frac{p}{(pn-qn')^2}.\frac{3P}{a^2\sqrt{(1-e^2)}}.\sin\overline{p(nt+\epsilon) - q(n't+\epsilon') + Q}.$$

Here the coefficient is divided *twice* by $pn - qn'$: and of course if that quantity is small, the coefficient is very much increased in magnitude. The numbers p and q must always be integers: if then two integers p and q can be found nearly in the proportion of the mean motions of m' and m, there will be a large term in the perturbation of m depending on the argument

$$\overline{p(nt+\epsilon)-q(n't+\epsilon')+Q}.$$

There is no difficulty in finding, among the planetary motions, proportions nearly coinciding with those of two numbers. The most remarkable is that of Jupiter and Saturn, where, the proportions of the mean motions being nearly 5 : 2, and the masses being large, although the term in question (as we shall shew hereafter) is a multiple of the cube of the eccentricities, yet the inequality depending on the term

$$\cos\overline{5(nt+\epsilon)-2(n't+\epsilon')+Q}$$

amounts to nearly fifty minutes. A singular instance has also been lately discovered in the perturbation of the Earth by Venus: eight times the mean motion of Venus is *very* nearly equal to thirteen times the mean motion of the Earth: and though (as we shall shew) this term

$$\cos\overline{13(nt+\epsilon)-8(n't+\epsilon')+Q}$$

must be multiplied by the fifth powers of the eccentricities, &c. which are very small, and though the disturbing mass is very small, this inequality is sensible.*

96. There are, in the value of $r\delta r$, terms of the same form (produced originally by the terms

$$\cos\overline{(p\pm 1)(nt+\epsilon)-q(n't+\epsilon')+Q})$$

which are *once* divided by

$$(pn\pm n-qn')^2-n^2, \quad \text{or} \quad (pn\pm 2n-qn')(pn-qn').$$

These terms then may be considerable in δr, but by no means so conspicuous as in $\delta\theta$. There are others introduced by the terms

$$\cos\overline{(p\pm 2)(nt+\epsilon)-q(n't+\epsilon')+Q}, \text{ \&c.}:$$

the same remark applies to them.

---

* The coefficient of the term $\cos\overline{13(nt+\epsilon)-8(n't+\epsilon')+Q}$ in R is in this instance multiplied in the integration by a number exceeding two millions.

97. Prop. 36. To describe the nature of the forces which produce the most striking inequalities in the radius vector, and to investigate their periodic times.

If τ be the time in which the term $P.\cos(pnt - qn't + Q)$ goes through all its possible changes of sign and magnitude and returns again to the same value, (which we shall call the period) it is evident that by adding τ to t, $pnt - qn't$ must have increased or decreased by 2π, and consequently

$$\tau = \frac{2\pi}{pn - qn'}, \quad \text{or} \quad = \frac{2\pi}{qn' - pn}.$$

Now as the disturbing force is expressed by means of R, we may conceive each term of R, (or some function of each term, having the same period) to represent a part of the disturbing force. The period of the force depending on the term

$$P.\cos(pnt - qn't + Q)$$

is therefore

$$\frac{2\pi}{pn - qn'}, \quad \text{or} \quad \frac{2\pi}{qn' - pn}.$$

Now the greatest terms (*cæteris paribus*) in δr or $r\delta r$ are those in which $pn - qn' \pm n$ is small (92); $\pm(pn - qn')$ therefore for these terms is nearly $= n$, and therefore the period of these terms is nearly $\frac{2\pi}{n}$, the periodic time of revolution. A force therefore which goes through all its periodic values nearly in the same time as the time of revolution will produce in the radius vector a considerable inequality. And the inequality will be expressed by a multiple of $\cos(pnt - qn't - Q)$: its period will be $\pm \frac{2\pi}{pn - qn'}$: and it will therefore go through all its values *nearly* in one revolution of the planet. This would amount to the same as altering the eccentricity and perihelion of the planet.

98. But because the period is not *exactly* equal to the revolution of the planet, the alteration of eccentricity and perihelion will not be constant but will vary slowly. For

suppose $pn - qn' - n$ to be small: δr will have a considerable term of the form $aN.\cos(pnt - qn't + Q)$: and adding this to the first variable term of r in (87), the principal inequality of r is

$$a\{-e\cos(nt+\epsilon-\varpi) + N.\cos(pnt-qn't+Q)\}$$

$$= a\{-e\cos(nt+\epsilon-\varpi)$$

$$+ N.\cos\overline{(pnt-qn't-nt+Q-\epsilon+\varpi)+(nt+\epsilon-\varpi)}\}$$

$$= -\cos(nt+\epsilon-\varpi)\{ae - aN.\cos(pnt-qn't-nt+Q-\epsilon+\varpi)\}$$

$$-\sin(nt+\epsilon-\varpi).aN.\sin(pnt-qn't-nt+Q-\epsilon+\varpi).$$

As in (66) and (68) this may be put under the form

$$-aE.\cos(nt+\epsilon-\varpi-K),$$

$$\text{where } E = e - N.\cos(pnt-qn't-nt+Q-\epsilon+\varpi)$$

$$K = \frac{N}{e}\sin(pnt-qn't-nt+Q-\epsilon+\varpi).$$

This is the same as the elliptic inequality in an orbit where E is the eccentricity and $\varpi + K$ the longitude of the perihelion. The alterations then both in the eccentricity and in the longitude of the perihelion are periodical: and their common period is $\frac{2\pi}{pn - qn' - n}$, which is very long.

99. The same term of R will also produce considerable terms in $r\delta r$ whose period is $\frac{2\pi}{pn - qn' - n}$: but they will be multiplied by e (see Art. 90). Hence there will be in δr no terms whose period is long that are comparable in magnitude to those whose period is nearly the same as the periodic time of the disturbed planet.

100. Prop. 37. To describe the nature of the forces which produce the most striking inequalities in the longitude, and to investigate their periodic times.

We have already remarked (94) that the terms of $\delta\theta$ produced by the considerable terms of $r\delta r$ will have a value only commensurate with their value in $r\delta r$, and consequently, though considerable, they cannot rise to first-rate importance. We shall therefore confine our attention to the term depending on $\cos(pnt - qn't + Q)$ where $pn - qn'$ is very small. The period of the force which produces this term is $\dfrac{2\pi}{pn - qn'}$, and is therefore very great. The period of the inequality is the same. It appears then that the terms which become very great are those which occupy a very long period, and which in consequence can hardly be discovered from observations that extend through a short time. Thus the inequality of Saturn produced by Jupiter (and a corresponding one of Jupiter) depending on

$$\cos \overline{5(nt + \epsilon) - 2(n't + \epsilon') + Q},$$

has a period exceeding 900 years, or 30 revolutions of the exterior planet. The inequality of the Earth alluded to in (94) has a period of about 240 years, or 240 revolutions of the exterior planet. There are many similar terms in the theory of the different planets, but no others so remarkable as these.

101. The increase of the coefficient from integration is inversely as $(pn - qn')^2$, or directly as $\left(\dfrac{2\pi}{pn - qn'}\right)^2$, or directly as the square of the period of the inequality. Of this, the following popular explanation may be offered. If we conceive a force to urge the planet in the same direction for a long time, or if we conceive the preponderance of forces acting in one direction over those acting in the opposite direction to be of the same kind for a long time, (which may be the case when $pn - qn'$ is small, because then the periodic times of the two planets are in the proportion of $p : q$ very nearly, and therefore after q revolutions of one planet or p revolutions of the other, they will again be in nearly the same relative situation at the same parts of their orbits, and thus the actions of the same kind are repeated over and over again for a long time), the velocity of the planet may be increased so that it describes a larger orbit than before, and its periodic time will therefore be increased,

and that by a quantity proportional to the time of action of these forces. During another equally long time, the periodic time may be diminished. During the former of these portions the planet will be dropping behind its mean place in longitude by an angle which is proportional to the product of the change in periodic time by the number of revolutions through which that change is of the same kind: and as each of these is proportional to the time expressed by $\frac{2\pi}{pn - qn'}$, the planet will have dropped behind its mean place by a quantity proportional to $\frac{1}{(pn - qn')^2}$. During the latter portion it will regain the same quantity. The whole inequality therefore, or the difference between its actual place and the place computed on the supposition of mean motion in longitude, will be proportional to $\frac{1}{(pn - q'n)^2}$. On this and similar points the reader is referred to the author's treatise entitled *Gravitation*.

PERTURBATIONS OF THE ELLIPTIC ELEMENTS.

102. Prop. 38. To explain algebraically the variation of parameters.

This artifice of solution is founded entirely upon the obvious fact, that any equation between two (or more) quantities (as x and y, or r and θ), may be changed into any other equation, provided that instead of the constant quantities we put functions of x and y properly chosen. For instance, we may change the equation $ax + by = c$ into the equation $(x - e)^2 + (y - f)^2 = c$, provided that instead of a we put $\frac{(x-e)^2}{x}$, and instead of b, $\frac{(y-f)^2}{y}$: or provided that instead of a we put $\frac{2 . (x-e)^2 + (y-f)^2}{2x}$, and instead of b, $\frac{(y-f)^2}{2y}$: or in an infinite number of ways. The constants a and b being sometimes called *parameters*, the artifice which puts variable quantities in their places is generally called the *variation of parameters.*

103. But it must not be supposed that we are often required to make such violent changes as that in the instance above given. In general, the principle is used when an equation, which is easily integrable, receives the addition of some small term which renders the form of the solution more complicated. It is then thought more simple and more intelligible to put the solution in a form exactly similar to that of the solution of the unaugmented equation, with the understanding that the parameters are now not constant but subject to a small variation. And it generally happens that, among the infinite number of ways of effecting this, we may choose one which will possess some other convenient property. Suppose, for instance, we had the equation

$$\frac{d^2 z}{dt^2} + n^2 z = 0.$$

The solution of this equation is $z = A\cos(nt - B)$, where A and B are constants, of the kind called arbitrary: that is, they are to be determined so as to suit some known properties of particular values of z and t. Now suppose we had the equation

$$\frac{d^2 z'}{dt^2} + n^2 z' + a \, . \cos(pt - Q) = 0$$

where a is small. The solution of this will contain only one term more than the former. But for many purposes it is more convenient that this solution should stand in the form $z' = A' . \cos(nt - B')$, where A' and B' have only small variations depending on the additional term: the use of the solution of the simpler equation being only to *suggest* a form for this solution. Among the infinite number of ways in which this can be done, it may be convenient to select that which shall give for $\frac{dz'}{dt}$ an expression of exactly the same form as the expression for $\frac{dz}{dt}$ deduced from the solution of the unaugmented equation. Assuming then $z' = A' . \cos(nt - B')$, we have

$$\frac{dz'}{dt} = -nA' . \sin(nt - B') + \cos(nt - B') . \frac{dA'}{dt}$$
$$+ A' . \sin(nt - B') . \frac{dB'}{dt} .$$

But $\frac{dz}{dt} = -nA . \sin(nt - B)$: and $\frac{dz'}{dt}$ is to have the same form: therefore we must have

$$\frac{dz'}{dt} = -nA' . \sin(nt - B').$$

Consequently the remaining terms of the complete expression for $\frac{dz'}{dt}$ must $= 0$; or

$$0 = \cos(nt - B') . \frac{dA'}{dt} + A' \sin(nt - B') . \frac{dB'}{dt} .$$

Differentiating $\frac{dz'}{dt}$, and substituting in the given equation,

$$0 = -\sin(nt - B') . \frac{dA'}{dt} + A' . \cos(nt - B') . \frac{dB'}{dt} + \frac{a}{n} . \cos(pt - Q).$$

From this equation and the last we find

$$\frac{dA'}{dt} = \frac{a}{n} . \sin(nt - B') . \cos(pt - Q),$$

$$\frac{dB'}{dt} = -\frac{a}{nA'} \cos(nt - B') . \cos(pt - Q).$$

The values of A' and B' will be readily found, if we suppose their difference from some constants A and B to be so small that in the terms multiplied by a we may put A for A', and B for B'. Then we have

$$z' = A' \cos(nt - B'),$$

$$\text{where } A' = A - \frac{a}{2n(n+p)} \cos(nt + pt - B - Q)$$
$$- \frac{a}{2n(n-p)} \cos(nt - pt - B + Q).$$

$$B' = B - \frac{a}{2n(n+p)A} \sin(nt + pt - B - Q)$$

$$- \frac{a}{2n(n-p)A} \sin(nt - pt - B + Q),$$

a form which is sometimes more convenient than that obtained by direct solution.

104. PROP. 39. To explain geometrically the variation of parameters.

This can be done most readily by reference to some instance: for example, that of the last Article. The equation

$$\frac{d^2 z}{dt^2} + n^2 z = 0,$$

is the equation corresponding to the motion of a simple pendulum whose length is $n^2 g$, in a cycloidal arc. The equation

$$\frac{d^2 z'}{dt^2} + n^2 z' + a \cos(pt - Q) = 0$$

represents the motion of the same pendulum, supposing that besides the force of gravity there is a disturbing force $a \cdot \cos(pt - Q)$ acting upon it in the direction of a tangent to the curve. The solution $z' = A' \cdot \cos(nt - B')$, if we look no further, merely asserts that the *place* of the pendulum at the time t may be expressed by means of the same formula as that which expresses the place in undisturbed vibration. Of this there is no doubt, as the same thing might be expressed in the same form with a hundred different values of the parameters. But if we differentiate it, we find (by the assumption of last Article) that the velocity, with the values of A' and B' that we have found, is $-nA' \sin(nt - B')$: that is, that the *velocity* of the body also is expressed by the same formula as if its vibration were not disturbed by the additional force. Consequently the *place* and *velocity* of the disturbed pendulum at the time t are the same as those of an undisturbed pendulum, whose arc of vibration on each side of the vertical is the value of A' at that instant, and which was at the extremity of its vibration at the time $\frac{B'}{n}$, giving B' the value corresponding to the same instant. Now we see a distinct and tangible meaning in the expression. If at any instant t the disturbing force should cease to act, the pendulum would go on vibrating so that its arc of vibration on each side would be the value which A' had at the instant t, and as if it had been at the extremity of a vibration at the time $\frac{B'}{n}$ where B' preserves constant the value which it had at the instant t. These two elements (the length of the arc of vibration, and the time at which the pendulum was at the extremity of the arc) may, if no disturbing force acts, be found from a knowledge of the place and the velocity of the pendulum at any instant: and therefore the variable elements in the disturbed motion (which, for the instant, are the same) may be inferred from the place and velocity of the disturbed pendulum, and may be considered as contained in them.

105. In exactly the same manner we may represent the motion of a disturbed planet by finding the values of the elliptic elements, which would represent the place of the planet at any time t, and which would also give the velocity and direction of motion of the planet at that time; the velocity being calculated from those elements as if they were the invariable elements of an undisturbed body. And if the disturbing force should at any instant cease to act, the planet would go on describing the ellipse whose elements are the values of the elements which correspond to that particular instant. And the elliptic instantaneous elements at any time may be inferred from the place and velocity and direction of motion of the planet at that time, and may be conceived as contained in them.

106. This method may be used for finding any of the inequalities produced by perturbation. But it is best adapted to those which result from changes in the elements whose period is very long. For though these changes may be considerable, so that in a century the orbit of a planet is

most materially changed, yet their value may not sensibly change from one year to another, so that a single revolution (so far as these long inequalities alone are concerned) may be *very exactly* represented by the ellipse whose elements are the values of the elements corresponding to any instant of that revolution. For the same reasons it is particularly well adapted to the discovery of those changes which continue constantly in the same direction, as for instance, the motion of the perihelion.

107. Prop. 40. To investigate the alteration of the semi-major axis in a disturbed orbit.

We shall, as before, put a, e, ϖ, for the invariable elements in an ellipse nearly coinciding with the curve described, and $a_{,}$, $e_{,}$, $\varpi_{,}$, for the variable elements which will represent the place and velocity of the planet in the curve described, provided the calculation be made for any instant as if they were invariable. Now upon examining the process above, it will be seen that the *place* and *velocity* or *first differential coefficient* in the real curve are the same as in this instantaneous ellipse: but there is no such condition respecting the *second differential coefficient*. We must then use the equations which have been once integrated. Now in an undisturbed elliptic orbit this equation is easily found, by the same process as in (79) and (80),

$$C = \left(\frac{dr_{,}}{dt}\right)^2 + r_{,}^2 \left(\frac{d\theta_{,}}{dt}\right)^2 - \frac{2\mu}{r_{,}}.$$

If we make this

$$= \left\{\frac{1}{r_{,}^4} \cdot \left(\frac{dr_{,}}{d\theta_{,}}\right)^2 + \frac{1}{r_{,}^2}\right\} r_{,}^4 \left(\frac{d\theta_{,}}{dt}\right)^2 - \frac{2\mu}{r_{,}}$$

$$= \left\{\left(\frac{du}{d\theta}\right)^2 + u^2\right\} h^2 - 2\mu u$$

as in (10) and (11), and put for u and h the values there given, the semi-major axis being $a_{,}$, we find $C = -\dfrac{\mu}{a_{,}}$. Thus in undisturbed motion in our fictitious orbit,

$$-\frac{\mu}{a_{\prime}} = \left(\frac{dr_{\prime}}{dt}\right)^2 + r_{\prime}^2\left(\frac{d\theta_{\prime}}{dt}\right)^2 - \frac{2\mu}{r_{\prime}}.$$

This equation may be regarded as an instance of what we have said (105), that the instantaneous elliptic elements may be inferred from the place and motion of the planet at any instant. For we here have $\frac{\mu}{a_{\prime}}$ absolutely expressed in terms of the polar co-ordinates and their differentials. In order then to express $\frac{\mu}{a_{\prime}}$ for every instant in terms of the disturbing forces, we have merely to express

$$\left(\frac{dr_{\prime}}{dt}\right)^2 + r_{\prime}^2\left(\frac{d\theta_{\prime}}{dt}\right)^2 - \frac{2\mu}{r_{\prime}}$$

in terms of those forces. This we can easily do by means of the equation of (80), which gives

$$\left(\frac{dr_{\prime}}{dt}\right)^2 + r_{\prime}^2\left(\frac{d\theta_{\prime}}{dt}\right)^2 - \frac{2\mu}{r_{\prime}} = C - 2\int_t \frac{d(R)}{dt} = C - 2n\int_t \frac{dR}{d\epsilon}.$$

$$\text{Consequently } -\frac{\mu}{a_{\prime}} = C - 2n\int_t \frac{dR}{d\epsilon}:$$

$$\text{and differentiating, } \frac{da_{\prime}}{dt} = -\frac{2na_{\prime}^2}{\mu}.\frac{dR}{d\epsilon},$$

$$\text{or nearly } = -\frac{2na^2}{\mu}.\frac{dR}{d\epsilon}.$$

108. Prop. 41. To investigate the alteration of the eccentricity in a disturbed orbit.

Let h be the area described in 1″ in the ellipse whose elements are a, e, ϖ: $h_{\prime}$ that in the disturbed orbit, or in the fictitious instantaneous ellipse: then $h_{\prime} = r_{\prime}^2\frac{d\theta_{\prime}}{dt}$.

$$\text{But by (82) } \frac{d}{dt}\left(r^2\frac{d\theta_{\prime}}{dt}\right) = -\frac{dR}{d\theta_{\prime}};$$

$$\text{hence } \frac{dh_{\prime}}{dt} = -\frac{dR}{d\theta_{\prime}} = -\frac{dR}{d\theta}, \text{ as in (86),}$$

in which equation it is supposed (82) that R is expressed in terms of r and θ.

109. This expression requires an alteration of form. For, as in (88), &c., we propose to express R entirely in terms of t, and therefore there will not be any explicit expression of R in terms of θ, from which $\frac{dR}{d\theta}$ can be found. We may thus find its value in terms of $\frac{dR}{d\epsilon}$ and $\frac{dR}{d\varpi}$. R, as in (86), is a function of r and θ: also r is expressed, see (87) and (88), by a series of the form

$$A + B\cos(nt + \epsilon - \varpi) + C.\cos(2nt + 2\epsilon - 2\varpi) + \&c.;$$

and $\theta = nt + \epsilon + M\sin(nt + \epsilon - \varpi) + N\sin(2nt + 2\epsilon - 2\varpi) + \&c.$

$$\text{Hence } \frac{dR}{d\varpi} = \frac{dR}{dr}\cdot\frac{dr}{d\varpi} + \frac{dR}{d\theta}\cdot\frac{d\theta}{d\varpi}$$

$$= \frac{dR}{dr}\{B\sin(nt + \epsilon - \varpi) + 2C.\sin(2nt + 2\epsilon - 2\varpi) + \&c.\}$$

$$+ \frac{dR}{d\theta}\{-M\cos(nt + \epsilon - \varpi) - 2N.\cos(2nt + 2\epsilon - 2\varpi) - \&c.\}.$$

$$\text{Similarly } \frac{dR}{d\epsilon} = \frac{dR}{dr}\cdot\frac{dr}{d\epsilon} + \frac{dR}{d\theta}\cdot\frac{d\theta}{d\epsilon}$$

$$= \frac{dR}{dr}.\{-B\sin(nt + \epsilon - \varpi) - 2C.\sin(2nt + 2\epsilon - 2\varpi) - \&c.\}$$

$$+ \frac{dR}{d\theta}.\{1 + M\cos(nt + e - \varpi) + 2N\cos(2nt + 2\epsilon - 2\varpi) + \&c.\}.$$

$$\text{Adding these, } \frac{dR}{d\varpi} + \frac{dR}{d\epsilon} = \frac{dR}{d\theta} *.$$

$$\text{Substituting, } \frac{dh_{\prime}}{dt} = -\left(\frac{dR}{d\varpi} + \frac{dR}{d\epsilon}\right).$$

---

* This is also true if the orbits of the two planets are inclined, provided that the original plane of the disturbed planet's orbit be taken for the plane of xy.

110. Now by (11) $h_{,} = \sqrt{\{\mu a_{,}(1 - e_{,}^2)\}}$:

whence $$\frac{dh_{,}}{dt} = \frac{1}{2}\sqrt{\left\{\frac{\mu(1-e_{,}^2)}{a_{,}}\right\}}\cdot\frac{da_{,}}{dt} - \sqrt{\left(\frac{\mu a_{,}}{1-e_{,}^2}\right)}\cdot e_{,}\frac{de_{,}}{dt},$$

and $$e_{,}\frac{de_{,}}{dt} = \frac{1-e_{,}^2}{2a_{,}}\cdot\frac{da_{,}}{dt} - \sqrt{\left(\frac{1-e_{,}^2}{\mu a_{,}}\right)}\cdot\frac{dh_{,}}{dt}.$$

Substituting the values of $\frac{da_{,}}{dt}$ and $\frac{dh_{,}}{dt}$

$$\frac{de_{,}}{dt} = \frac{-na}{\mu e}\cdot(1-e^2)\frac{dR}{d\epsilon} + \frac{na\sqrt{(1-e^2)}}{\mu e}\cdot\left(\frac{dR}{d\epsilon} + \frac{dR}{d\varpi}\right).$$

111. PROP. 42. To investigate the alteration of the longitude of perihelion in a disturbed orbit.

The radius vector $r_{,}$ or $\frac{a_{,}(1-e_{,}^2)}{1+e_{,}\cos(\theta_{,}-\varpi_{,})}$ is a function of $a_{,}$, $e_{,}$, and $\varpi_{,}$, as well as of $\theta_{,}$. And to represent correctly the place of the body in the orbit really described, $a_{,}$, $e_{,}$, and $\varpi_{,}$, must all be considered variable; and therefore to find the correct value of $\frac{dr_{,}}{dt}$, we must differentiate considering $a_{,}$, $e_{,}$, $\varpi_{,}$, all as functions of t. The correct value of $\frac{dr_{,}}{dt}$ in the real orbit is therefore

$$\frac{dr_{,}}{d\theta_{,}}\cdot\frac{d\theta_{,}}{dt} + \frac{dr_{,}}{da_{,}}\cdot\frac{da_{,}}{dt} + \frac{dr_{,}}{de_{,}}\cdot\frac{de_{,}}{dt} + \frac{dr_{,}}{d\varpi_{,}}\cdot\frac{d\varpi_{,}}{dt}.$$

But if the disturbing force should cease to act, the body would describe the ellipse (105) of which the elements are $a_{,}$, $e_{,}$, $\varpi_{,}$, and therefore in that ellipse $\frac{dr_{,}}{dt} = \frac{dr_{,}}{d\theta_{,}}\cdot\frac{d\theta_{,}}{dt}$. And by the suppositions of (105), the velocities and directions of motion in the two cases are to be the same, and consequently $\frac{dr_{,}}{dt}$, $\frac{dr_{,}}{d\theta_{,}}$, and $\frac{d\theta_{,}}{dt}$, are to be the same in both. Making the two expressions equal,

$$0 = \frac{dr_{,}}{da_{,}}\cdot\frac{da_{,}}{dt} + \frac{dr_{,}}{de_{,}}\cdot\frac{de_{,}}{dt} + \frac{dr_{,}}{d\varpi_{,}}\cdot\frac{d\varpi_{,}}{dt},$$

$$\text{or } 0 = \frac{1 - e_{,}^{2}}{1 + e_{,} \cos(\theta_{,} - \varpi_{,})} \cdot \frac{da_{,}}{dt} - a_{,} \cdot \frac{2e_{,} + (1 + e_{,}^{2}) . \cos(\theta_{,} - \varpi_{,})}{(1 + e_{,} \cos \overline{\theta_{,} - \varpi_{,}})^{2}} \cdot \frac{de'}{dt}$$

$$- a_{,} \frac{(1 - e_{,}^{2}) . e_{,} \sin(\theta_{,} - \varpi_{,})}{(1 + e_{,} \cos \overline{\theta_{,} - \varpi_{,}})^{2}} \cdot \frac{d\varpi_{,}}{dt};$$

whence, putting a, e, &c. for $a_{,}$, $e_{,}$, &c.

$$\frac{d\varpi_{,}}{dt} = \frac{1}{e \sin(\theta - \varpi)} \left\{ \frac{1 + e . \cos(\theta - \varpi)}{a} \cdot \frac{da_{,}}{dt} \right.$$

$$\left. - \frac{2e + (1 + e^{2}) . \cos(\theta - \varpi)}{1 - e^{2}} \cdot \frac{de_{,}}{dt} \right\}$$

$$= \frac{na}{\mu e \sin(\theta - \varpi)} \left\{ \frac{1 - e^{2}}{e} \cos(\theta - \varpi) . \frac{dR}{d\epsilon} \right.$$

$$\left. - \frac{2e + (1 + e^{2}) . \cos(\theta - \varpi)}{e \sqrt{(1 - e^{2})}} \left(\frac{dR}{d\epsilon} + \frac{dR}{d\varpi} \right) \right\}.$$

We shall soon find a method of simplifying this expression.

112. Prop. 43. To find the values of $\frac{d\theta}{de}$ and $\frac{dr}{de}$ in an elliptic orbit.

$$\text{By (12) } \frac{dt}{d\theta} = \frac{a^{\frac{3}{2}} (1 - e^{2})^{\frac{3}{2}}}{\sqrt{(\mu)} (1 + e \cos \overline{\theta - \varpi})^{2}}.$$

For e put $e + \delta e$, and expand as far as the first power of δe:

$$\text{then } \frac{dt}{d\theta} = \frac{a^{\frac{3}{2}} (1 - e^{2})^{\frac{3}{2}}}{\sqrt{(\mu)} (1 + e \cos \overline{\theta - \varpi})^{2}}$$

$$- \frac{a^{\frac{3}{2}} \sqrt{(1 - e^{2})}}{\sqrt{(\mu)}} \cdot \frac{3e + (2 + e^{2}) . \cos(\theta - \varpi)}{(1 + e \cos \overline{\theta - \varpi})^{3}} \delta e.$$

$$\text{Integrating, } t = \int_{\theta} \frac{a^{\frac{3}{2}} (1 - e^{2})^{\frac{3}{2}}}{\sqrt{(\mu)} (1 + e \cos \overline{\theta - \varpi})^{2}}$$

$$- \frac{a^{\frac{3}{2}} \sqrt{(1 - e^{2})}}{\sqrt{(\mu)}} \cdot \frac{\sin(\theta - \varpi) (2 + e \cos \overline{\theta - \varpi})}{(1 + e \cos \overline{\theta - \varpi})^{2}} . \delta e,$$

the constant, in the integration of the term multiplying δe, being made $=0$, because, whatever be the value of e, we always make the equation of the center $=0$ when $nt+\epsilon-\varpi=0$ (19), or when $\theta-\varpi=0$: and the same must therefore hold for δe.

$$\text{And } t+\frac{a^{\frac{3}{2}}\sqrt{(1-e^2)}}{\sqrt{(\mu)}}\cdot\frac{\sin(\theta-\varpi)(2+e\cos\overline{\theta-\varpi})}{(1+e\cos\overline{\theta-\varpi})^2}\delta e$$

$$=\int_\theta\frac{a^{\frac{3}{2}}(1-e^2)^{\frac{3}{2}}}{\sqrt{(\mu)}(1+e\cos\overline{\theta-\varpi})^2}.$$

The second side will be some function of θ which we will call $\phi(\theta)$. For convenience, put t' for the second term on the first side, then

$$\phi(\theta)=t+t',$$

and consequently

$$\phi(\theta+\theta')=t+t'+\phi'(\theta).\theta',$$

θ' being supposed small. Here the longitude $\theta+\theta'$ corresponds to the time $t+t'+\phi'(\theta).\theta'$, on the same suppositions as those on which the investigation preceding has been made, namely, that the eccentricity is $e+\delta e$. Make

$$t'+\phi'(\theta).\theta'=0, \qquad \text{or } \theta'=-\frac{t'}{\phi'(\theta)};$$

then the longitude $\theta-\dfrac{t'}{\phi'(\theta)}$ corresponds to the time t; or the longitude at the time $t=$ longitude as usually expressed

$$+\frac{a^{\frac{3}{2}}\sqrt{1-e^2}}{\sqrt{\mu}}\cdot\frac{\sin(\theta-\varpi)(2+e\cos\overline{\theta-\varpi})}{(1+e\cos\overline{\theta-\varpi})^2}\delta e.\frac{\sqrt{\mu}.(1+e\cos\overline{\theta-\varpi})^2}{a^{\frac{3}{2}}(1-e^2)^{\frac{3}{2}}}$$

$=$ longitude as usually expressed

$$+\frac{\sin(\theta-\varpi)(2+e\cos\overline{\theta-\varpi})}{1-e^2}\delta e,$$

and hence by the definition of the differential coefficient,

$$\frac{d\theta}{de}=\frac{\sin(\theta-\varpi)(2+e\cos\overline{\theta-\varpi})}{1-e^2}.$$

113. For $\frac{dr}{de}$ we must observe that the value of r at a given time depends on e, partly because it contains e explicitly, and partly because it contains θ which depends on e. Thus

$$\frac{d(r)}{de} = \frac{dr}{de} + \frac{dr}{d\theta} \cdot \frac{d\theta}{de} = -a \cdot \frac{2e + (1 + e^2) \cdot \cos(\theta - \varpi)}{(1 + e\cos . \overline{\theta - \varpi})^2}$$

$$+ \frac{a(1 - e^2) e \sin(\theta - \varpi)}{(1 + e\cos\overline{\theta - \varpi})^2} \cdot \frac{\sin(\theta - \varpi)(2 + e\cos\overline{\theta - \varpi})}{1 - e^2}$$

$$= -a\cos(\theta - \varpi).$$

114. Prop. 44. To find an expression for $\frac{dR}{de}$ in terms of $\frac{dR}{d\epsilon}$ and $\frac{dR}{d\varpi}$, R being supposed expanded in terms of t.

Since R is originally a function of r and θ,

$$\frac{dR}{de} = \frac{dR}{dr} \cdot \frac{dr}{de} + \frac{dR}{d\theta} \cdot \frac{d\theta}{de}.$$

We have already (109) found

$$\frac{dR}{d\theta} = \frac{dR}{d\epsilon} + \frac{dR}{d\varpi}:$$

and to find $\frac{dR}{dr}$ we shall proceed thus.

$$\frac{d(R)}{dt} = \frac{dR}{dr} \cdot \frac{dr}{dt} + \frac{dR}{d\theta} \cdot \frac{d\theta}{dt}.$$

But by (88) $\frac{d(R)}{dt} = n\frac{dR}{d\epsilon}$;

also if r be expressed in terms of θ without t, r is a function of t only because θ is a function of t, and therefore

$$\frac{dr}{dt} = \frac{dr}{d\theta} \cdot \frac{d\theta}{dt}.$$

Hence we have

$$n\frac{dR}{d\epsilon} = \left(\frac{dR}{dr}\cdot\frac{dr}{d\theta} + \frac{dR}{d\theta}\right)\cdot\frac{d\theta}{dt}$$

$$= \left(\frac{dR}{dr}\cdot\frac{a(1-e^2)e.\sin(\theta-\varpi)}{(1+e\cos\overline{\theta-\varpi})^2} + \frac{dR}{d\epsilon} + \frac{dR}{d\varpi}\right)\frac{\sqrt{(\mu)}(1+e\cos\overline{\theta-\varpi})^2}{a^{\frac{3}{2}}(1-e^2)^{\frac{3}{2}}}.$$

Observing that $\frac{\sqrt{(\mu)}}{a^{\frac{3}{2}}} = n$, this becomes

$$\frac{dR}{d\epsilon} = \frac{dR}{dr}\cdot\frac{ae\sin(\theta-\varpi)}{\sqrt{(1-e)^2}} + \left(\frac{dR}{d\epsilon} + \frac{dR}{d\varpi}\right)\cdot\frac{(1+e\cos\overline{\theta-\varpi})^2}{(1-e^2)^{\frac{3}{2}}},$$

whence

$$\frac{dR}{dr} = \frac{\sqrt{(1-e^2)}}{ae\sin(\theta-\varpi)}\left\{\frac{dR}{d\epsilon} - \left(\frac{dR}{d\epsilon} + \frac{dR}{d\varpi}\right)\cdot\frac{(1+e\cos\overline{\theta-\varpi})^2}{(1-e^2)^{\frac{3}{2}}}\right\}.$$

Putting this value in the expression for $\frac{dR}{de}$, and the values of $\frac{dr}{de}$, $\frac{d\theta}{de}$, from (112) and (113), $\frac{dR}{de} =$

$$-\frac{\sqrt{(1-e^2)}}{e\sin(\theta-\varpi)}\left\{\frac{dR}{d\epsilon} - \left(\frac{dR}{d\epsilon} + \frac{dR}{d\varpi}\right)\cdot\frac{(1+e\cos\overline{\theta-\varpi})^2}{(1-e^2)^{\frac{3}{2}}}\right\}\cos(\theta-\varpi)$$

$$+\left(\frac{dR}{d\epsilon} + \frac{dR}{d\varpi}\right)\cdot\frac{\sin(\theta-\varpi)(2+e\cos\overline{\theta-\varpi})}{1-e^2}$$

$$= -\frac{\sqrt{(1-e^2)}\,.\cos(\theta-\varpi)}{e\sin(\theta-\varpi)}\cdot\frac{dR}{d\epsilon}$$

$$+\frac{(1+e^2)\cos(\theta-\varpi)+2e}{(1-e^2)e\sin(\theta-\varpi)}\left(\frac{dR}{d\epsilon} + \frac{dR}{d\varpi}\right).$$

115. Comparing this with the value of $\frac{d\varpi_{,}}{dt}$ in (111)

$$\frac{d\varpi_{,}}{dt} = -\frac{na\sqrt{(1-e^2)}}{\mu e}\cdot\frac{dR}{de}.$$

116. We have now obtained the variations of the elements $a_{,}$, $e_{,}$, and $\varpi_{,}$, upon which the dimensions and position

of the imaginary elliptic orbit at any instant depend. Another parameter however is necessary to determine the body's place in the instantaneous orbit, namely the epoch, or the mean longitude when $t = 0$, which we have called ϵ. We now proceed to find its variation.

117. Prop. 45. To find the variation of the epoch in a disturbed orbit.

By assuming that the planet moves for an instant in a part of an elliptic orbit, we assume that all the expressions for the longitude and the radius vector and their first differentials with respect to t, whether expressed in terms of θ or of t, are the same as if the elements of the elliptic motion were for an instant invariable. Yet it is plain that the differentials will be correctly found by considering those elements as variable, since they *must* be varied to represent the motion from any point to any distant point. And all that applies to both $r_{\prime}$ and $\theta_{\prime}$ will apply to any function of $r_{\prime}$ and $\theta_{\prime}$, as for instance R, which depends on m only by being a function of $r_{\prime}$ and $\theta_{\prime}$. Now suppose $r_{\prime}$ and $\theta_{\prime}$ expanded as in (87): they will be functions of $a_{\prime}$, $e_{\prime}$, $\varpi_{\prime}$, and $n_{\prime}t+\epsilon_{\prime}$, where it must be remarked that $n_{\prime}$ or $\frac{\sqrt{(\mu)}}{a_{\prime}^{\frac{3}{2}}}$ is variable if $a_{\prime}$ is variable. Consequently R, supposed to be expanded in the same way, is a function of $a_{\prime}$, $e_{\prime}$, $\varpi_{\prime}$, and $n_{\prime}t + \epsilon_{\prime}$. Hence $\frac{d(R)}{dt}$ (observing that $n_{\prime}t$ and $\epsilon_{\prime}$ always go together) is

$$\frac{dR}{da_{\prime}} \cdot \frac{da_{\prime}}{dt} + \frac{dR}{de_{\prime}} \cdot \frac{de_{\prime}}{dt} + \frac{dR}{d\varpi_{\prime}} \cdot \frac{d\varpi_{\prime}}{dt} + \frac{dR}{d\epsilon_{\prime}} \cdot \frac{d(n_{\prime}t + \epsilon_{\prime})}{dt},$$

$$\text{or nearly } \frac{dR}{da} \cdot \frac{da_{\prime}}{dt} + \frac{dR}{de} \cdot \frac{de_{\prime}}{dt} + \frac{dR}{d\varpi} \cdot \frac{d\varpi_{\prime}}{dt}$$

$$+ \frac{dR}{d\epsilon}\left(n_{\prime} + t\frac{dn_{\prime}}{dt} + \frac{d\epsilon_{\prime}}{dt}\right).$$

But if the elements were for an instant invariable, $\frac{d(R)}{dt}$ would be $\frac{dR}{d\epsilon} n_{\prime}$. Making these equal, we get this equation

$$0 = \frac{dR}{da} \cdot \frac{da_{\prime}}{dt} + \frac{dR}{de} \cdot \frac{de_{\prime}}{dt} + \frac{dR}{d\varpi} \cdot \frac{d\varpi_{\prime}}{dt} + \frac{dR}{d\epsilon}\left(t\frac{dn_{\prime}}{dt} + \frac{d\epsilon_{\prime}}{dt}\right).$$

Substituting the values of $\frac{da_{\prime}}{dt}$, $\frac{de_{\prime}}{dt}$, $\frac{d\varpi_{\prime}}{dt}$, from (107), (110), and (115), and putting for $\frac{dn_{\prime}}{dt}$ its value $-\frac{3n_{\prime}}{2a_{\prime}} \cdot \frac{da_{\prime}}{dt}$,

$$\text{or } \frac{3n^2 a}{\mu} \cdot \frac{dR}{d\epsilon},$$

and dividing the whole by $\frac{dR}{d\epsilon}$,

$$\frac{d\epsilon_{\prime}}{dt} = -\frac{3n^2 a}{\mu} \cdot \frac{dR}{d\epsilon} t + \frac{2na^2}{\mu} \cdot \frac{dR}{da}$$

$$-\frac{na}{\mu e}\{\sqrt{(1-e^2)} - (1-e^2)\}\frac{dR}{de}.$$

118. PROP. 46. To calculate the longitude of a disturbed planet.

For convenience of reference, we will collect here the expressions for the variations of the elements.

$$\frac{da_{\prime}}{dt} = -\frac{2na^2}{\mu} \cdot \frac{dR}{d\epsilon} \quad (107)$$

$$\frac{dn_{\prime}}{dt} = \frac{3n^2 a}{\mu} \cdot \frac{dR}{d\epsilon}$$

$$\frac{de_{\prime}}{dt} = -\frac{na}{\mu e}(1-e^2)\frac{dR}{d\epsilon} + \frac{na(1-e^2)^{\frac{1}{2}}}{\mu e}\left(\frac{dR}{d\epsilon} + \frac{dR}{d\varpi}\right) \quad (110)$$

$$\frac{d\varpi_{\prime}}{dt} = -\frac{na}{\mu e}(1-e^2)^{\frac{1}{2}} \cdot \frac{dR}{de} \quad (115)$$

$$\frac{d\epsilon_{\prime}}{dt} = -\frac{3n^2 a}{\mu} \cdot \frac{dR}{d\epsilon} t + \frac{2na^2}{\mu} \cdot \frac{dR}{da} - \frac{na}{\mu\epsilon}(\sqrt{1-e^2} - \overline{1-e^2})\frac{dR}{de} \quad (117).$$

When R is developed, the values of $a_{\prime}$, $n_{\prime}$, $e_{\prime}$, $\varpi_{\prime}$, and $\epsilon_{\prime}$, are to be found from these expressions: and they are then to be substituted in the series

$$\theta_{,} = n_{,}t + \epsilon_{,} + \left(2e_{,} - \frac{e_{,}^{3}}{4} + \&c.\right)\sin(n_{,}t + \epsilon_{,} - \varpi_{,})$$

$$+ \left(\frac{5e_{,}^{2}}{4} + \&c.\right)\sin(2n_{,}t + 2\epsilon_{,} - 2\varpi_{,}) + \&c.$$

which applies to an elliptic orbit of which they are the elements (87).

119. Ex. Suppose one term in the development of R to be

$$P \,.\, e^{3} e'^{2} \,.\, \cos \overline{13(nt+\epsilon) - 8(n't+\epsilon') - 3\varpi - 2\varpi'},$$

where P is a function of a and a'. (We shall for abbreviation call this $P e^{3} e'^{2} \cos T$).

Here $\dfrac{dR}{d\epsilon} = -13 \,.\, P e^{3} e'^{2} \sin T,$

$$\frac{dR}{d\varpi} = 3 P e^{3} e'^{2} \sin T,$$

$$\text{whence } \frac{dR}{d\epsilon} + \frac{dR}{d\varpi} = -10 P e^{3} e'^{2} \sin T,$$

$$\frac{dR}{de} = 3 P e^{2} e'^{2} \cos T,$$

$$\frac{dR}{da} = \frac{dP}{da} \,.\, e^{3} e'^{2} \cos T.$$

Hence $\dfrac{da_{,}}{dt} = \dfrac{26 n a^{2}}{\mu} P e^{3} e'^{2} \sin T$:

$$\text{integrating, } a_{,} = a - \frac{26 n a^{2}}{\mu(13n - 8n')} P e^{3} e'^{2} \cos T.$$

$$\frac{dn_{,}}{dt} = -\frac{39 n^{2} a}{\mu} P e^{3} e'^{2} \sin T:$$

$$n_{,} = n + \frac{39 n^{2} a}{\mu(13n - 8n')} P e^{3} e'^{2} \cos T.$$

$$\frac{de_{,}}{dt} = \frac{13na}{\mu}(1 - e^2) P e^2 e'^2 \sin T$$

$$- \frac{10na(1 - e^2)^{\frac{1}{2}}}{\mu} P e^2 e'^2 \sin T,$$

$$e_{,} = e - \frac{13na(1 - e^2)}{\mu(13n - 8n')} P e^2 e'^2 \cos T$$

$$+ \frac{10na(1 - e^2)^{\frac{1}{2}}}{\mu(13n - 8n')} P e^2 e'^2 \cos T.$$

$$\frac{d\varpi_{,}}{dt} = - \frac{3na(1 - e^2)^{\frac{1}{2}}}{\mu} P e e'^2 . \cos T:$$

$$\varpi_{,} = \varpi - \frac{3na(1 - e^2)^{\frac{1}{2}}}{\mu(13n - 8n')} P e e'^2 \sin T.$$

$$\frac{d\epsilon_{,}}{dt} = \frac{39n^2 a}{\mu} P e^3 e'^2 . t . \sin T + \frac{2na^2}{\mu} . \frac{dP}{da} e^3 e'^2 . \cos T$$

$$- \frac{3na}{\mu}(\sqrt{1 - e^2} - \overline{1 - e^2}) P e e'^2 . \cos T:$$

$$\epsilon_{,} = \epsilon - \frac{39n^2 a}{\mu(13n - 8n')} P e^3 e'^2 . t \cos T + \frac{39n^2 a}{\mu(13n - 8n')^2} P e^3 e'^2 \sin T$$

$$+ \frac{2na^2}{\mu(13n - 8n')} . \frac{dP}{da} e^3 e'^2 \sin T$$

$$- \frac{3na}{\mu(13n - 8n')}(\sqrt{1 - e^2} - \overline{1 - e^2}) P e e'^2 \sin T.$$

Hence, expanding $\sqrt{1 - e^2}$, and retaining in the substitutions no power of e beyond e^3, $n_{,}t + \epsilon_{,} = nt + \epsilon$

$$+ \left\{ \frac{39n^2 a P}{\mu(13n - 8n')^2} + \frac{2na^2}{\mu(13n - 8n')} . \frac{dP}{da} - \frac{3naP}{2\mu(13n - 8n')} \right\} e^3 e'^2 \sin T,$$

$$n_{,}t + \epsilon_{,} - \varpi_{,} = nt + \epsilon - \varpi$$

$$+ \left\{ \left(\frac{39n^2 a P}{\mu(13n - 8n')^2} + \frac{2na^2}{\mu(13n - 8n')} \frac{dP}{da} - \frac{3naP}{\mu(13n - 8n')} \right) e^3 e'^2 \right.$$

$$\left. + \frac{3na}{\mu(13n - 8n')} P e e'^2 \right\} \sin T,$$

(which we shall call $nt + \epsilon - \varpi + x$)

$$e_{,} = e - \frac{3naP}{\mu(13n - 8n')} e^2 e'^2 \cos T.$$

Substituting these in the series for $\theta_{,}$, and observing that

$$\sin(nt + \epsilon - \varpi + x) = \sin(nt + \epsilon - \varpi) + x\cos(nt + \epsilon - \varpi),$$

and neglecting products which produce higher powers than $e^3 e'^2$,

$$\theta_{,} = nt + \epsilon + (2e - \frac{e^3}{4} + \&c.)\sin(nt + \epsilon - \varpi)$$

$$+ \left(\frac{5e^2}{4} + \&c.\right)\sin(2nt + 2\epsilon - 2\varpi) + \&c.$$

$$+ \left\{\frac{39n^2 aP}{\mu(13n-8n')^2} + \frac{2na^2}{\mu(13n-8n')}\frac{dP}{da} - \frac{3naP}{2\mu(13n-8n')}\right\} e^3 e'^2 \sin T$$

$$+ \frac{3naP}{\mu(13n-8n')} e^2 e'^2 . \{\sin \overline{T - (nt + \epsilon - \varpi)} + \sin \overline{T + (nt + \epsilon - \varpi)}\}$$

$$+ \frac{3naP}{\mu(13n-8n')} e^2 e'^2 . \{\sin \overline{T - (nt + \epsilon - \varpi)} - \sin \overline{T + (nt + \epsilon - \varpi)}\}.$$

The first two lines are the longitude calculated with invariable elements. The additional terms are

$$\left\{\frac{39n^2 aP}{\mu(13n - 8n')^2} + \frac{2na^2}{\mu(13n - 8n')} \cdot \frac{dP}{da} - \frac{3naP}{2\mu(13n - 8n')}\right\}$$

$$\times e^3 e'^2 \sin \overline{13(nt + \epsilon) - 8(n't + \epsilon') - 3\varpi - 2\varpi'}$$

$$+ \frac{6naP}{\mu(13n - 8n')} e^2 e'^2 \sin . \overline{12(nt + \epsilon) - 8(n't + \epsilon') - 2\varpi - 2\varpi'}$$

constituting two distinct inequalities. The value of these has been adverted to in (100) &c. supposing $13n - 8n'$ small. We may remark that in this case the first inequality is twice divided by $13n - 8n'$: the second, though only once divided by $13n - 8n'$, is multiplied only by $e^2 e'^2$ instead of $e^3 e'^2$: the comparative magnitude of the terms depends then on the proportion of $\frac{13n - 8n'}{n}$ and e.

120. PROP. 47. To calculate the radius vector of a disturbed planet.

This must be done by substituting for $a_{\prime}$, $e_{\prime}$, &c. in the expression

$$r_{\prime} = a_{\prime}\left\{1 + \frac{e_{\prime}^2}{2} + \&c. + (e_{\prime} + \&c.)\cos(n_{\prime}t + \epsilon_{\prime} - \varpi_{\prime}) - \&c.\right\}.$$

Proceeding in the same manner as in the last Article, and using the same instance, we find for the additional terms in $r_{\prime}$

$$-\frac{26na^2}{\mu(13n - 8n')} Pe^3e'^2 \cos \overline{13(nt+\epsilon) - 8(n't+\epsilon') - 3\varpi - 2\varpi'}$$

$$-\frac{3na^2}{\mu(13n - 8n')} Pe^2e'^2 . \cos \overline{12(nt+\epsilon) - 8(n't+\epsilon') - 2\varpi - 2\varpi'}.$$

The latter of these is generally the more important.

DEVELOPMENT OF R.

121. PROP. 48. Supposing $\dfrac{1}{\sqrt{(a^2 - 2aa'\cos\omega + a'^2)}}$ expanded in a series of the form

$$\tfrac{1}{2}C^{(0)} + C^{(1)}\cos\omega + C^{(2)}\cos 2\omega + \&c.:$$

to calculate numerically the values of $C^{(0)}$ and $C^{(1)}$.

Case 1. Let a and a' be very unequal: suppose for instance that a' is much less than a. Then the fraction

$$= \frac{1}{a}\left(1 - 2\frac{a'}{a}\cos\omega + \frac{a'^2}{a^2}\right)^{-\frac{1}{2}}.$$

For $2\cos\omega$ put $\epsilon^{\omega\sqrt{-1}} + \epsilon^{-\omega\sqrt{-1}}$ (ϵ being here the base of Napierian logarithms) then the fraction

$$= \frac{1}{a}\left(1 - \frac{a'}{a}\epsilon^{\omega\sqrt{-1}}\right)^{-\frac{1}{2}} . \left(1 - \frac{a'}{a}\epsilon^{-\omega\sqrt{-1}}\right)^{-\frac{1}{2}}$$

$$= \frac{1}{a}\left\{1 + \frac{1a'}{2a}\epsilon^{\omega\sqrt{-1}} + \frac{1.3a'^2}{2.4a^2}\epsilon^{2\omega\sqrt{-1}} + \frac{1.3.5a'^3}{2.4.6a^3}\epsilon^{3\omega\sqrt{-1}} + \&c.\right\}$$

$$\times \left\{1 + \frac{1a'}{2a}\epsilon^{-\omega\sqrt{-1}} + \frac{1.3a'^2}{2.4a^2}\epsilon^{-2\omega\sqrt{-1}} + \frac{1.3.5a'^3}{2.4.6a^3}\epsilon^{-3\omega\sqrt{-1}} + \&c.\right\}$$

$$= \frac{1}{a}\left\{1 + \left(\frac{1}{2}\right)^2\frac{a'^2}{a^2} + \left(\frac{1.3}{2.4}\right)^2\frac{a'^4}{a^4} + \left(\frac{1.3.5}{2.4.6}\right)^2\frac{a'^6}{a^6} + \&c.\right\}$$

$$+ \frac{1}{a}\left\{\frac{1}{2}\frac{a'}{a} + \frac{1}{2}\cdot\frac{1.3}{2.4}\cdot\frac{a'^3}{a^3} + \frac{1.3}{2.4}\cdot\frac{1.3.5}{2.4.6}\cdot\frac{a'^5}{a^5} + \&c.\right\} \times$$

$$(\epsilon^{\omega\sqrt{-1}} + \epsilon^{-\omega\sqrt{-1}})$$

$$+ (\&c.)\,(\epsilon^{2\omega\sqrt{-1}} + \epsilon^{-2\omega\sqrt{-1}}) + \&c.$$

$$= \frac{1}{a}\left\{1 + \left(\frac{1}{2}\right)^2\cdot\frac{a'^2}{a^2} + \left(\frac{1.3}{2.4}\right)^2\cdot\frac{a'^4}{a^4} + \left(\frac{1.3.5}{2.4.6}\right)^2\cdot\frac{a'^6}{a^6} + \&c.\right\}$$

$$+ \frac{2}{a}\left\{\frac{1}{2}\cdot\frac{a'}{a} + \frac{1}{2}\cdot\frac{1.3}{2.4}\cdot\frac{a'^3}{a^3} + \frac{1.3}{2.4}\cdot\frac{1.3.5}{2.4.6}\cdot\frac{a'^5}{a^5} + \&c.\right\}\cos\omega + \&c.$$

As $\frac{a'}{a}$ is small, these series converge very rapidly.

122. *Case* 2. Suppose that a' is not very much less than a. Put α for $\frac{a'}{a}$ and $\pi - 2\psi$ for ω: then

$$\frac{1}{a\sqrt{(1+2\alpha\cos 2\psi + \alpha^2)}} = \tfrac{1}{2}C^{(0)} - C^{(1)}\cos 2\psi + C^{(2)}\cos 4\psi - \&c.$$

Integrating both sides with respect to ψ from $\psi = 0$ to $\psi = \frac{\pi}{2}$,

$$\text{we get } \frac{\pi}{4}C^{(0)} = \int_\psi \frac{1}{a\sqrt{(1+2\alpha\cos 2\psi + \alpha^2)}}:$$

and the problem is now reduced to finding the value of this definite integral. Now let

$$\sin\psi' = \frac{\sin 2\psi}{\sqrt{(1+2\alpha\cos 2\psi+\alpha^2)}};$$

$$\cos\psi' = \frac{\cos 2\psi + \alpha}{\sqrt{(1+2\alpha\cos 2\psi + \alpha^2)}}: \quad \tan\psi' = \frac{\sin 2\psi}{\cos 2\psi + \alpha}.$$

It is plain that as ψ increases from 0 till $\cos 2\psi = -a$, ψ' increases from 0 to 90°; and as ψ increases to 90°, ψ' increases to 180°.

By differentiating the equation $\tan\psi'(\cos 2\psi + a) = \sin 2\psi$, we get

$$\frac{\cos 2\psi + a}{\cos^2\psi'} \cdot \frac{d\psi'}{d\psi} - 2\tan\psi' . \sin 2\psi = 2\cos 2\psi;$$

$$\text{or} \quad \frac{\cos 2\psi + a}{\cos\psi'} \cdot \frac{d\psi'}{d\psi} = 2(\cos 2\psi \cos\psi' + \sin 2\psi \sin\psi').$$

On substituting the values of $\cos\psi'$ and $\sin\psi'$, this becomes

$$\sqrt{(1 + 2a\cos 2\psi + a^2)} \cdot \frac{d\psi'}{d\psi} = 2\frac{1 + a\cos 2\psi}{\sqrt{(1 + 2a\cos 2\psi + a^2)}}$$

$$= 2\sqrt{\left(1 - \frac{a^2\sin^2 2\psi}{1 + 2a\cos 2\psi + a^2}\right)} = 2\sqrt{(1 - a^2\sin^2\psi')},$$

$$\text{whence} \quad \frac{1}{\sqrt{(1 + 2a\cos 2\psi + a^2)}} = \frac{1}{2\sqrt{(1 - a^2\sin^2\psi')}} \cdot \frac{d\psi'}{d\psi}.$$

$$\text{hence} \ \frac{\pi}{4} C^{(0)} = \frac{1}{2a}\int_\psi \frac{1}{\sqrt{(1 - a^2\sin^2\psi')}} \cdot \frac{d\psi'}{d\psi} \left(\text{from } \psi = 0 \text{ to } \psi = \frac{\pi}{2}\right)$$

$$= \frac{1}{2a}\int_{\psi'} \frac{1}{\sqrt{(1 - a^2\sin^2\psi')}} \quad (\text{from } \psi' = 0 \text{ to } \psi' = \pi)$$

$$= \frac{1}{a}\int_{\psi'} \frac{1}{\sqrt{(1 - a^2\sin^2\psi')}} \left(\text{from } \psi' = 0 \text{ to } \psi' = \frac{\pi}{2}\right).$$

$$\text{Let} \quad \frac{1}{\sqrt{(1 - a^2\sin^2\psi')}} = \frac{P}{\sqrt{\{(1 + a')^2 - 4a'\sin^2\psi'\}}}:$$

$$\text{then } P = 1 + a': \text{ and } \frac{4a'}{(1 + a')^2} = a^2, \text{ or } a' = \frac{1 - \sqrt{(1 - a^2)}}{1 + \sqrt{(1 - a^2)}}:$$

$$\text{and thus } \frac{\pi}{4} C^{(0)} = \frac{1 + a'}{a}\int_{\psi'} \frac{1}{\sqrt{(1 + 2a'\cos 2\psi' + a'^2}}$$

from $\psi' = 0$ to $\psi' = \frac{\pi}{2}$; where a' is smaller than a. If we proceed, forming ψ'' from ψ', ψ''' from ψ'', &c. a'' from

a', a''' from a'', &c. by the same law, it is plain that we shall get $\frac{\pi}{4} C^{(0)}$

$$= \frac{1}{a} \cdot (1 + a') \cdot (1 + a'') \cdot (1 + a''') \ldots\ldots (1 + a^{(n)}) \times$$

$$\int_{\psi^{(n)}} \frac{1}{\sqrt{(1 + 2\, a^{(n)} \cos 2\psi^{(n)} + a^{(n)2})}}.$$

And as a', a'', a''', &c. diminish very rapidly, we may soon find one of them $a^{(n)}$ so small that the term under the integral sign will not differ sensibly from $\int_{\psi^{(n)}} . 1 = \frac{\pi}{2}$.

$$\text{Then } C^{(0)} = \frac{2}{a} (1 + a') \cdot (1 + a'') \cdot (1 + a''') \cdot \&c.$$

The calculation of this is very easy:

for if $a = \sin\phi$, $a' = \dfrac{1 - \cos\phi}{1 + \cos\phi} = \tan^2\dfrac{\phi}{2}$; $\therefore$ $(1 + a') = \sec^2\dfrac{\phi}{2}$.

And if a' or $\tan^2\dfrac{\phi}{2} = \sin\phi'$, $(1 + a'')$ is found similarly to be $= \sec^2\dfrac{\phi'}{2}$: &c. Thus the rule is simply this:

Make $\sin\phi = a$: $\sin\phi' = \tan^2\dfrac{\phi}{2}$: $\sin\phi'' = \tan^2\dfrac{\phi'}{2}$: &c.:

then $C^{(0)} = \dfrac{2}{a} \sec^2\dfrac{\phi}{2} . \sec^2\dfrac{\phi'}{2} . \&c.$ as far as the secant differs sensibly from 1.

123. $C^{(1)}$ may be thus calculated. Multiply both sides of the equation at the beginning of (122) by $\cos 2\psi$ and integrate:

then $\dfrac{\pi}{4} C^{(1)} = -\dfrac{1}{a} \displaystyle\int_{\psi} \frac{\cos 2\psi}{\sqrt{\{1 + 2a\cos 2\psi + a^2\}}}$ from $\psi = 0$ to $\psi = \dfrac{\pi}{2}$.

By making the same assumption as in (122) and by solving the equation

$$\frac{\cos 2\psi + a}{\sqrt{(1 + 2a\cos 2\psi + a^2)}} = \cos\psi',$$

which gives

$$\cos 2\psi = -a\sin^2\psi' + \cos\psi'\sqrt{(1 - a^2\sin^2\psi')},$$

this may be converted into

$$\frac{1}{a}\int_{\psi'}\left\{-\frac{1}{2}\cos\psi' + \frac{a(2\sin^2\psi' - 1)}{4\sqrt{(1 - a^2\sin^2\psi')}} + \frac{a}{4}\cdot\frac{1}{\sqrt{(1 - a^2\sin^2\psi')}}\right\}$$

$$= -\frac{1}{a}\cdot\frac{a\cdot(1 + a')}{2}\int_{\psi'}\cdot\frac{\cos 2\psi'}{\sqrt{(1 + 2a'\cos 2\psi' + a'^2)}}\begin{Bmatrix}\psi' = 0\\ \psi' = \frac{\pi}{2}\end{Bmatrix}$$

$$+ \frac{\pi}{8}\cdot aC^{(0)}.$$

Continuing the process we find

$$\frac{\pi}{4}C^{(1)} = \frac{\pi}{4}C^{(0)}\left\{\frac{a}{2} + \frac{aa'}{2.2} + \frac{aa'a''}{2.2.2} + \&c.\right\},$$

$$\text{or}\quad C^{(1)} = C^{(0)}\left\{\frac{\sin\phi}{2} + \frac{\sin\phi}{2}\cdot\frac{\sin\phi'}{2} + \&c.\right\}$$

which is very easily calculated.

124. Prop. 49. Given the numerical values of two consecutive coefficients $C^{(k-1)}$, $C^{(k)}$, to find the numerical value of the next $C^{(k+1)}$.

$$\text{Since}\quad \frac{1}{(a^2 - 2aa'\cos\omega + a'^2)^{\frac{1}{2}}} = \tfrac{1}{2}C^{(0)} + C^{(1)}\cos\omega + \&c.$$

$$+ C^{(k-1)}\cos(k-1)\omega + C^{(k)}\cos k\omega + C^{(k+1)}\cos(k+1)\omega + \&c.;$$

differentiating the logarithm of each side with respect to ω

$$\frac{aa'\sin\omega}{(a^2 + a'^2) - 2aa'\cos\omega}$$

$$\frac{C^{(1)}\sin\omega + \&c. + \overline{k-1}.C^{(k-1)}\sin(k-1)\omega + k.C^{(k)}\sin k\omega + \overline{k+1}.C^{(k+1)}\sin(k+1)\omega + \&c.}{\frac{1}{2}C^{(0)} + C^{(1)}\cos\omega + \&c. + C^{(k-1)}\cos(k-1)\omega + C^{(k)}\cos k\omega + C^{(k+1)}\cos(k+1)\omega + \&c.}$$

Multiplying cross-wise, and taking the coefficient of $\sin k\omega$,

$$(a^2 + a'^2) \cdot k \cdot C^{(k)} - aa' \cdot (k-1) \cdot C^{(k-1)} - aa' \cdot (k+1) \cdot C^{(k+1)}$$

$$= \tfrac{1}{2}\, aa'\, C^{(k-1)} - \tfrac{1}{2}\, aa'\, C^{(k+1)},$$

$$\text{whence } C^{(k+1)} = \frac{2k}{2k+1} \cdot \frac{a^2 + a'^2}{aa'}\, C^{(k)} - \frac{2k-1}{2k+1}\, C^{(k-1)}.$$

125. This formula includes (as will be seen on trial) $C^{(0)}$, $C^{(1)}$, $C^{(2)}$. And as the two first are calculated, all the others may be calculated from them by this formula.

126. In the same manner,

$$\text{if } \frac{1}{(a^2 - 2aa'\cos\omega + a'^2)^{\frac{3}{2}}} = \tfrac{1}{2}\, D^{(0)} + D^{(1)} \cos\omega + \&c.$$

$$D^{(k+1)} = \frac{2k}{2k-1} \cdot \frac{a^2 + a'^2}{aa'}\, D^{(k)} - \frac{2k+1}{2k-1}\, D^{(k-1)}.$$

127. Prop. 50. To express $D^{(k)}$ in terms of $C^{(k-1)}$ and $C^{(k)}$.

Since

$$\frac{1}{(a^2 - 2aa'\cos\omega + a'^2)^{\frac{1}{2}}} = (a^2 - 2aa'\cos\omega + a'^2) \frac{1}{(a^2 - 2aa'\cos\omega + a'^2)^{\frac{3}{2}}}$$

$$\text{or } \tfrac{1}{2}\, C^{(0)} + C^{(1)} \cos\omega + \&c.$$

$$= (a^2 - 2aa'\cos\omega + a'^2) \left\{ \tfrac{1}{2}\, D^{(0)} + D^{(1)} \cos\omega + \&c. \right\}$$

we have, comparing the coefficients of $\cos k\omega$,

$$C^{(k)} = (a^2 + a'^2)\, D^{(k)} - aa' D^{(k-1)} - aa' D^{(k+1)}.$$

Putting for $D^{(k+1)}$ the value in (126),

$$C^{(k)} = -\frac{1}{2k-1} (a^2 + a'^2)\, D^{(k)} + \frac{2}{2k-1}\, aa' D^{(k-1)}.$$

Again, comparing the coefficients of $\cos \overline{(k-1)\omega}$

$$C^{(k-1)} = (a^2 + a'^2)\, D^{(k-1)} - aa' D^{(k-2)} - aa' D^{(k)}.$$

But the formula of (126) becomes, on putting $k-1$ for k

$$D^{(k)} = \frac{2k-2}{2k-3} \cdot \frac{a^2+a'^2}{a a'} \cdot D^{(k-1)} - \frac{2k-1}{2k-3} D^{(k-2)}.$$

Whence $D^{(k-2)} = \dfrac{2k-2}{2k-1} \cdot \dfrac{a^2+a'^2}{a a'} D^{(k-1)} - \dfrac{2k-3}{2k-1} D^{(k)}$,

and substituting,

$$C^{(k-1} = -\frac{2}{2k-1} a a' D^{(k)} + \frac{1}{2k-1} (a^2+a'^2) D^{(k-1)}$$

Eliminating $D^{(k-1)}$ from the expressions for $C^{(k)}$ and $C^{(k-1)}$

$$D^{(k)} = (2k-1) \frac{2aa'}{(a^2-a'^2)^2} C^{(k-1)} - (2k-1) \cdot \frac{a^2+a'^2}{(a^2-a'^2)^2} C^{(k)}.$$

In the same manner we might find the terms in the expansion

$$\text{of } \frac{1}{(a^2 - 2aa'\cos\omega + a'^2)^{\frac{5}{2}}}, \text{ \&c.}$$

128. Prop. 51. To express $\dfrac{d \,.\, C^{(k)}}{da}$, $\dfrac{d \,.\, C^{(k)}}{da'}$, &c. in terms of quantities already found.

These will be found by differentiating $\dfrac{1}{(a^2-2aa'\cos\omega+a'^2)^{\frac{1}{2}}}$ and taking the coefficient of $\cos k\omega$ in the expansion of this differential. Now with respect to a the differential coefficient is $\dfrac{-a + a'\cos\omega}{(a^2-2aa'\cos\omega+a'^2)^{\frac{3}{2}}}$ and the coefficient of $\cos k\omega$ is

$$-a D^{(k)} + \frac{a'}{2} D^{(k-1)} + \frac{a'}{2} D^{(k+1)}.$$

As these are already calculated, $\dfrac{d \,.\, C^{(k)}}{da}$ is found.

$\dfrac{d \,.\, C^{(k)}}{da'}$ may then be most readily calculated in this manner.

Since $C^{(k)}$ is a quantity of -1 dimensions in a and a', we have by a well known theorem,

$$-C^{(k)} = a\frac{dC^{(k)}}{da} + a'\frac{dC^{(k)}}{da'}.$$

$$\text{Consequently } \frac{dC^{(k)}}{da'} = -\frac{1}{a'}C^{(k)} - \frac{a}{a'}\cdot\frac{dC^{(k)}}{da}.$$

$\dfrac{d^2.C^{(k)}}{da^2}$ will be obtained by differentiating

$$-aD^{(k)} + \frac{a'}{2}D^{(k-1)} + \frac{a'}{2}D^{(k+1)},$$

the values of $\dfrac{d.D^{(k)}}{da}$, &c. having been previously expressed by the terms of the expansion of $\dfrac{1}{(a^2-2aa'\cos\omega+a'^2)^{\frac{5}{2}}}$, &c. Then $\dfrac{d^2.C^{(k)}}{da.da'}$ will be found by observing that $\dfrac{dC^{(k)}}{da}$ is of -2 dimensions in a and a', and therefore

$$-2\frac{dC^{(k)}}{da} = a\frac{d^2C^{(k)}}{da^2} + a'\frac{d^2C^{(k)}}{dada'}.$$

Similarly $\dfrac{d^2.C^{(k)}}{da'^2}$ is found from the equation

$$-2\frac{dC^{(k)}}{da'} = a\frac{d^2C^{(k)}}{dada'} + a'\frac{d^2C^{(k)}}{da'^2}.$$

And thus we may proceed to any number of differential coefficients.

129. Prop. 52. To expand $\dfrac{1}{\sqrt{\{r^2-2rr'\cos(\theta-\theta')+r'^2\}}}$ in terms of t.

We have seen (87) that

$$r = a + a\left\{\frac{e^2}{2} + \&c. - (e + \&c.)\cos(nt + \epsilon - \varpi)\right.$$

$$\left. - \left(\frac{e^2}{2} + \&c.\right)\cos(2nt + 2\epsilon - 2\varpi) + \&c.\right\} = a + v:$$

and similarly $r' = a' + v'$, where the expression for v' is similar to that for v, putting a', e', n', ϵ', ϖ', for a, e, n, ϵ, ϖ. The expansion required will therefore be effected if we first expand

$$\frac{1}{\sqrt{\{a^2 - 2aa'\cos(\theta - \theta') + a'^2\}}},$$

and then in every part put $a + v$ for a and $a' + v'$ for a'. Let $\omega = \theta - \theta'$: then by Prop. 49 and 50

$$\frac{1}{\sqrt{\{a^2 - 2aa'\cos(\theta - \theta') + a'^2\}}}$$

may be expanded in the form

$$\tfrac{1}{2}C^{(0)} + C^{(1)}\cos(\theta - \theta') + C^{(2)}\cos 2(\theta - \theta') + C^{(3)}\cos 3(\theta - \theta') + \&c.,$$

where $C^{(0)}$, $C^{(1)}$, &c. are functions of a and a'. If then in each term we put $a + v$ for a and $a' + v'$ for a', the series will become the sum of a multitude of series, in this form

$$\begin{aligned}
\frac{1}{\sqrt{\{r^2 - 2rr'\cos(\theta - \theta') + r'^2\}}}
&= \tfrac{1}{2}C^{(0)} + \tfrac{1}{2}\frac{dC^{(0)}}{da}v + \tfrac{1}{4}\frac{d^2C^{(0)}}{da^2}v^2 \\
&\quad + \tfrac{1}{2}\frac{dC^{(0)}}{da'}v' + \tfrac{1}{2}\frac{d^2C^{(0)}}{da\,da'}vv' \\
&\quad + \tfrac{1}{4}\frac{d^2C^{(0)}}{da'^2}v'^2 + \&c. \\
&\quad + \cos(\theta - \theta') \times \left\{\begin{aligned} & C^{(1)} + \frac{dC^{(1)}}{da}v + \tfrac{1}{2}\frac{d^2C^{(1)}}{da^2}v^2 \\ & + \frac{dC^{(1)}}{da'}v' + \frac{d^2C^{(1)}}{da\,da'}vv' \\ & + \tfrac{1}{2}\frac{d^2C^{(1)}}{da'^2}v'^2 + \&c. \end{aligned}\right. \\
&\quad + \cos 2(\theta - \theta') \times \left\{\begin{aligned} & C^{(2)} + \frac{dC^{(2)}}{da}v \\ & + \frac{dC^{(2)}}{da'}v' + \&c. \end{aligned}\right. \\
&\quad + \&c.
\end{aligned}$$

and since v and v' are expressed entirely in simple cosines of multiples of $(nt+\epsilon-\varpi)$ and $(n't+\epsilon'-\varpi')$, and since every power of a cosine may be expressed by simple cosines of multiple arcs, and every product of cosines may be expressed by simple cosines of the sums and differences of the arcs, it follows that the multiplier of $\cos k(\theta-\theta')$ may be expressed entirely by simple cosines of arcs arising from the combination of multiples of $nt+\epsilon-\varpi$ and $n't+\epsilon'-\varpi'$. The coefficients, as they depend only on $C^{(k)}$, $\dfrac{dC^{(k)}}{da}$, &c. may be calculated numerically from the preceding propositions.

130. The expression for θ in (87) is

$$nt+\epsilon+(2e-\frac{e^2}{4}+\&c.)\sin(nt+\epsilon-\varpi)+\&c.:$$

and a similar expression holds for θ':

$$\text{consequently } k(\theta-\theta')=k(nt+\epsilon)-k(n't+\epsilon')$$

$$k\{(2e-\frac{e^3}{4}+\&c.)\sin(nt+\epsilon-\varpi)-\left(2e'-\frac{e'^3}{4}+\&c.\right)\sin(n't+\epsilon'-\varpi')+\&c.\}$$

$$=k(nt+\epsilon)-k(n't+\epsilon')+z.$$

The cosine of $k(\theta-\theta')$ is $\cos\overline{k(nt+\epsilon)-k(n't+\epsilon')}\,.\cos z$

$$-\sin\overline{k(nt+\epsilon)-k(n't+\epsilon')}\,.\sin z:$$

$$\text{and } \cos z=1-\frac{z^2}{1.2}+\frac{z^4}{1.2.3.4}-\&c.:$$

$$\sin z=z-\frac{z^3}{1.2.3}+\frac{z^5}{1.2.3.4.5}-\&c.$$

The latter expressions when expanded will produce a series of simple cosines of combinations of multiples of $(nt+\epsilon-\varpi)$ and $(n't+\epsilon'-\varpi')$: and consequently $\cos k(\theta-\theta')$, when expanded, will produce a series of cosines of combinations of

$$k(nt+\epsilon)-k(n't+\epsilon'),$$

multiples of $nt+\epsilon-\varpi$,

and multiples of $n't+\epsilon'-\varpi'$.

And when this is multiplied by the series which in (129) stands as its multiplier, the product will be a series to which the same remarks may be applied, in which the coefficient of every term is, by the methods above given, numerically calculable. The multiples of $(nt+\epsilon)-(n't+\epsilon')$, $nt+\epsilon-\varpi$, and $n't+\epsilon'-\varpi'$, will, as is plain, be arranged in every possible way of addition and subtraction in the argument of each cosine.

131. The value of R (86) being

$$\frac{m'r \cdot \cos(\theta-\theta')}{r'^2} - \frac{m'}{\sqrt{\{r^2-2rr'\cos(\theta-\theta')+r'^2\}}};$$

we are now able to expand it completely (as the first term will be expanded by multiplying together the series for r, $\cos(\theta-\theta')$, and $\left.\frac{1}{r'^2}\right)$. It is only necessary now to explain the expansion of $r\frac{dR}{dr}$. In the introduction of this expression $\{$see (82) &c.$\}$ it is to be remarked that R was to be differentiated with respect to r as if r had no dependance on θ. Consequently $r\frac{dR}{dr}$ means the quantity which $a \cdot \frac{d\text{R}}{da}$ becomes (R being the value of R when a is put for r) on putting $a+v$ for a, without altering θ. If then in R the coefficient of $\cos k(\theta-\theta')$ is $C^{(k)}$, it will be in $a\frac{d\text{R}}{da}$, $a\frac{dC^{(k)}}{da}$; and in $r\frac{dR}{dr}$,

$$a\frac{dC^{(k)}}{da}+\frac{d}{da}\left(a\frac{d \cdot C^{(k)}}{da}\right) \cdot v$$

$$+\frac{d}{da'}\left(a\frac{d \cdot C^{(k)}}{da}\right)v' + \text{\&c.}$$

$$= a \cdot \frac{dC^{(k)}}{da}+\left(a\frac{d^2 \cdot C^{(k)}}{da^2}+\frac{d \cdot C^{(k)}}{da}\right)v$$

$$+a\frac{d^2C^{(k)}}{da\,da'}v', + \text{\&c.}$$

and the coefficient of every term can be calculated.

132. PROP. 53. In the expansion of R, $r\dfrac{dR}{dr}$, $\dfrac{dR}{da}$, $\dfrac{dR}{d\epsilon}$, and $\dfrac{dR}{d\varpi}$, the principal coefficient of a term

$$\cos\overline{(pn - qn')t + Q}$$

will contain powers of e and e' to $p - q$ dimensions*.

(1) If we examine the expressions for v and v' it is immediately seen that a term $\cos g\overline{(nt + \epsilon - \varpi)}$ has for coefficient a multiple of e^g, and that a term $\cos h\overline{(n't + \epsilon' - \varpi')}$ has for coefficient a multiple of e'^h, &c.

(2) In any power of v, each term is either a power h of $e^g . \cos g\overline{(nt + \epsilon - \varpi)}$ or a product of $e^g . \cos g\overline{(nt + \epsilon - \varpi)}$ by $e^h . \cos h\overline{(nt + \epsilon - \varpi)}$. The *first* of these will produce $e^{gh} . \{A \cos gh\overline{(nt + \epsilon - \varpi)} + B \cos g . (h-2)\overline{(nt + \epsilon - \varpi)} + \&c.\}$ where $e^{gh} . B$ is not the principal term in the coefficient of $\cos g . (h-2)\overline{(nt + \epsilon - \varpi)}$, since lower powers of e will be found multiplying that cosine, in the expansion of the power $h - 2$ of $e^g . \cos g\overline{(nt + \epsilon - \varpi)}$: but where e^{gh} is the lowest power in the coefficient of $\cos gh\overline{(nt + \epsilon - \varpi)}$ since it has been produced by the simplest combination which could form that cosine. The *second* will produce

$$e^{g+h}\{\cos\overline{(g + h)(nt + \epsilon - \varpi)} + \cos\overline{(g - h)(nt + \epsilon - \varpi)}\}.$$

The first term is produced by the simplest combination of arcs which could form it; the second is not, for it might be produced by the product of

$$e^{g-2h} . \cos\overline{(g - 2h)(nt + \epsilon - \varpi)} \text{ and } e^h \cos h\overline{(nt + \epsilon - \varpi)},$$

which would have for coefficient e^{g-h}. In all these terms then we have this general rule: the principal coefficient of $\cos x\overline{(nt + \epsilon - \varpi)}$ is a multiple of e^x.

(3) A similar proposition applies to the terms of the expansion of any power of v'.

---

* This is also true if the orbits be inclined: the inclination being supposed of the same order as the eccentricities.

(4) In the product of a power of v by a power of v', the product of two such terms as

$$e^x . \cos \overline{x(nt+\epsilon-\varpi)} \quad \text{and} \quad e'^{x'} \cos \overline{x'(n't+\epsilon'-\varpi')}$$

will produce

$$e^x e'^{x'} \cos \overline{x(nt+\epsilon-\varpi)+x'(n't+\epsilon'-\varpi')}$$

$$\text{and} \quad e^x e'^{x'} \cos \overline{x(nt+\epsilon-\varpi)-x'(n't+\epsilon'-\varpi')}.$$

Tracing in the same way the effects of multiplying these by another cosine we shall get this general rule: the principal coefficient of $\cos \overline{x(nt+\epsilon-\varpi) \pm x'(n't+\epsilon'-\varpi')}$ is a multiple of $e^x e'^{x'}$.

(5) Suppose now we consider how in the expansion of R such a term as $\cos \overline{(13n-8n')t+Q}$ is to be produced. It might be produced by

$$7(\overline{nt+\epsilon}-\overline{n't+\epsilon'}),\ 6(nt+\epsilon-\varpi), \text{ and } (n't+\epsilon'-\varpi').$$

This (by what has gone before) would have for coefficient a multiple of $e^6 e'$; of the seventh order. But it might be produced by

$$8(\overline{nt+\epsilon}-\overline{n't+\epsilon'}),\ 5(nt+\epsilon-\varpi)$$

$$9(\overline{nt+\epsilon}-\overline{n't+\epsilon'}),\ 4(nt+\epsilon-\varpi),\ (n't+\epsilon'-\varpi')$$

$$10(\overline{nt+\epsilon}-\overline{n't+\epsilon'}),\ 3(nt+\epsilon-\varpi),\ 2(n't+\epsilon'-\varpi')$$

$$11(\overline{nt+\epsilon}-\overline{n't+\epsilon'}),\ 2(nt+\epsilon-\varpi),\ 3(n't+\epsilon'-\varpi')$$

$$12(\overline{nt+\epsilon}-\overline{n't+\epsilon'}),\ (nt+\epsilon-\varpi),\ 4(n't+\epsilon'-\varpi')$$

$$13(\overline{nt+\epsilon}-\overline{n't+\epsilon'}),\ 5(n't+\epsilon'-\varpi');$$

and the coefficients of these combinations would be multiples respectively of e^5, e^4e', $e^3e'^2$, $e^2e'^3$, ee'^4, e'^5: all of the fifth order: and $13-8=5$. By trying any other combinations it will be seen that their coefficients are of the seventh, ninth, or some higher order, but that none can be formed with coefficients of a lower order. And as the same reasoning applies in all other cases, we easily arrive at this general conclusion, that

the principal coefficient of a term $\cos \overline{(pn-qn')t+Q}$ in the expansion of R is of the order $p-q$.

133. This remark applies to the terms

$$r\frac{dR}{dr},\quad \frac{dR}{da},\quad \frac{dR}{d\epsilon},\quad \text{and}\quad \frac{dR}{d\varpi},$$

since the powers of e and e' are not altered by these differentiations. But in $\frac{dR}{de}$ one power of e is lost in the differentiation, and therefore

the coefficient of a term $\cos\overline{(pn-qn')t+Q}$ in $\frac{dR}{de}$ is of the order $p-q-1$.

134. PROP. 54. The principal part of the coefficient of a term $\frac{\sin}{\cos}\overline{(pn-qn')t+Q}$, either in the longitude or in the radius vector, is of the order $p-q$.

By referring to (119) it will be seen that the coefficient of the term depending on $\overline{(13n-8n')t+Q}$ is of the fifth order, and that this is a consequence of the coefficient of the corresponding term in R (depending on that argument) being of the fifth order: and in the same way it would be seen that the form of one term in the longitude is of the same form as the term of R by which it was produced, and that its coefficient is of the same order as the coefficient of the term in R. But besides this there is a term depending on $\overline{(12n-8n')t+S}$, of the fourth order: and thus it would be seen that from any other term of R there is derived a term in which $p-q$ is less by 1 than in the term of R, and in which the order is also lowered by 1. Thus it is seen that the law which we have mentioned is strictly preserved in all the terms of the longitude.

It might be supposed that the loss of one power of e in $\frac{dR}{de}$ by differentiation, and that of another power in its application, $\left\{\text{for } \frac{d\varpi_{,}}{dt}=-\frac{na\sqrt{1-e^2}}{\mu e}\frac{dR}{de}\ (115)\right\}$, would have interrupted this law: but it does not. For after the substitution of the integral of this term in $nt+\epsilon-\varpi+x$ (119) and the expansion of the sine, it is multiplied by e; and the sine and cosine are then combined in such a manner with sine and cosine of $nt+\epsilon-\varpi$, that a term such as $\overline{(13n-8n')t+Q}$ is changed to such a one as $\overline{(12n-8n')t+S}$; and the law that we have mentioned is still preserved.

And by referring to (120) it will be seen that the same remarks apply in all respects to the terms in the radius vector.

135. We have shewn in (94) and in (119) that the terms which produce the most important perturbations in the longitude and radius vector are those in which $pn-qn'$ is small, or in which p and q must be nearly as $n':n$. But from this proposition it appears that they will have coefficients of the order $p-q$, and therefore they will not be considerable except $p-q$ be small. Consequently p and q must both be small. The problem of finding the terms which rise to importance is therefore reduced to this; to find two small integer numbers p and q very nearly in the proportion of n' to n. And it is by discovering terms of this kind that all the most important inequalities depending on the eccentricities, &c. have been found.

136. Prop. 55. To point out the relation between the mutual disturbances of two planets.

Suppose m and m' to be the masses of the two planets: R and R' the function on which the disturbance of each depends. Then

$$R=\frac{m'r.\cos(\theta'-\theta)}{r'^2}-\frac{m'}{\sqrt{\{r'^2-2rr'\cos(\theta'-\theta)+r^2\}}},$$

$$R'=\frac{mr'.\cos(\theta-\theta')}{r^2}-\frac{m}{\sqrt{\{r^2-2r'r\cos(\theta-\theta')+r'^2\}}}.$$

Since $\cos(\theta-\theta')=\cos(\theta'-\theta)$, the last terms of these expressions differ only in having m' and m respectively in their numerators.

137. Now the first terms can produce, when developed in terms of t, only cosines of combinations of $nt+\epsilon-(n't+\epsilon')$ with multiples of $nt+\epsilon-\varpi$ and of $n't+\epsilon'-\varpi'$. Hence they cannot produce the principal terms in any of the important inequalities of the form $\cos\overline{(pn-qn')t+Q}$, except p or $q=1$. For instance, they could produce the term

$$\cos.\overline{13(nt+\epsilon)-8(n't+\epsilon')+Q}$$

only by the product of

$$\cos\overline{nt+\epsilon-(n't+\epsilon')},\ e^{12}.\cos\overline{12(nt+\epsilon-\varpi)},\ e'^{7}\cos\overline{7(n't+\epsilon'-\varpi')},$$

which would be of the 19th order: whereas there are other terms of the 5th order with the same argument. And the terms in which p or q or both are $=1$, are added to terms resulting from the expansion of the other fraction in such a manner that no simple relation can be assigned.

138. If then we confine ourselves to the terms where neither p nor $q=1$, and in which $pn-qn'$ is small, and if we take only the most important part of their results, we can find a simple relation between the perturbations of the two planets. For let one term of R be the sum of the series of terms

$$m'\{Me^k e'^l.\cos\overline{(pn-qn')t+p\epsilon-q\epsilon'-k\varpi-l\varpi'}$$
$$+Ne^o e'^s\cos\overline{(pn-qn')t+p\epsilon-q\epsilon'-o\varpi-s\varpi'}$$
$$+\&c.\}$$

where $p-q=k+l=o+s=\&c$: which sum admits of being collected in the form $P\cos\overline{(pn-qn')t+p\epsilon-q\epsilon'+Q}$: then, proceeding as in (119), we find for the principal inequality in the longitude of m,

$$+\frac{m'}{\mu}.aP.\frac{3n^2p}{(pn-qn')^2}\sin\overline{(pn-qn')t+p\epsilon-q\epsilon'+Q}.$$

Similarly for the principal inequality in the longitude of m',

$$-\frac{m}{\mu}.a'P.\frac{3n'^2q}{(pn-qn')^2}\sin\overline{(pn-qn')t+p\epsilon-q\epsilon'+Q}.$$

The proportion of these is $-\frac{m'an^2p}{ma'n'^2q}$. But as pn and qn' are very nearly equal, $\frac{np}{n'q}=1$ nearly, and the proportion is $-\frac{m'an}{ma'n'}$. Now $n=\frac{\sqrt{(\mu)}}{a^{\frac{3}{2}}}$; $n'=\frac{\sqrt{(\mu)}}{a'^{\frac{3}{2}}}$: and μ (the sum of the masses of the sun and planet) may be considered the same for both. Hence $\frac{n}{n'}=\frac{a'^{\frac{3}{2}}}{a^{\frac{3}{2}}}$: and therefore

$$\frac{\text{coefficient of long inequality of } m}{\text{coefficient of corresponding inequality of } m'}=-\frac{m'a'^{\frac{1}{2}}}{ma^{\frac{1}{2}}}.$$

From the sign it appears that while in consequence of this inequality one planet is *before* its mean place, the other is *behind* its mean place.

139. The same term of R will furnish other inequalities in the longitude of m and m', as in (119). The proportion might be simply assigned if they were produced by only a single term with the coefficient $Me^ke'^l$. But from the manner in which the inequalities are derived, the proportion will be different for each different term, and therefore no simple proportion of the aggregates can be assigned.

140. PROP. 56. To examine the parts of R which are independent of t.

We shall proceed in our examination only to the second order of excentricities.

(1) If we examine the term

$$\frac{m'r.\cos(\theta-\theta')}{r'^2},\text{ or } m'.\frac{a(1-e\cos\overline{nt+\epsilon-\varpi}+\&c.)}{a'^2(1-e'\cos\overline{n't+\epsilon'-\varpi'}+\&c.)^2}$$
$$\times\{\cos\overline{nt+\epsilon-(n't+\epsilon')}+\&c.\},$$

we shall easily see that it will contain one term

$$-\frac{m'a}{a'^2}.2ee'.\cos(nt+\epsilon-\varpi).\cos(n't+\epsilon'-\varpi').\cos\overline{nt+\epsilon-(n't+\epsilon')},$$

arising from the small terms of $\frac{r}{r'^2}$.

The product of the two first cosines will produce one term

$$\cos \overline{nt + \epsilon - \varpi - (n't + \epsilon' - \varpi')},$$

and hence there will be in R the term

$$-\frac{m'a}{a'^2} ee'. \cos \overline{nt + \epsilon - \varpi - (n't + \epsilon' - \varpi')} . \cos \overline{nt + \epsilon - (n't + \epsilon')}$$

one part of the development of which is

$$-\frac{m'a}{a'^2} . \frac{ee'}{2} . \cos (\varpi - \varpi').$$

Again, if we observe that, in the expansion of $\cos (\theta - \theta')$, (130), the term of the second order is

$$-\cos \overline{nt + \epsilon - (n't + \epsilon')} . \frac{z^2}{2}$$

which contains one term

$$\cos \overline{nt + \epsilon - (n't + \epsilon')} . \tfrac{1}{2} . 4ee' . \sin \overline{nt + \epsilon - \varpi} . \sin \overline{n't + \epsilon' - \varpi'}$$

we shall see that the product of the last factors will have one term

$$ee' . \cos \overline{nt + \epsilon - \varpi - (n't + \epsilon' - \varpi')}$$

which multiplied by the first, and by $\frac{m'a}{a'^2}$, will introduce the term $+\frac{m'a}{a'^2} . \frac{ee'}{2} . \cos (\varpi - \varpi')$.

This term destroys that which we have just found. There are other terms involving higher powers of e and e', which we shall not consider.

(2) If we examine the development of

$$\frac{1}{\sqrt{\{r^2 - 2rr' \cos (\theta - \theta') + r'^2\}}} \text{ in (129)}$$

we shall find in $-\frac{m'}{\sqrt{\{r^2 - 2rr' \cos (\theta - \theta') + r'^2\}}}$ these constant terms

$$-\frac{m'C^{(0)}}{2}-\frac{m'ae^2}{4}\cdot\frac{dC^{(0)}}{da}-\frac{m'a'e'^2}{4}\cdot\frac{dC^{(0)}}{da'}$$

$$-\frac{m'a^2e^2}{8}\cdot\frac{d^2C^{(0)}}{da^2}-\frac{m'a'^2e'^2}{8}\cdot\frac{d^2C^{(0)}}{da'^2}$$

(where the two last terms are introduced by $\sin^2\overline{nt+\epsilon-\varpi}$ &c.)

$$-m'C^{(1)}.ee'\cos(\varpi-\varpi')$$

$$-m'.\frac{ee'}{2}\left(a'\frac{dC^{(1)}}{da'}+a\frac{dC^{(1)}}{da}\right)\cos(\varpi-\varpi')$$

$$-m'aa'\frac{ee'}{4}\cdot\frac{d^2C^{(1)}}{dada'}\cos(\varpi-\varpi')$$

(by expanding $\cos\overline{\theta-\theta'}$)

and other smaller terms.

By the use of the formulæ of (127) and (128) these collected terms may be put under the form

$$-\frac{m'C^{(0)}}{2}-m'.\frac{D^{(1)}aa'}{8}(e^2+e'^2)+m'\frac{D^{(2)}aa'}{4}ee'\cos(\varpi-\varpi').$$

141. PROP. 57. To examine the terms which produce permanent variations in any of the elements.

The terms of R which depend on t are always in the form of a cosine with a constant coefficient. These terms then {as will be seen on referring to (118)} can produce only periodical variations in the elements. It might at first sight appear that there is in $\epsilon_{,}$ a term multiplied by t, but {as in (119)} it will be seen that there is always in $n_{,}t$ a similar term which destroys it. It will be sufficient therefore to examine the constant terms found above.

(1) With regard to these terms, $\frac{dR}{d\epsilon}=0$ (since ϵ is not found in them): therefore there is no constant term in $\frac{da_{,}}{dt}$, or no permanent variation in the axis major. Similarly there is no permanent variation in $n_{,}$.

(2) The value of $\frac{dR}{d\varpi}$ is

$$-m'\frac{D^{(2)}aa'}{4}ee'.\sin(\varpi-\varpi'):$$

hence the constant term of $\frac{de_{,}}{dt}$ is

$$-\frac{m'}{\mu}.\frac{na^2a'D^{(2)}}{4}e'_{,}.\sin(\varpi-\varpi').$$

Similarly the constant term of $\frac{de'_{,}}{dt}$ is

$$+\frac{m}{\mu}.\frac{n'aa'^2D^{(2)}}{4}e_{,}.\sin(\varpi-\varpi').$$

The eccentricities change therefore in opposite directions: it is easily seen that

$$mn'a'e_{,}\frac{de_{,}}{dt}+m'nae'_{,}\frac{de'_{,}}{dt}=0,$$

or that $mn'a'e^2_{,}+m'nae'^2_{,}$ is constant.

(3) The value of $\frac{dR}{de}$ is

$$-m'\frac{D^{(1)}aa'e}{4}+m'\frac{D^{(2)}aa'}{4}e'\cos(\varpi-\varpi'),$$

whence $\frac{d\varpi_{,}}{dt}=\frac{m'}{\mu}.\frac{na^2a'}{4}(D^{(1)}-D^{(2)}\frac{e'}{e}.\cos\overline{\varpi-\varpi'}).$

$D^{(1)}$ is greater than $D^{(2)}$; and hence, unless $\frac{e'}{e}$ be large, and $\cos(\varpi-\varpi')$ positive, the longitude of perihelion will increase, or the apse will progress.

$\frac{d\varpi'_{,}}{dt}$ similarly $=\frac{m}{\mu}.\frac{n'aa'^2}{4}(D^{(1)}-D^{(2)}\frac{e}{e'}\cos\overline{\varpi-\varpi'}).$

(4) The principal part of $\frac{dR}{da}$ is $-\frac{m'}{2}.\frac{dC^{(0)}}{da}$,

$$\text{whence } \frac{d\epsilon_{,}}{dt} = -\frac{m'}{\mu} na^2 \frac{dC^{(0)}}{da}.$$

If the orbit of m be exterior to that of m', or if a be $> a'$, it appears from the expansion of (121) that $\frac{dC^{(0)}}{da}$ is negative, and therefore $\frac{d\epsilon_{,}}{dt}$ is positive. Let it $= B$: then $\epsilon_{,} = \epsilon + Bt$: hence the mean longitude or $n_{,}t + \epsilon_{,} = (n + B) . t + \epsilon$. This shews that the mean motion is greater than that in an undisturbed orbit with the same axis major. {We have in (91) arrived at a result equivalent to this.} If a were less than a', $C^{(0)}$ would be expanded by powers of $\frac{a}{a'}$, $\frac{dC^{(0)}}{da}$ would be positive, and the mean motion would be less than in an undisturbed orbit.

142. It is almost unnecessary to remark that the disturbing effect of several planets will be the sum of their separate disturbances. This always holds when the equations for the perturbations are linear, as in the planetary theory.

143. We have stated that though we ought in strictness to substitute in every term of R the *variable* values of the elements, yet it is generally sufficient to use *constant* values. But this is hardly accurate enough in the calculation of some equations of long period, where $e_{,}$ and $\varpi_{,}$ alter considerably in the period. In general however it will be sufficient to substitute for them $e + Et$ and $\varpi + \Pi t$, {where E and Π are the values of $\frac{de}{dt}$ and $\frac{d\varpi}{dt}$ found in (141)} in the values of

$$\frac{dR}{d\epsilon}, \quad \frac{dR}{d\varpi}, \quad \frac{dR}{de},$$

and then proceed with the integration.

144. PROP. 58. To investigate the secular variations of the eccentricity and longitude of perihelion.

We have now sufficiently explained the method of calculating those variations of the elements whose period is not very long. For these we either neglect or estimate approximately the secular variations of the elements. Similarly to calculate accurately the secular variations, we may neglect the variations of short period. Putting $e\cos\varpi = u$, $e\sin\varpi = v$, and $e'\cos\varpi' = u'$, $e'\sin\varpi' = v'$, the equations of (141) for the variations of the eccentricity and perihelion are easily changed to

$$\frac{du}{dt} = \frac{m'}{\mu} \cdot \frac{na^2a'}{4}(-D^{(1)}v + D^{(2)}v'),$$

$$\frac{dv}{dt} = \frac{m'}{\mu} \cdot \frac{na^2a'}{4}(D^{(1)}u - D^{(2)}u'),$$

$$\frac{du'}{dt} = \frac{m}{\mu} \cdot \frac{n'a'^2a}{4}(-D^{(1)}v' + D^{(2)}v),$$

$$\frac{dv'}{dt} = \frac{m}{\mu} \cdot \frac{n'a'^2a}{4}(D^{(1)}u' - D^{(2)}u).$$

Solving these as simultaneous equations, we find

$$u = p_1\cos(a_1t - q_1) + p_2\cos(a_2t - q_2),$$

$$v = p_1\sin(a_1t - q_1) + p_2\sin(a_2t - q_2),$$

$$u' = \frac{p_1}{x_1}\cos(a_1t - q_1) + \frac{p_2}{x_2}\cos(a_2t - q_2),$$

$$v' = \frac{p_1}{x_1}\sin(a_1t - q_1) + \frac{p_2}{x_2}\sin(a_2t - q_2),$$

where p_1, p_2, q_1, q_2, are arbitrary constants, to be determined so as to make the expressions for u, v, &c. represent the known values of u, v, &c. at any one time: the two values of a_1 and a_2 are

$$\frac{aa'}{8\mu}\{(mn'a' + m'na)D^{(1)}$$

$$\mp\sqrt{(\overline{m'na - mn'a'}|^2 . D^{(1)2} + 4m'na . mn'a' . D^{(2)2})}\},$$

and the two values of x_1 and x_2

$$\frac{1}{2mn'a'D^{(2)}}\{(mn'a' - m'na)D^{(1)}$$

$$\pm\sqrt{(\overline{m'na - mn'a'}|^2 D^{(1)2} + 4m'na . mn'a' . D^{(2)2})}\}.$$

Then

$$e=\sqrt{(u^2+v^2)}: \quad \tan\varpi=\frac{v}{u}: \quad e'=\sqrt{(u'^2+v'^2)}: \quad \tan\varpi'=\frac{v'}{u'}.$$

A similar process is to be used when the secular variations produced by the mutual perturbations of several planets are to be calculated.

On examining the geometrical meaning* of these expressions it is easily seen that the motion of perihelion is on the whole progressive; and that its mean period is $\frac{2\pi}{a_1}$ if p_1 be greater than p_2, and $\frac{2\pi}{a_2}$ if p_2 be greater than p_1. Similarly the motion of perihelion of m' is on the whole progressive; and its mean period is $\frac{2\pi}{a_1}$ if $\frac{p_1}{x_1}$ be greater than $\frac{p_2}{x_2}$, or $\frac{2\pi}{a_2}$ if $\frac{p_2}{x_2}$ be greater than $\frac{p_1}{x_1}$.

145. We shall conclude with shortly describing the process to be followed when the orbits of the planets are inclined. If we put

$$R=-\frac{m'}{\sqrt{\{(x'-x)^2+(y'-y)^2+(z'-z)^2\}}}+\frac{m'(x'x+y'y+z'z)}{(x'^2+y'^2+z'^2)^{\frac{3}{2}}}$$

equations similar to those of (77) may be found. If we take for xy the plane of the disturbed planet's orbit at a given time, z is only the effect of the disturbing force and may therefore be neglected in R. And all the equations for the perturbations of radius vector and longitude, or for the variation of the major axis, &c. hold equally well: the only difference in the process is, that for R we must expand

$$-\frac{m'}{\sqrt{\{(x'-x)^2+(y'-y)^2+z'^2\}}}+\frac{m'(x'x+y'y)}{(x'^2+y'^2+z'^2)^{\frac{3}{2}}}.$$

* The center of each orbit describes an Epitrochoid round the focus which coincides with the Sun's center. This will be seen on comparing the expressions for u and v (which are rectangular co-ordinates, referred to the focus) with the expressions for x and y in Peacock's *Examples*, p. 193. The circles must be supposed to revolve uniformly.

The perturbation in latitude may be calculated from the equation

$$\frac{d^2 z}{dt^2} = -\frac{\mu}{r^3} z + \frac{m' z'}{\{(x'-x)^2 + (y'-y)^2 + z'^2\}^{\frac{3}{2}}}$$

$$- \frac{m' z'}{(x'^2 + y'^2 + z'^2)^{\frac{3}{2}}},$$

which being integrated as in the note to (87), the tan. latitude or $\frac{z}{r}$ is found.

The latitude may also be found by investigating the variations of the longitude of the node and of the inclinations. For this, as well as for the peculiarities of the theory of Jupiter's satellites, and other more recondite parts of the Problem of Three Bodies, we refer to the Encyclopædia Metropolitana Art. *Physical Astronomy*, to Woodhouse's Physical Astronomy, to Lubbock's Essays, to the Mécanique Céleste of Laplace, or to the Théorie Analytique of Pontécoulant.

FIGURE OF THE EARTH.

PRELIMINARY PROPOSITIONS.

1. Prop. 1. The sections of two similar and concentric and similarly situated spheroids, made by the same plane, are similar and concentric ellipses, similarly situated.

Let BK, CG, (fig. 1) be the sections, which will be ellipses (Hustler's *Conic Sections*, p. 62; Hymers's *Geometry of Three Dimensions*, Art. 103). Join W, the center of the spheroid, with N, the center of the ellipse BK, and let ABE, HCD be sections by any plane through NW: these will evidently be similar ellipses, whose common center is W. And since BK is bisected in N, BN is an ordinate to the semidiameter AW in the ellipse ABE, and is therefore parallel to the tangent at A, and therefore to that at H (as the ellipses are similar); hence, CN is an ordinate to WH, in the smaller ellipse. Let WD and WE be the semi-conjugate diameters: then,

$$NC^2 = \frac{WD^2}{WH^2}\cdot(WH^2 - WN^2); \quad NB^2 = \frac{WE^2}{WA^2}\cdot(WA^2 - WN^2);$$

and, as the ellipses DCH, EBA, are similar,

$$\frac{WD^2}{WH^2} = \frac{WE^2}{WA^2}; \quad \therefore \frac{NC^2}{NB^2} = \frac{WH^2 - WN^2}{WA^2 - WN^2},$$

a constant value for the same sections; hence, the sections are similar, and similarly situated, with respect to N: but N is the center of the ellipse BK; therefore it is also the center of CG.

2. Prop. 2. Let AGB, CSK, (fig. 2 and $2a$) be two ellipses, concentric, similar, and similarly situated; through C, the extremity of the axis of the smaller ellipse, draw ECO

perpendicular to that axis, meeting the larger ellipse in E and O; draw EF parallel to the same axis; then, if the angles GEF, FEH, KCD, DCL, be all equal, $EG + EH$ will be equal to $CK + CL$.

3. For draw the diameter $QTNV$ bisecting EG in R; and draw OP parallel to EG.

Since EG is bisected by QV, EG is an ordinate to that diameter; so also is OP, which is parallel to EG. And because the two ellipses are similar, the tangent at Q is parallel to that at S; hence CK being parallel to EG, or parallel to the tangent at Q, is also parallel to the tangent at S; it is therefore an ordinate to the diameter SN, and is bisected by it in T. Now, because OP is parallel to EG, and EF perpendicular to EO, the angle COP is the complement of GEF; and CEH is also the complement of HEF; but $GEF=HEF$; $\therefore$ CEH is equal to COP. And the point O has the same situation in the semi-ellipse AOB, which E has in AEB. Hence, $OP=EH$. But as $CE=CO$, and ER, CT, OV, are parallel, if through T the line rTv be drawn parallel to ECO, Rr will $= Vv$, and $Er=Ov=CT$. Hence

$$ER + OV = Er - Rr + Ov + Vv = 2CT;$$

doubling both sides,

$$EG + OP, \text{ or } EG + EH = 2CK = CK + CL.$$

4. Prop. 3. If the angles made by EG and EH with EF be increased till the point G falls on the other side of E, as in fig. 3, then

$$EH - EG = CK + CL.$$

Making the same construction, we find

$$OV - ER = Ov + Vv - (Rr - Er) = 2CT,$$

$$\text{whence } OP - EG, \text{ or } EH - EG, = 2CK = CK + CL.$$

5. Prop. 4. The vertical solid angle of a homogeneous pyramid being given, its attraction upon a particle placed

at the vertex, is proportional to its length, $\left\{\text{the force to each particle} \propto \frac{\text{mass}}{(\text{distance})^2}\right\}$. For conceive the pyramid to be divided into an indefinitely great number of strata of the same thickness, by sections perpendicular to its axis: the homologous sides of these sections will be as the distance from the vertex; therefore the areas of the sections will be as the square of that distance; and therefore the mass included between two sections will be ultimately as the square of that distance. But the attraction on a particle at the vertex is as the mass directly, and as the square of the distance inversely. Hence, the attraction of every stratum is the same; and, consequently, the whole attraction of the pyramid will be proportional to the number of strata, that is, to its length.

In the same manner it appears, that the attraction of a frustrum of the same pyramid upon a particle placed at the vertex of the pyramid, is as the length of the frustrum.

6. Prop. 5. If the base of a pyramid, whose vertical solid angle is small, be given, the attraction on a point in the vertex $\propto \frac{\text{base}}{\text{length}}$.

For if b be the base, l the length, x the distance of any section from the vertex, the area of this section $= \frac{bx^2}{l^2}$; hence, the mass included between the section at the distance x, and that at the distance $x + \delta x$, ultimately $= \frac{bx^2 \delta x}{l^2}$; and its attraction $= \frac{b \delta x}{l^2}$; putting u for the attraction, $\frac{du}{dx} = \frac{b}{l^2}$; $\therefore u = \frac{bx}{l^2}$, which for the whole pyramid $= \frac{b}{l}$.

7. If we put k for the density of the matter, that is, if the attraction of the matter in the volume M, on a point at the distance D, produce an accelerating force $= \frac{Mk}{D^2}$,

then the attraction of one stratum ultimately $= \frac{kb\delta x}{l^2}$, whence $u = \frac{kbx}{l^2}$, or for the whole pyramid $= \frac{kb}{l}$.

ON THE ATTRACTION OF AN OBLATE SPHEROID.

8. PROP. 6. To find the attraction of an oblate spheroid on a particle placed at its pole.

Let B, (fig. 4), be the pole of the spheroid, BD the axis; let the spheroid be divided into wedges, by planes passing through BD, two of which are BPD, BQD, making with each other the very small angle ω; in these planes draw BP, BQ making with BD the angle θ, and Bp, Bq, making with BP, BQ the very small angle $\delta\theta$; and suppose the wedge divided into pyramids similar to BPq. Let x be the abscissa of P, measured along the axis of the spheroid; y its ordinate; let $BP = r$. If through qp a section $pqts$ be drawn perpendicular to the axis of the pyramid; since qp ultimately $= y\omega$, and $qt = r\delta\theta$, the area of this section $= ry\omega\delta\theta$; therefore by (7), the attraction of the pyramid $= \frac{kry\omega\delta\theta}{r} = ky\omega\delta\theta$. This is in the direction BP: but as the whole attraction of the spheroid will evidently be in the direction BD, we must resolve the attraction of the pyramid into one parallel to BD, and one perpendicular to BD: the former will be effective, but the latter will be counteracted by forces in the opposite direction. The effective part

$$= ky\omega\delta\theta \cdot \frac{x}{r} = kr \cdot \sin\theta \cdot \cos\theta \cdot \omega \cdot \delta\theta.$$

Let a be the equatoreal radius of the spheroid, b the semiaxis;

$$\text{then } y^2 = \frac{a^2}{b^2}(2bx - x^2).$$

Putting for x and y their values $r\cos\theta$ and $r\sin\theta$, this becomes

$$r^2\sin^2\theta = \frac{a^2}{b^2}(2br\cos\theta - r^2\cos^2\theta),$$

$$\text{whence } r = \frac{2b\cos\theta}{1 - \dfrac{a^2-b^2}{a^2}\sin^2\theta} = \frac{2b\cos\theta}{1 - e^2\sin^2\theta},$$

putting e for the eccentricity of the generating ellipse. Hence, the attraction of the pyramid ultimately

$$= k\omega \, . \frac{2b\cos^2\theta \, . \sin\theta \, . \delta\theta}{1 - e^2\sin^2\theta},$$

and if w be the attraction of the wedge,

$$\frac{dw}{d\theta} = 2kb\omega \, . \frac{\cos^2\theta \, . \sin\theta}{1 - e^2\sin^2\theta}.$$

Let $\cos\theta = z$; then

$$\frac{dw}{dz} = -2kb\omega \, . \frac{z^2}{1 - e^2 + e^2z^2};$$

integrating

$$w = -\frac{2kb\omega}{e^2}\left\{z - \frac{\sqrt{(1-e^2)}}{e} \, . \tan^{-1}\frac{ez}{\sqrt{(1-e^2)}}\right\}.$$

Taking this from $\theta = 0$, to $\theta = \frac{\pi}{2}$, or from $z = 1$, to $z = 0$,

$$w = 2kb\omega\left\{\frac{1}{e^2} - \frac{\sqrt{(1-e^2)}}{e^3}\tan^{-1}\frac{e}{\sqrt{(1-e^2)}}\right\}.$$

This is the attraction of a wedge whose angle is ω; and since the attraction of every wedge with an equal angle must be the same, the attraction of the whole spheroid will be found by putting 2π in the place of ω; that is, the attraction

$$= 4\pi \, . kb\left\{\frac{1}{e^2} - \frac{\sqrt{(1-e^2)}}{e^3}\tan^{-1}\frac{e}{\sqrt{(1-e^2)}}\right\}$$

$$= 4\pi \, . kb\left\{\frac{1}{e^2} - \frac{\sqrt{(1-e^2)}}{e^3}\sin^{-1}e\right\}.$$

9. If the spheroid differs very little from a sphere, we may put $a = b(1+\epsilon)$; ϵ is then called the ellipticity of the spheroid. Then $e^2 = 2\epsilon$, nearly. Hence, the attraction of the pyramid

$$= 2bk\omega \frac{\cos^2\theta \sin\theta\,\delta\theta}{1 - 2\epsilon\sin^2\theta}$$

$$= 2bk\omega(\cos^2\theta \,.\, \sin\theta \,.\, \delta\theta + 2\epsilon \,.\, \cos^2\theta \sin^3\theta \,.\, \delta\theta), \text{ nearly}$$

$$= 2bk\omega(\cos^2\theta + 2\epsilon\cos^2\theta - 2\epsilon\cos^4\theta)\sin\theta \,.\, \delta\theta;$$

whence the attraction of the wedge

$$= 2bk\omega\left(-\frac{\cos^3\theta}{3} - \frac{2\epsilon}{3}\cos^3\theta + \frac{2\epsilon}{5}\cos^5\theta\right);$$

and taking this integral from $\theta = 0$, to $\theta = \frac{\pi}{2}$, the attraction of the wedge

$$= 2bk\omega\left(\frac{1}{3} + \frac{2\epsilon}{3} - \frac{2\epsilon}{5}\right) = \frac{2}{3}bk\omega\left(1 + \frac{4\epsilon}{5}\right).$$

Then the attraction of the whole spheroid is found as before, by putting 2π for ω, and is, therefore,

$$= \frac{4\pi}{3}kb\left(1 + \frac{4\epsilon}{5}\right).$$

10. Prop. 7. To find the attraction of an oblate spheroid on a particle at its equator.

Let ARM, fig. 5, be the equator of the spheroid; AZ a perpendicular to it from the attracted point A; suppose the spheroid divided into wedges, by planes passing through AZ; let two of these planes, very near each other, be APR, Apr; and drawing the diameter AM, let $MAR = \phi$, $MAr = \phi + \delta\phi$. Then suppose the lines AP, Ap, to be drawn in the planes APR, Apr, making with AR, Ar, equal angles θ, and AQ, Aq to be very near them, making with AR, Ar, equal angles $\theta + \delta\theta$. If through q we draw a plane $qtsw$ perpendicular to the axis of the small pyramid APq,

$$wq \text{ ultimately } = r\cos\theta \,.\, \delta\phi, \quad qt = r\delta\theta,$$

and the base of the pyramid

$$= wq \,.\, qt = r^2 \,.\, \cos\theta \,.\, \delta\phi \,.\, \delta\theta\,;$$

therefore by (7), its attraction in the direction AP

$$= \frac{k \,.\, r^2 \,.\, \cos\theta \,.\, \delta\phi \,.\, \delta\theta}{r} = k \,.\, r \,.\, \cos\theta \,.\, \delta\phi \,.\, \delta\theta.$$

Draw PN perpendicular to the plane of the equator: NO perpendicular to AM; let $AO = x$, $ON = y$, $NP = z$. If we resolve the attraction of the pyramid into two parts, one in the direction AN, and the other in NP, the latter of these will be counteracted by the attraction of another pyramid in the same wedge, equally inclined to AR but on the opposite side; and the former

$$= k \,.\, r \,.\, \cos\theta \,.\, \delta\phi \,.\, \delta\theta \,.\, \cos\theta = k \,.\, r \,.\, \cos^2\theta \,.\, \delta\phi \,.\, \delta\theta.$$

This is in the direction AN; if we resolve it into two in the directions AO, ON, the latter of these will be counteracted by the attraction of an equal pyramid, in another wedge which makes the same angle with AM; and the former

$$= k \,.\, r \,.\, \cos^2\theta \,.\, \delta\phi \,.\, \delta\theta \,.\, \cos\phi = k \,.\, r \,.\, \cos\phi \,.\, \cos^2\theta \,.\, \delta\phi \,.\, \delta\theta.$$

Now the equation to the spheroid is

$$PN^2 = \frac{b^2}{a^2}(AC^2 - CO^2 - ON^2)$$

$$= \frac{b^2}{a^2}(AC^2 - \overline{AC - AO}^2 - ON^2),$$

$$\text{or } z^2 = \frac{b^2}{a^2}(2ax - x^2 - y^2)\,;$$

putting for x, y, and z, their values

$$r \,.\, \cos\theta \,.\, \cos\phi, \qquad r \,.\, \cos\theta \,.\, \sin\phi, \qquad r \,.\, \sin\theta,$$

it becomes

$$r^2 \,.\, \sin^2\theta = \frac{b^2}{a^2}(2ar \,.\, \cos\theta \,.\, \cos\phi - r^2 \,.\, \cos^2\theta \,.\, \cos^2\phi - r^2 \,.\, \cos^2\theta \,.\, \sin^2\phi);$$

$$\therefore\ r = \frac{2b^2}{a} \,.\, \frac{\cos\theta \,.\, \cos\phi}{(1 - e^2) \,.\, \cos^2\theta + \sin^2\theta}.$$

Hence, the effective attraction of the pyramid

$$= \frac{2kb^2}{a} \cdot \frac{\cos^2\phi \,.\, \cos^3\theta \,.\, \delta\phi \,.\, \delta\theta}{(1-e^2) \,.\, \cos^2\theta + \sin^2\theta}.$$

Let w be the effective attraction of the wedge;

$$\frac{dw}{d\theta} = \frac{2kb^2}{a} \cos^2\phi \,.\, \delta\phi \,.\, \frac{\cos^3\theta}{(1-e^2) \,.\, \cos^2\theta + \sin^2\theta}.$$

To integrate this, let $\sin\theta = v$;

$$\frac{dw}{dv} = \frac{2kb^2}{a} \cos^2\phi \,.\, \delta\phi \,.\, \frac{1-v^2}{(1-e^2)+e^2v^2};$$

$$\therefore\ w = \frac{2kb^2}{a} \cos^2\phi \,.\, \delta\phi \,.\left\{-\frac{v}{e^2} + \frac{1}{e^3\sqrt{(1-e^2)}} \,.\, \tan^{-1}\frac{ev}{\sqrt{(1-e^2)}}\right\}.$$

Taking this from $\theta = -\frac{\pi}{2}$, to $\theta = +\frac{\pi}{2}$, or from $v = -1$, to $v = +1$,

$$w = \frac{2kb^2}{a} \cos^2\phi \,.\, \delta\phi \,.\left\{\frac{2}{e^3\sqrt{(1-e^2)}} \,.\, \tan^{-1}\frac{e}{\sqrt{(1-e^2)}} - \frac{2}{e^2}\right\}$$

$$= 4kb \cos^2\phi \,.\, \delta\phi \,.\left\{\frac{1}{e^3}\sin^{-1}e - \frac{\sqrt{(1-e^2)}}{e^2}\right\}.$$

Hence, if u be the attraction of the spheroid,

$$\frac{du}{d\phi} = 4kb\cos^2\phi\left\{\frac{1}{e^3}\sin^{-1}e - \frac{\sqrt{(1-e^2)}}{e^2}\right\};$$

integrating from $\phi = -\frac{\pi}{2}$, to $\phi = +\frac{\pi}{2}$,

$$u = 2kb\pi\left\{\frac{1}{e^3}\sin^{-1}e - \frac{\sqrt{(1-e^2)}}{e^2}\right\}.$$

11. If the spheroid differ little from a sphere, putting, as before, $a = b(1+\epsilon)$, we find

$$e^2 = 2\epsilon, \qquad \frac{b^2}{a} = b(1-\epsilon),$$

and $\frac{dw}{dv} = 2kb(1-\epsilon)\cos^2\phi \,.\, \delta\phi \,.\, \frac{1-v^2}{1-2\epsilon+2\epsilon v^2}$

$= 2kb \,.\, (1-\epsilon) \,.\, \cos^2\phi \,.\, \delta\phi \,.\, (1 - v^2 + 2\epsilon - 4\epsilon v^2 + 2\epsilon v^4)$, nearly,

and $w = 2kb(1-\epsilon) \,.\, \cos^2\phi \,.\, \delta\phi \,.\, \left(v - \frac{v^3}{3} + 2\epsilon v - \frac{4\epsilon v^3}{3} + \frac{2\epsilon v^5}{5}\right)$,

which, from $v = -1$, to $v = +1$

$$= 4kb(1-\epsilon)\cos^2\phi \,.\, \delta\phi \,.\, \left(\frac{2}{3} + \frac{16\epsilon}{15}\right)$$

$$= \frac{8kb}{3}(1-\epsilon)\cos^2\phi \,.\, \delta\phi \,.\, (1 + \frac{8}{5}\epsilon)$$

$$= \frac{8kb}{3}\cos^2\phi \,.\, \delta\phi\,(1 + \frac{3}{5}\epsilon);$$

$$\therefore \frac{du}{d\phi} = \frac{8kb}{3}\cos^2\phi\,(1 + \frac{3}{5}\epsilon);$$

therefore integrating from $\phi = -\frac{\pi}{2}$ to $\phi = +\frac{\pi}{2}$,

$$u = \frac{4kb\pi}{3}(1 + \frac{3}{5}\epsilon).$$

12. PROP. 8. If E, (fig. 6), be any point on the surface of a spheroid, and EC be drawn perpendicular to the plane of the equator, and a spheroid be described concentric, similar, and similarly situated to the given spheroid, touching EC at C; then the attraction of the given spheroid on E, in a direction parallel to the radius CW, is equal to the attraction of the smaller spheroid in the same direction on the point C.

13. Suppose both spheroids divided into wedges by planes passing through EC: let $EGHB$, $CKDL$, be the sections of both, made by one plane; and $ERSb$, $CTdV$, the sections made by another plane very near the former. Draw EF parallel to CD, and let the angles GEF, HEF, KCD, LCD, be all equal. Let the angles gEF, hEF, kCD, lCD be also equal to each other, and very nearly

equal to the former; and suppose the wedge divided into pyramids by planes passing through these lines perpendicular to the plane AEB. Then the angles GEg, HEh, KCk, LCl, are equal. And therefore since the axes of the pyramids GEr, HEs, KCt, LCv, are equally inclined to EC, the edge of the wedge, their solid angles will be equal. Consequently, by Prop. 4., their attractions in the directions of their axes will be as their lengths. And the attraction of each in the direction EF or CD will be as its length multiplied by the cosine of the angle which its axis makes with EF or CD; or since this angle is the same for all, the attraction of each in the direction EF or CD will be as its length. Hence the sum of the attractions of EGr and EHs in direction EF : sum of attractions of CKt and CLv in direction CD :: $EG+EH$: $CK+CL$. But, by Prop. 1., AEB and CKD are similar and concentric ellipses; therefore by Prop. 2.,

$$EG + EH = CK + CL;$$

or the attractions of GEr and HEs in direction EF = attractions of KCt, LCv, in direction CD.

If the point G had fallen on the other side of E, we should have had, by Prop. 3.,

$$EH - EG = CK + CL,$$

and therefore the difference of the attractions of the two pyramids, whose vertices are at E = the sum of the attractions of those whose vertices are at C; but since in this case the resolved part of the attraction of EGr is in a direction opposite to EF, we may still say, that the attractions of GEr, HEs, in direction EF = attractions of KCt, LCv, in direction CD. And the same is true for all other corresponding pairs of pyramids.

Now since the angle $GEg=KCk$, and $HEh=LCl$, by taking the same number of pairs of pyramids, we shall at the same time have taken the whole double wedge $AEBO$, and the whole wedge $CKDL$; and for every corresponding pair of pyramids, the attraction in direction CD or EF

is the same; therefore the attraction of E by the double wedge $AEBO$ in direction EF, is the same as the attraction of C by the wedge $CKDL$, in direction CD.

Now resolve each of these attractions into two, one perpendicular to the axis of the spheroid, another at right angles to this. Since EC is parallel to the axis of the spheroid, the perpendiculars upon the axis from E and C will be parallel; but EF and CD are parallel: therefore the angle made by EF with the perpendicular from E, is equal to the angle made by CD with the perpendicular from C. Consequently, the resolved parts of the equal attractions, in directions perpendicular to the axis of the spheroid, will also be equal. Now the double wedge $AEBO$, and the wedge $CKDL$, are formed by the same planes; and therefore the number of wedges into which the two spheroids can be cut is the same; and since the attractions of each corresponding pair of wedges, in direction perpendicular to the axis, are the same, the attractions of the whole spheroids in that direction will be the same; or the attraction of the larger spheroid on E, in direction perpendicular to the axis, is equal to the attraction of the smaller spheroid on C, in the same direction.

14. Prop. 9. To find the attraction on E, in a direction perpendicular to the axis of the spheroid.

By the last proposition, this is equal to the attraction of the spheroid CN in the same direction on the point C. And by Prop. 7, the attraction of a spheroid, whose polar and equatoreal radii are b and a, on a point in its equator

$$= 2kb\pi\left\{\frac{1}{e^3}\sin^{-1}e - \frac{\sqrt{(1-e^2)}}{e^2}\right\}$$

$$= 2ka\pi\left\{\frac{\sqrt{(1-e^2)}}{e^3}\sin^{-1}e - \frac{1-e^2}{e^2}\right\}.$$

This will represent the attraction of the spheroid CN on the point C, if $a = CW$.

Consequently, the attraction on E, in the direction perpendicular to the axis of the spheroid

$$= CW . 2k\pi . \left\{\frac{\sqrt{(1-e^2)}}{e^3}\sin^{-1} e - \frac{1-e^2}{e^2}\right\},$$

where e is the eccentricity of the smaller spheroid, which is the same as that of the larger, as the spheroids are similar. In the same spheroid, it will be observed, this attraction is proportional to CW.

15. Prop. 10. If from any point E, (fig. 7) on the surface of a spheroid, a line EX be drawn perpendicular to the axis, and a spheroid XY be described concentric, similar, and similarly situated, to the given spheroid, touching that line at its pole X; then the attraction of the given spheroid on the point E, in a direction parallel to its axis, is equal to the attraction of the smaller spheroid on a point at its pole X.

The demonstration of this is, in all respects, similar to the demonstration of Prop. 8. The spheroids must be divided into wedges, by planes passing through the line EX; and the sections of the spheroids, made by one plane, will be similar and concentric ellipses.

16. Prop. 11. To find the attraction of E, in a direction parallel to the axis of the spheroid.

By Prop. 10, this is equal to the attraction of the spheroid XY, on a particle at X. But, by Prop. 6., the attraction of the spheroid XY, on a particle at X

$$= 4\pi k . WX . \left\{\frac{1}{e^2} - \frac{\sqrt{(1-e^2)}}{e^3}\sin^{-1} e\right\};$$

therefore the attraction of the larger spheroid on E, in a direction parallel to its axis, is

$$4\pi k . EC . \left\{\frac{1}{e^2} - \frac{\sqrt{(1-e^2)}}{e^3}\sin^{-1} e\right\}.$$

In the same spheroid this is proportional to EC.

17. PROP. 12. If a point E, (fig. 8) be placed at the interior surface of a shell, bounded by similar and concentric spheroidical surfaces, the attraction of the whole shell will be 0.

For suppose the small pyramids EF, EG, to be formed by the same planes passing through E; let a plane pass through the axis of the pyramids, and through the common center of the spheroids: through H the point of bisection of EM, draw $WHKL$. Since EM is bisected in H, the tangent at K is parallel to EM; and since the ellipses are similar and concentric, the tangent at L is parallel to that at K; it is therefore parallel to FG; and FG therefore is bisected in H, or $FH = HG$. But $EH = HM$; $\therefore EF = MG$. Now the attractions of the pyramid FE, and the frustrum MG, upon a point at E, are proportional to their lengths EF and MG, by Prop. 4.; they are, therefore, equal; and they are in opposite directions; therefore they destroy each other. Now the whole shell may be divided into pairs of pyramids, in each of which it may be shewn that the attraction is 0; therefore the attraction of the whole shell = 0.

18. PROP. 13. To find the attraction of a spheroid on a point within it.

From E the point, (fig. 8), draw EC perpendicular to the plane of the equator. By Prop. 12, the attraction of the shell external to the spheroid EKM is 0; and by Prop. 9., the attraction of the spheroid EKM on E, in the direction perpendicular to the axis of the spheroid, is

$$CW.2k\pi.\left\{\frac{\sqrt{(1-e^2)}}{e^3}\sin^{-1}e - \frac{1-e^2}{e^2}\right\}.$$

By Prop. 11, the attraction in direction parallel to the axis, is

$$EC.4k\pi\left\{\frac{1}{e^2} - \frac{\sqrt{(1-e^2)}}{e^3}\sin^{-1}e\right\}.$$

19. The results of all these propositions may be thus stated.

The attraction on any particle of a spheroid, perpendicular to the axis, equals its distance from the axis × Q, where

$$Q = 2k\pi\left\{\frac{\sqrt{(1-e^2)}}{e^3}\sin^{-1}e - \frac{1-e^2}{e^2}\right\}.$$

The attraction perpendicular to the plane of the equator equals its distance from the plane of the equator × P, where

$$P = 4k\pi\left\{\frac{1}{e^2} - \frac{\sqrt{(1-e^2)}}{e^3}\sin^{-1}e\right\}.$$

If the spheroid differ little from a sphere,

$$Q = \frac{4k\pi}{3}\left(1 - \frac{2}{5}\epsilon\right), \quad P = \frac{4k\pi}{3}\left(1 + \frac{4}{5}\epsilon\right).$$

These expressions may be found by expanding the values just given: or by making the attractions at the pole and equator coincide with those found in (11) and (9).

APPLICATION OF THESE THEOREMS TO THE FIGURE OF THE EARTH.

20. In investigating the figure of the Earth, we shall suppose that the Earth was originally a homogeneous fluid mass, every particle attracting every other particle with an accelerating force, proportional to the mass of the attracting particle directly, and the square of the distance of the attracted particle inversely. This mass we suppose to revolve about an axis in $23^h. 56^m. 4^s$.

21. Now if the Earth had no motion of rotation, it would evidently assume a spherical figure. For the mutual attraction of the particles would collect the whole into one mass; and if any one part were then protuberant above the rest, the direction of gravity would not be perpendicular to its surface, and it would not remain in that form, but would run down. The form then must be such as would leave no part protuberant above the rest; that is, it must be spherical.

22. But, in consequence of the Earth's rotation, every particle has a centrifugal force, or a tendency to recede from the axis of rotation. The effect of this, it is plain, will be to enlarge the Earth at its equator, and to flatten it at the poles. It is our object now to shew, that the figure which the Earth would assume is accurately that of an oblate spheroid.

23. To prove this we shall shew that, upon giving a proper value to the ellipticity, the whole force which acts upon any point at the surface is perpendicular to the surface; and that, if two canals of any form be made in the fluid, terminated at any points in the surface, and leading to the same point in the interior, the pressure on this point is the same from the fluid in both canals.

24. Let T be the time of revolution: the centrifugal force on the particle E, fig. 6, 7, and 8, is

$$\frac{4\pi^2}{T^2}EX, \text{ or } \frac{4\pi^2}{T^2}CW,$$

and is in the direction XE. Adding this to the forces mentioned in (19), we have, the whole force acting upon the point E in direction $EX = \left(Q - \frac{4\pi^2}{T^2}\right) . CW$: that in direction $EC = P . EC$. The forces therefore upon any point, estimated in directions perpendicular to the axis, and perpendicular to the equator, are still proportional to the distances from the axis, and from the plane of the equator.

25. Prop. 14. In order that the fluid in a canal from the equator to the center, and the fluid in a canal from the pole to the center, may produce the same pressure on a particle at the center, the whole force at the pole must be to that at the equator as the radius of the equator to the radius of the pole.

Let two sections of the polar canal be made by planes at the distances p and $p + \delta p$ from the center. The pressure which is produced by the fluid included between these, estimated by its action upon a given surface, is proportional to the length of the included column multiplied by the accelerating force that acts on it, and may therefore be represented by $Pp.\delta p$*. Let u be the whole pressure; since, upon increasing the distance by δp, the pressure is diminished by $Pp.\delta p$, we have

$$\frac{du}{dp} = -Pp; \quad \therefore \; u = C - \frac{Pp^2}{2}.$$

But the pressure at the surface $= 0$, or $C - \frac{Pb^2}{2} = 0$;

$$\therefore \; u = \frac{P}{2}(b^2 - p^2);$$

hence, the pressure at the center $= \frac{Pb^2}{2}$. Similarly the pressure at the center, produced by the equatoreal column,

$$= \left(Q - \frac{4\pi^2}{T^2}\right)\frac{a^2}{2}.$$

---

* The accelerating force is represented by the *acceleration* which it would cause in the matter upon which it acts, if that matter were allowed to move freely under its action: it is measured by the velocity in feet per second which its action during one second would cause: and cannot be correctly represented in any other way.

The pressure caused by a quantity of fluid or other matter under the action of an accelerating force is a pressure of the same kind as that which we commonly call weight. It is connected with the accelerating force by means of the third law of motion, which teaches that the pressure corresponding to a certain acceleration of a certain quantity of matter is proportional to the product of the acceleration by the quantity of matter.

If we estimate the quantity of matter by the weight in pounds of that matter at a certain point of the Earth's surface, under the action of gravity at that place, and if we put g for the acceleration which gravity would produce in 1″ at the same place; then we mean by this that the pressure W estimated in pounds corresponds to the accelerating force g acting on the mass W. Substituting these in the general theorem,

$$\text{pressure} = C \times \text{acceleration} \times \text{quantity of matter},$$

we have $$W = C \times g \times W,$$

and therefore $C = \frac{1}{g}$. Hence in any other instance of accelerating force,

$$\text{pressure in pounds} = \frac{\text{accelerating force}}{g} \times \text{weight in pounds},$$

where the weight in pounds and the value of g are both referred to the same locality.

We shall frequently omit the divisor g in merely representing by a proportional quantity the pressure which accelerating forces cause.

When the pressures are equal (which, from the nature of fluids, is necessary for equilibrium),

$$Pb^2 = \left(Q - \frac{4\pi^2}{T^2}\right) a^2,$$

$$\text{or } P.b : \left(Q - \frac{4\pi^2}{T^2}\right) a :: a : b.$$

But Pb = force at the pole; $\left(Q - \frac{4\pi^2}{T^2}\right) a$ = the whole force at the equator; therefore the force at the pole : force at the equator :: equatoreal radius : polar radius.

26. Prop. 15. When this proportion holds, the whole force at any point on the surface is perpendicular to the surface.

Let E (fig. 9) be the point on the surface: take EC to represent the force in the direction EC, and CN to represent that in direction EX: then, EN will represent the magnitude and direction of the whole force at E. Now,

$$EC : CN :: P.EC : \left(Q - \frac{4\pi^2}{T^2}\right).CW, \text{ by (24)},$$

$$\text{or } :: a^2.EC : b^2.CW,$$

by the demonstration of Prop. 14;

$$\text{or } EC : CN :: EC : \frac{b^2}{a^2} CW.$$

Hence, $CN = \frac{b^2}{a^2} CW$ = subnormal (by *Conic Sections*); therefore, EN is the normal; that is, the whole force is perpendicular to the surface. It appears also that the whole force is represented in magnitude by the normal.

27. Prop. 16. When the same proportion holds, if to any point within the spheroid canals of any form be drawn, terminated any where in the surface; the pressure on that point, found by adding the pressures of successive portions of any of the canals, will be the same for every canal.

Let E (fig. 11) be the point, $EOoF$ a canal; take O and o, two points very near each other; draw ON, on, perpendicular to the plane of the equator, and OM, om, perpendicular to the axis; draw Ol perpendicular to no, and Ok perpendicular to mo: let $MO = x$, $NO = y$, $EO = s$; $mo = x + \delta x$, $no = y + \delta y$, $Eo = s + \delta s$. The accelerating force in the direction $ON = P.ON = P.y$: the resolved part of this in the direction of the canal

$$= P.y.\cos Oon = P.y.\frac{\delta y}{\delta s} = \text{ultimately } P.y.\frac{dy}{ds}.$$

And the accelerating force in the direction OM

$$= \left(Q - \frac{4\pi^2}{T^2}\right).OM = \left(Q - \frac{4\pi^2}{T^2}\right).x;$$

the resolved part, in the direction of the canal, is found by multiplying it by $\cos Ook$, that is, by $\frac{ok}{Oo}$ or $\frac{\delta x}{\delta s}$, or ultimately by $\frac{dx}{ds}$: hence, the whole accelerating force in that direction

$$= Py\frac{dy}{ds} + \left(Q - \frac{4\pi^2}{T^2}\right).x.\frac{dx}{ds}:$$

and the pressure produced by the action of this force on the fluid in Oo (see the Note to Art. 25) may be represented by

$$\delta s.\left(Py\frac{dy}{ds} + \overline{Q - \frac{4\pi^2}{T^2}}.x\frac{dx}{ds}\right).$$

This, if we put p for the pressure, is the decrement of p, produced on increasing s by δs: hence, $\frac{dp}{ds} = \frac{\delta p}{\delta s}$ ultimately

$$= -Py\frac{dy}{ds} - \overline{Q - \frac{4\pi^2}{T^2}}.x\frac{dx}{ds}.$$

$$\text{Integrating, } p = C - \frac{Py^2}{2} - \overline{Q - \frac{4\pi^2}{T^2}}.\frac{x^2}{2}.$$

Let the values of x and y, at the point E, be f and g; and where the canal meets the surface, let the values of x and y be v and w: then, observing that the pressure at the surface is $= 0$, we find the pressure at E

$$= \frac{Pw^2}{2} + \overline{Q - \frac{4\pi^2}{T^2}} \cdot \frac{v^2}{2} - \frac{Pg^2}{2} - \overline{Q - \frac{4\pi^2}{T^2}} \cdot \frac{f^2}{2}.$$

But, by the equation to the generating ellipse,

$$w^2 = \frac{b^2}{a^2}(a^2 - v^2); \qquad \therefore \quad \frac{Pw^2}{2} = \frac{Pb^2}{2} - \frac{Pb^2}{a^2} \cdot \frac{v^2}{2}:$$

and, by the demonstration of Prop. 14,

$$\frac{Pb^2}{a^2} = Q - \frac{4\pi^2}{T^2}.$$

Hence, the pressure at E

$$= \frac{Pb^2}{2} - \frac{Pg^2}{2} - \overline{Q - \frac{4\pi^2}{T^2}} \cdot \frac{f^2}{2}.$$

This expression, it may be remarked, is independent of the form of the canal, and of the place at which it terminates in the surface; and therefore we should have found the same for the pressure produced by the fluid in any other canal, as GE. For all canals therefore leading to the same point, the pressure on that point is the same.

28. From Prop. 15, we find, that if a fluid mass have the form of an oblate spheroid, there will be no tendency to disturb the particles at the surface; and from Prop. 16. it appears that, as the pressure on every particle is equal in all directions, none of the interior particles will have any tendency to motion. Every part therefore will be at rest: and therefore the oblate spheroid is the form of equilibrium, if the force at the pole : whole force at the equator :: equatoreal axis : polar axis.

29. Prop. 17. To find the proportion of the axes of the spheroid which is in equilibrium.

The force at the pole, by Prop. 6,

$$= 4\pi . kb\left\{\frac{1}{e^2} - \frac{\sqrt{(1-e^2)}}{e^3}\sin^{-1}e\right\}.$$

The attraction at the equator, by Prop. 7,

$$= 2\pi . kb\left\{\frac{1}{e^3}\sin^{-1}e - \frac{\sqrt{(1-e^2)}}{e^2}\right\};$$

the centrifugal force there

$$= \frac{4\pi^2}{T^2}a = \frac{4\pi^2}{T^2}.\frac{b}{\sqrt{(1-e^2)}}:$$

hence, the whole force at the equator

$$= 2\pi . kb\left\{\frac{1}{e^3}\sin^{-1}e - \frac{\sqrt{(1-e^2)}}{e^2}\right\} - \frac{4\pi^2}{T^2}.\frac{b}{\sqrt{(1-e^2)}}.$$

These must be in the proportion of $a : b$, or $1 : \sqrt{(1-e^2)}$; that is,

$$2k\left\{\frac{1}{e^2} - \frac{\sqrt{(1-e^2)}}{e^3}\sin^{-1}e\right\} : k\left\{\frac{1}{e^3}\sin^{-1}e - \frac{\sqrt{(1-e^2)}}{e^2}\right\} - \frac{2\pi}{T^2}.\frac{1}{\sqrt{(1-e^2)}}$$

$$:: 1 : \sqrt{(1-e^2)}.$$

Let $q = \dfrac{2\pi}{kT^2}$: substituting and reducing,

$$\frac{3(1-e^2)}{e^2} - \frac{(3-2e^2)\sqrt{(1-e^2)}}{e^3}\sin^{-1}e + q = 0.$$

The solution of this equation will give e, when q is known.

30. As this is a transcendental equation, it can be solved only by approximation; but some properties of its roots may be found thus. The left side of the equation is positive, when $e = 0$, and when $e = 1$, which are the extreme values of e that can be admitted; and it has therefore no roots, or an even number. And by constructing a curve, as fig. 10, in which the abscissa is e, and the ordinate is proportionate to the value of the first side of the equation,

we find that the curve cannot cut the axis in more than two points, and therefore there can be but two forms of the oblate spheroid, which are figures of equilibrium. If in one of these forms e be small, in the other it will be very nearly $= 1$.

31. To find these forms on the supposition that the centrifugal force is small, or q small, we will begin with supposing e small. In this case it will be most convenient to use the formulæ of (9) and (11). Then,

$$\frac{4\pi}{3}bk\left(1+\frac{4\epsilon}{5}\right) : \frac{4\pi}{3}bk\left(1+\frac{3\epsilon}{5}\right) - 2\pi kqb\,(1+\epsilon) :: 1+\epsilon : 1;$$

or, neglecting $q\epsilon$, which is the product of two small quantities,

$$1+\frac{4\epsilon}{5} : 1+\frac{3\epsilon}{5}-\frac{3q}{2} :: 1+\epsilon : 1.$$

Hence, $1+\frac{4\epsilon}{5} = 1+\frac{8\epsilon}{5}-\frac{3q}{2}$, or $\epsilon = \frac{5}{4}\cdot\frac{3}{2}q.$

32. It is convenient to express the ellipticity in terms of the proportion of the centrifugal force at the equator to gravity. This proportion is

$$\frac{2\pi kbq}{\frac{4\pi}{3}bk\left(1+\frac{3\epsilon}{5}\right)-2\pi kq} = \frac{2\pi kbq}{\frac{4\pi}{3}bk} \text{ nearly } = \frac{3}{2}q:$$

let this $= m$. Then $\epsilon = \frac{5m}{4}$.

33. To find the other form of equilibrium, we observe that e is nearly $= 1$, and therefore $\frac{b}{a}$ is small. We must therefore expand the terms of the equation in powers of $\frac{b}{a}$.

For e^2 put $1-\frac{b^2}{a^2}$, or $1-c^2$, where $c=\frac{b}{a}$, and it becomes

$$\frac{3c^2}{1-c^2} - \frac{(1+2c^2)c}{(1-c^2)^{\frac{3}{2}}}\cos^{-1}c + q = 0.$$

Taking only the first power of c, we have

$$-c\,.\,\frac{\pi}{2}+q=0, \qquad \text{or } \frac{b}{a}=c=\frac{2q}{\pi}.$$

34. In the Earth it is found that $m=\frac{1}{289}$, or $q=\frac{1}{434}$: hence, supposing the ellipticity small, $\epsilon=\frac{1}{230}$: or, supposing the eccentricity nearly $=1$, $c=\frac{1}{681}$. That is, the Earth would be an oblate spheroid, with axes either in the proportion of 230 : 231, or in the proportion of 1 : 681. It is found by measurement, that the ratio of the axes is nearly 300 : 301: hence, the Earth is not homogeneous.

35. In the spheroid of small ellipticity, the proportion of gravity at the pole to that at the equator is the same as the ratio of the axes, or is the ratio 231 : 230, supposing the Earth homogeneous. By observation, it is found to be about 188 : 187.

36. Resuming the consideration of the general equation of (29), and the construction of (30), it is easily seen that, upon giving to q a certain value, the curve will touch the line of abscissæ: and upon increasing q the curve will not meet the line of abscissæ at all. In the former case, then, there is but one form of equilibrium, and, in the latter, equilibrium is not possible. To find e', the value of e, which gives but one form, we may observe, that two roots of the equation have become coincident, or are equal: if then we take the differential coefficient of the first side of the equation, it must have one of the equal roots: or the same value of e will make it $=0$. This gives

$$-\frac{9}{e'^3}+\frac{2}{e'}+\left(\frac{9}{e'^4}-\frac{8}{e'^2}\right)\cdot\frac{\sin^{-1}e'}{\sqrt{(1-e'^2)}}=0:$$

solving this equation by approximation, $e'=,92995$, whence $\frac{b'}{a'}=,36769$, $\frac{a'}{b'}=2,7197$. Substituting this value of e in the general equation, $q' = ,224671$. And, if T' be the time of revolution, since $q' = \frac{2\pi}{kT'^2}$, we have

$$T' = \sqrt{\left(\frac{2\pi}{kq'}\right)} = \frac{5,2883}{\sqrt{(k)}}.$$

37. In a fluid whose density is the same as the density of the Earth, supposed homogeneous,

$$\frac{1}{434} = \frac{2\pi}{k(23^h.56'.4'')^2}, \text{ by (34)}.$$

Dividing this by the equation $q' = \frac{2\pi}{kT'^2}$, we have

$$\frac{1}{434.q'} = \left(\frac{T'}{23^h.56'.4''}\right)^2,$$

$$\text{or } T' = 23^h.56'.4''.\sqrt{\left(\frac{1}{434.q'}\right)} = 2^h.25'.26''.$$

A spheroid then cannot remain in equilibrium, if it revolve in a shorter time than $2^h.25'.26''$, its density being the same as that of the Earth.

38. The expression for the ellipticity in (31), gives us the means of comparing the ellipticities of different planets, supposed homogeneous. For the ellipticity

$$= \frac{15}{8}q = \frac{15}{4}.\frac{\pi}{kT^2}.$$

If then we can in any manner compare the masses of the planets, (which can be done immediately with those that have satellites,) and if we know the ratio of their diameters, the ratio of their densities will be known: and, knowing also the ratio of their times of revolution, their ellipticities will be inversely as their densities × the square of the times of revolution.

For an investigation of the remarkable case of equilibrium of a revolving ellipsoid when its three axes are unequal, the reader is referred to Mr Ivory's paper in the *Philosophical Transactions* for 1838, page 57.

ON THE FIGURE OF THE EARTH, SUPPOSING IT HETEROGENEOUS.

39. The results which we have deduced relative to the figure of the Earth, supposing it homogeneous, do not agree with observation: the homogeneity of the Earth is also, *a priori*, very improbable. We shall now proceed to shew that, supposing the Earth heterogeneous, a spheroidal form, of ellipticity different from that which we have found, will be a form of equilibrium. As before, we shall suppose that the Earth was originally a fluid mass: and we shall consider the density of different parts to be different, either from its original constitution, or from the difference of pressure produced by the weight of the superincumbent mass. From the difficulty of the investigation, we are obliged to suppose the ellipticity small, and to reject all quantities depending on its square and higher powers.

40. Prop. 18. If the base of a prism be very small, to find its attraction in the direction of its axis on a point any where without it.

Let BC, (fig. 12) be the given prism: A the given point: draw AD perpendicular to the axis, produced if necessary; take two sections perpendicular to the axis, passing through the points E and F, which are very near each other: join AB, AE, AC. Let $AD = a$, $DB = b$, $DC = c$, the section of the prism $= S$, its density $= \rho$: $DE = x$, $DF = x + \delta x$. The mass included between the two sections through E and $F = S \cdot \delta x$; therefore its attraction on $A = \dfrac{\rho \cdot S \cdot \delta x}{(AE)^2}$ ultimately; therefore the resolved part in the direction of the prism's axis

$$= \frac{\rho \cdot S \cdot \delta x}{(AE)^2} \cdot \frac{DE}{AE} = \frac{\rho \cdot S \cdot DE \cdot \delta x}{(AE)^3}:$$

hence, if u be the whole attraction, $\dfrac{du}{dx} =$ ultimate value of $\dfrac{\delta u}{\delta x}$

$$= \frac{\rho \cdot S \cdot x}{(a^2 + x^2)^{\frac{3}{2}}}; \text{ integrating, } u = -\frac{\rho \cdot S}{(a^2 + x^2)^{\frac{1}{2}}} + C.$$

Determining C so as to make $u=0$ when $x=DB=b$, we find the whole attraction

$$=\rho\,.\,S\,.\left\{\frac{1}{\sqrt{(a^2+b^2)}}-\frac{1}{\sqrt{(a^2+c^2)}}\right\}=\rho\,.\,S\,.\left(\frac{1}{AB}-\frac{1}{AC}\right).$$

41. PROP. 19. To find the expression upon whose integration depends the attraction of an ellipsoid on a point any where within or without it.

Let CD, CE, CF, (fig. 13) be the three semi-axes of the ellipsoid, $=a$, b, c, respectively: take these directions for the directions of x, y, and z: then, the equation to the surface of the ellipsoid is

$$\frac{x^2}{a^2}+\frac{y^2}{b^2}+\frac{z^2}{c^2}=1.$$

Let the co-ordinates of the point A be f, g, h. Suppose the ellipsoid divided into slices by planes, as PMQ, pmq, parallel to the plane of yz: and suppose these slices divided into prisms by planes, as PNQ, pnq, parallel to the plane of xz. Let

$$CM=x,\quad Mm=\delta x,\quad MN=y,\quad mn=y+\delta y,\quad PN=z.$$

Then the parallelogram $Nn=\delta x\,.\,\delta y$. By the last Proposition therefore the attraction of the prism Pq on A, in a direction parallel to z, is

$$\rho\,.\,\delta x\,.\,\delta y\left(\frac{1}{AP}-\frac{1}{AQ}\right)$$

$$.\,\delta x\,.\,\delta y\left(\frac{1}{\sqrt{\{\overline{f-x}|^2+\overline{g-y}|^2+\overline{h-z}|^2\}}}-\frac{1}{\sqrt{\{\overline{f-x}|^2+\overline{g-y}|^2+\overline{h+z}|^2\}}}\right).$$

Let $x=ar$, $y=bs$; then, the attraction of the prism

$$=\rho\,.\,a\,.\,b\,.\,\delta r\,.\,\delta s\left(\frac{1}{AP}-\frac{1}{AQ}\right).$$

The attraction of the slice therefore will

$$=\rho\,.\,a\,.\,b\,.\,\delta r\,.\int_s\left(\frac{1}{AP}-\frac{1}{AQ}\right),$$

taken from G to H. To find the values of y, corresponding to those points, we must make $z=0$, in the equation to the surface: then

$$\frac{x^2}{a^2}+\frac{y^2}{b^2}=1; \quad \therefore \; y=\pm\frac{b}{a}\sqrt{(a^2-x^2)}, \quad \text{and} \quad s=\pm\sqrt{(1-r^2)}.$$

The attraction of the slice therefore

$$=\rho . a . b . \delta r \times \int_s \left(\frac{1}{AP}-\frac{1}{AQ}\right),$$

taken from $s=-\sqrt{(1-r^2)}$ to $s=+\sqrt{(1-r^2)}$.

Let this $=v\delta r$: then the attraction of the ellipsoid $=\int_r v$, taken from $x=-a$ to $x=+a$, or from $r=-1$, to $r=+1$. The attraction therefore of the ellipsoid $=\rho . a . b \times$

$$\int_r . \int_s . \left(\frac{1}{\sqrt{\{\overline{f-ar}|^2+\overline{g-bs}|^2+\overline{h-ct}|^2\}}}-\frac{1}{\sqrt{\{\overline{f-ar}|^2+\overline{g-bs}|^2+\overline{h+ct}|^2\}}}\right),$$

where $ct=z$; the first integral being taken from $s=-\sqrt{(1-r^2)}$ to $s=+\sqrt{(1-r^2)}$, and the second from $r=-1$ to $r=+1$.

42. Prop. 20. If a spheroid, (fig. 13), whose semi-axes are a, b, c, attract a point A without it whose co-ordinates are $l\alpha$, $m\beta$, $n\gamma$, where $l^2+m^2+n^2=1$.

And if a spheroid (fig. 14) of the same density, whose semi-axes are α, β, γ, attract a point A' within it*, whose co-ordinates are la, mb, nc;

---

* That A' is within the spheroid attracting it is easily shewn. For since $l^2+m^2+n^2=1$, its co-ordinates satisfy the equation

$$\frac{x^2}{a^2}+\frac{y^2}{b^2}+\frac{z^2}{c^2}=1,$$

and it is therefore in the surface of an ellipsoid, concentric to the ellipsoid which attracts it, and whose semi-axes are a, b, c. In the same manner, A is in the surface of an ellipsoid concentric to the given ellipsoid, whose semi-axes are α, β, γ. Now, since

$$\alpha^2-a^2=\beta^2-b^2=\gamma^2-c^2,$$

the surfaces of the ellipsoids do not cut each other: and the point A being without the given ellipsoid, α must be $>a$, $\beta>b$, $\gamma>c$. The ellipsoid therefore whose semi-axes are a, b, c, is entirely within the other, supposing them concentric; and A' therefore is within the ellipsoid, whose semi-axes are α, β, γ.

And if $a^2 - c^2 = \alpha^2 - \gamma^2$, $b^2 - c^2 = \beta^2 - \gamma^2$.

Then, the attraction on A parallel to c, is to the attraction on A' parallel to γ, as ab to $\alpha\beta$.

43. By the last Proposition, the attraction on A parallel to c

$$= \rho . a . b . \textstyle\int_r \int_s \left\{ \dfrac{1}{\sqrt{(\overline{l\alpha - ar}|^2 + \overline{m\beta - bs}|^2 + \overline{n\gamma - ct}|^2)}} \right.$$

$$\left. - \frac{1}{\sqrt{(\overline{l\alpha - ar}|^2 + \overline{m\beta - bs}|^2 + \overline{n\gamma + ct}|^2)}} \right\}.$$

The square of the denominator of the first fraction

$$= l^2\alpha^2 - 2a\alpha lr + a^2r^2$$

$$+ m^2\beta^2 - 2b\beta ms + b^2s^2$$

$$+ n^2\gamma^2 - 2c\gamma nt + c^2t^2.$$

But $l^2 + m^2 + n^2 = 1$: and the equation to the ellipsoid, or

$$\frac{x^2}{a^2} + \frac{y^2}{b^2} + \frac{z^2}{c^2} = 1$$

gives us $r^2 + s^2 + t^2 = 1$:

eliminating, therefore, n^2 and t^2, the denominator is

$$l^2(\alpha^2 - \gamma^2) + m^2(\beta^2 - \gamma^2) + \gamma^2$$

$$- 2\{a\alpha lr + b\beta ms + c\gamma n\sqrt{(1 - r^2 - s^2)}\}$$

$$+ r^2(a^2 - c^2) + s^2(b^2 - c^2) + c^2.$$

The second denominator differs from this only in the sign of

$$2c\gamma n\sqrt{(1 - r^2 - s^2)}.$$

Now, the attraction on A' parallel to γ, found in the same way

$$= \rho\alpha\beta \times$$

$$\textstyle\int_r \int_s \left\{ \dfrac{1}{\sqrt{(\overline{la - \alpha r}|^2 + \overline{mb - \beta s}|^2 + \overline{nc - \gamma t}|^2)}} \right.$$

$$\left. - \frac{1}{\sqrt{(\overline{la - \alpha r}|^2 + \overline{mb - \beta s}|^2 + \overline{nc + \gamma t}|^2)}} \right\}.$$

The square of the first denominator, expanded as above,

$$= l^2 (a - c^2) + m^2 (b^2 - c^2) + c^2$$
$$- 2\{\alpha a l r + \beta b m s + \gamma c n \sqrt{(1 - r^2 - s^2)}\}$$
$$+ r^2 (\alpha^2 - \gamma^2) + s^2 (\beta^2 - \gamma^2) + \gamma^2.$$

But, by supposition,

$$a^2 - c^2 = \alpha^2 - \gamma^2, \quad b^2 - c^2 = \beta^2 - \gamma^2;$$

hence, this is precisely equal to the square of the denominator of the first factor above: or the first fraction here = the first fraction above. Similarly, the second fraction here = the second fraction above. Hence, the whole expression under the sign of integration is the same in both. And the limits of integration are the same; therefore the integrals will be the same. In the first, the integral is multiplied by $\rho a b$, and in the second, by $\rho \alpha \beta$: hence, the attraction of A parallel to c is to that of A' parallel to γ as ab to $\alpha\beta$.

44. In the same way, the attraction of A parallel to a, is to that of A' parallel to α, as bc to $\beta\gamma$: and the attraction of A parallel to b, is to that of A' parallel to β, as ac to $\alpha\gamma$. If the ellipsoid become a spheroid, by making $a = b$, then $\alpha = \beta$, and the forces parallel to c and γ, are as a^2 to α^2: those parallel to a and α are as ac to $\alpha\gamma$: those parallel to b and β as ac to $\alpha\gamma$.

45. Prop. 21. To find the attraction of an oblate spheroid, whose ellipticity is small, on a point without it.

Let c, the semi-axis of revolution of the spheroid, coincide with the axis of z: let f, g, h, be the co-ordinates of the attracted point. And, as in the last Proposition, let $f = l\alpha$, $g = m\alpha$, $h = n\gamma$, where

$$l^2 + m^2 + n^2 = 1, \text{ and } \alpha^2 - \gamma^2 = a^2 - c^2.$$

$$\text{Let } a = c(1 + e), \quad \alpha = \gamma(1 + \epsilon):$$

then, rejecting e^2, &c., the last equation becomes

$$2\gamma^2 \epsilon = 2c^2 e, \quad \text{or} \quad \epsilon = \frac{c^2}{\gamma^2} e.$$

And from the equation

$$l^2 + m^2 + n^2 = 1,$$

or $\dfrac{f^2+g^2}{a^2} + \dfrac{h^2}{\gamma^2} = 1$, or $\dfrac{1}{\gamma^2}\{f^2 + g^2 + h^2 - 2\epsilon(f^2+g^2)\} = 1$,

we get $\gamma^2 = f^2 + g^2 + h^2 - 2\epsilon(f^2+g^2)$;

$$\text{but } \epsilon = \frac{c^2}{\gamma^2}e = \frac{c^2}{f^2+g^2+h^2}e;$$

$$\therefore\ \gamma^2 = f^2 + g^2 + h^2 - 2e\,.\frac{c^2(f^2+g^2)}{f^2+g^2+h^2},$$

$$\text{and } \gamma = \sqrt{(f^2+g^2+h^2)} - e\,.\frac{c^2(f^2+g^2)}{(f^2+g^2+h^2)^{\frac{3}{2}}}.$$

Also, since $a^2 - \gamma^2 = a^2 - c^2 = 2c^2e$,

$$a^2 = \gamma^2 + 2c^2e = f^2 + g^2 + h^2 + 2e\frac{c^2h^2}{f^2+g^2+h^2};$$

$$\therefore\ a = \sqrt{(f^2+g^2+h^2)} + e\frac{c^2h^2}{(f^2+g^2+h^2)^{\frac{3}{2}}}.$$

46. Now, the co-ordinates of A' (Prop. 20.) $= la, mb, nc$: or $= \dfrac{fa}{a}, \dfrac{ga}{a}, \dfrac{hc}{\gamma}$. The distance of this point, therefore, from the axis of the spheroid

$$= \sqrt{\left(\frac{f^2a^2}{a^2} + \frac{g^2a^2}{a^2}\right)} = \frac{a}{a}\sqrt{(f^2+g^2)}.$$

And, as this point is within the spheroid whose semi-axes are a and γ, its attraction towards the axis, by (19),

$$= \frac{4\pi}{3}\rho\left(1 - \frac{2}{5}\epsilon\right).\frac{a}{a}\sqrt{(f^2+g^2)}:$$

hence, its attraction in the direction of x

$$= \frac{f}{\sqrt{(f^2+g^2)}}\cdot\frac{4\pi}{3}\rho\left(1 - \frac{2}{5}\epsilon\right).\frac{a}{a}\sqrt{(f^2+g^2)}$$

$$= \frac{4\pi}{3}\rho\,.\left(1 - \frac{2}{5}\epsilon\right).\frac{a}{a}.f.$$

Similarly, its attraction in the direction of y

$$= \frac{4\pi}{3}\rho . (1 - \frac{2}{5}\epsilon) . \frac{a}{a} . g.$$

And, by (19), its attraction in the direction of z

$$= \frac{4\pi}{3}\rho (1 + \frac{4}{5}\epsilon) . \frac{c}{\gamma} . h.$$

47. By Prop. 20, therefore, the attraction on A

in the direction of $x = \frac{ac}{a\gamma} . \frac{4\pi}{3}\rho (1 - \frac{2}{5}\epsilon) . \frac{a}{a} . f$

$$= \frac{4\pi}{3} . \rho . \frac{a^2 c}{a^2\gamma} . (1 - \frac{2}{5}\epsilon) f;$$

that in the direction of $y = \frac{ac}{a\gamma} . \frac{4\pi}{3}\rho (1 - \frac{2}{5}\epsilon) . \frac{a}{a} . g$

$$= \frac{4\pi}{3}\rho . \frac{a^2 c}{a^2\gamma} . (1 - \frac{2}{5}\epsilon) g;$$

that in the direction of $z = \frac{a^2}{a^2} . \frac{4\pi}{3}\rho (1 + \frac{4}{5}\epsilon) . \frac{c}{\gamma} . h$

$$= \frac{4\pi}{3}\rho . \frac{a^2 c}{a^2\gamma} (1 + \frac{4}{5}\epsilon) h.$$

Now $a^2 c = c^3 (1 + 2e)$: and, from the expressions above,

$$a^2\gamma = (f^2 + g^2 + h^2)^{\frac{3}{2}} + e\frac{c^2 (2h^2 - f^2 - g^2)}{(f^2 + g^2 + h^2)^{\frac{1}{2}}};$$

$$\therefore \frac{a^2 c}{a^2\gamma} = \frac{c^3}{(f^2 + g^2 + h^2)^{\frac{3}{2}}} \left\{1 + 2e - e\frac{c^2 (2h^2 - f^2 - g^2)}{(f^2 + g^2 + h^2)^2}\right\}.$$

$$\text{Also, } 1 - \frac{2}{5}\epsilon = 1 - e . \frac{2}{5} . \frac{c^2}{f^2 + g^2 + h^2},$$

(putting $f^2 + g^2 + h^2$ for γ^2).

$$\text{And } 1 + \frac{4}{5}\epsilon = 1 + e . \frac{4}{5} . \frac{c^2}{f^2 + g^2 + h^2}.$$

Substituting these, we find, at length, the attraction of A in the direction of x

$$= \frac{4\pi}{3}\rho \cdot \frac{c^3}{(f^2+g^2+h^2)^{\frac{3}{2}}} \cdot \left\{1 + 2e - e\frac{c^2(12h^2 - 3f^2 - 3g^2)}{5(f^2+g^2+h^2)^2}\right\} f;$$

that in the direction of y

$$= \frac{4\pi}{3}\rho \cdot \frac{c^3}{(f^2+g^2+h^2)^{\frac{3}{2}}} \cdot \left\{1 + 2e - e\frac{c^2(12h^2 - 3f^2 - 3g^2)}{5(f^2+g^2+h^2)^2}\right\} g;$$

that in the direction of z

$$= \frac{4\pi}{3}\rho \cdot \frac{c^3}{(f^2+g^2+h^2)^{\frac{3}{2}}} \cdot \left\{1 + 2e - e\frac{c^2(6h^2 - 9f^2 - 9g^2)}{5(f^2+g^2+h^2)^2}\right\} h.$$

48. Prop. 22. To find the attraction of an oblate spheroid on a point without it; the spheroid being heterogeneous; and all the surfaces passing through points at which the density is the same, being spheroidal, of variable ellipticity.

Let EF, (fig. 15) be the spheroid; let $E'F'$ be a spheroidal surface, at every one of whose points the density is the same, and $E''F''$ a spheroidal surface very near the former, of different ellipticity, at all of whose points the density is the same, but differing from that at the surface $E'F'$. Let $CF' = c$, $CF'' = c + \delta c$. Since the ellipticity varies when the semi-axis of the spheroid is varied, e must be a function of c. Now the density of all the matter included between $E'F'$ and $E''F''$ is not uniform; but by diminishing δc, it may be made to approximate as nearly as we please to uniformity. Conceive, now, for the moment, the interior matter of the spheroid $E'F'$ to be of the same density as that at its surface, or to be equal to ρ; let its attraction in the direction of $x = \rho . A$. Then the attraction of the spheroid $E''F''$ in the same direction, will be the value which A receives when $c + \delta c$ is put for c, and when, instead of e we put the value of the ellipticity in the spheroid $E''F''$. But if we consider e as a function of c, this is included in considering the variation which it receives in consequence of the variation of c.

Hence A will be changed to $A + \frac{dA}{dc}\delta c + \&c.$; and therefore the attraction of the spheroid $E''F''$

$$= \rho\left(A + \frac{dA}{dc}\delta c + \&c.\right).$$

The difference of the attraction of the two spheroids, or the attraction of the shell included between them, is therefore ultimately $= \rho \cdot \frac{dA}{dc}\delta c$. If, then, u be the attraction of the heterogeneous spheroid, whose polar semi-axis $= c$, we find $\frac{du}{dc}$ = ultimate value of $\frac{\delta u}{\delta c} = \rho \cdot \frac{dA}{dc}$;

$$\therefore\ u = \int_c \rho \cdot \frac{dA}{dc};$$

where in the differentiation e must be considered as a function of c; and in the integration, ρ and e must both be considered as functions of c.

49. Now,

$$A = \frac{4\pi}{3}\left\{\frac{1}{(f^2+g^2+h^2)^{\frac{3}{2}}}c^3(1+2e) - \frac{12h^2-3f^2-3g^2}{5(f^2+g^2+h^3)^{\frac{7}{2}}}c^5 e\right\}f;$$

and when we differentiate this with respect to c, since f, g, and h, are perfectly independent of c, the only variable terms will be $c^3(1+2e)$ and $c^5 e$. Let

$$\int_c \rho \cdot \frac{d(c^3.\overline{1+2e})}{dc} = \phi(c); \quad \int_c \rho \cdot \frac{d(c^5 e)}{dc} = \psi(c);$$

both integrals being made to vanish when $c = 0$. Then by the expression found in the last article, the attraction in the direction of x

$$= \frac{4\pi}{3} \cdot \left\{\frac{1}{(f^2+g^2+h^2)^{\frac{3}{2}}} \cdot \phi(c) - \frac{12h^2-3f^2-3g^2}{5(f^2+g^2+h^2)^{\frac{7}{2}}}\psi(c)\right\} \cdot f.$$

Similarly from the values in (47), we find the attraction in the direction of y

$$= \frac{4\pi}{3}\left\{\frac{1}{(f^2+g^2+h^2)^{\frac{3}{2}}}\cdot\phi(c) - \frac{12h^2-3f^2-3g^2}{5(f^2+g^2+h^2)^{\frac{7}{2}}}\psi(c)\right\}.g.$$

And the attraction in the direction of z

$$= \frac{4\pi}{3}\left\{\frac{1}{(f^2+g^2+h^2)^{3}}\cdot\phi(c) - \frac{6h^2-9f^2-9g^2}{5(f^2+g^2+h^2)^{\frac{7}{2}}}\psi(c)\right\}.h.$$

It will be necessary to remark, that the second term in each expression is small, when the ellipticity is small, but the first is not small.

50. Prop. 23. To find the attraction of a heterogeneous shell on a point within it; the points of equal density being situated in spheroidal surfaces, and the interior and exterior surfaces being spheroids, in every part of which the density is the same.

Let f, g, h be the co-ordinates of the attracted point A', (fig. 15). Take $E'F'$, $E''F''$ as in the last Proposition, and suppose the whole spheroid $E'CF'$ to have the uniform density ρ. Then by (19), the attraction of the spheroid $E'CF'$ on the point A', directed towards the axis

$$= \frac{4\pi}{3}\rho\sqrt{f^2+g^2}.\left(1-\frac{2}{5}e\right);$$

whence the attraction in the direction of x

$$= \frac{4\pi}{3}\rho.f\left(1-\frac{2}{5}e\right);$$

that in the direction of y

$$= \frac{4\pi}{3}\rho.g\left(1-\frac{2}{5}e\right).$$

And by (19), the attraction in the direction of z

$$= \frac{4\pi}{3}\rho.h\left(1+\frac{4}{5}e\right).$$

Then by exactly the same reasoning as that in Prop. 22, it may be shewn, that the attraction of the heterogeneous shell will be found by differentiating these expressions, (after taking away ρ) with respect to c, multiplying then by ρ, and integrating with regard to c. Let $\int_c \rho \cdot \frac{de}{dc} = \chi(c)$, the integral being made to vanish when $c = 0$. If we suppose the polar semi-axis of the interior surface to be c, that of the exterior to be c, then $\int_c \rho \frac{de}{dc}$ for the shell, between these limits, $= \chi(\mathrm{c}) - \chi(c)$. Hence it will easily be seen, that the force in the direction of x

$$= -\frac{4\pi}{3} \cdot \frac{2}{5} \{\chi(\mathrm{c}) - \chi(c)\} \cdot f,$$

that in the direction of y

$$= -\frac{4\pi}{3} \cdot \frac{2}{5} \{\chi(\mathrm{c}) - \chi(c)\} \cdot g,$$

that in the direction of z

$$= +\frac{4\pi}{3} \cdot \frac{4}{5} \{\chi(\mathrm{c}) - \chi(c)\} \cdot h.$$

51. Prop. 24. To find the attraction of a heterogeneous spheroid of the same kind as that in Prop. 22, upon any point within it.

Let γ be the polar semi-axis of that spheroidal surface which passes through the given point, and through all other points at which the density is the same. And let c be the semi-axis of the exterior surface. Then the given point is external to all the spheroids whose semi-axes are less than γ, and to these, therefore, the integration in Prop. 22. must be applied. Taking these integrals, then, from $c = 0$ to $c = \gamma$, we find for the forces in direction of x, y and z,

$$\frac{4\pi}{3} \left\{ \frac{1}{(f^2 + g^2 + h^2)^{\frac{3}{2}}} \phi(\gamma) - \frac{12h^2 - 3f^2 - 3g^2}{5(f^2 + g^2 + h^2)^{\frac{7}{2}}} \psi(\gamma) \right\} f,$$

$$\frac{4\pi}{3} \left\{ \frac{1}{(f^2 + g^2 + h^2)^{\frac{3}{2}}} \phi(\gamma) - \frac{12h^2 - 3f^2 - 3g^2}{5(f^2 + g^2 + h^2)^{\frac{7}{2}}} \psi(\gamma) \right\} g,$$

$$\frac{4\pi}{3} \left\{ \frac{1}{(f^2 + g^2 + h^2)^{\frac{3}{2}}} \phi(\gamma) - \frac{6h^2 - 9f^2 - 9g^2}{5(f^2 + g^2 + h^2)^{\frac{7}{2}}} \psi(\gamma) \right\} h.$$

Again, the given point is interior to all the spheroidal surfaces whose polar semi-axes are greater than γ, and less than c;

and, therefore, the expressions of Prop. 23. must be taken between these limits. These give us for the forces in the directions of x, y and z,

$$-\frac{4\pi}{3}\cdot\frac{2}{5}\{\chi(c)-\chi(\gamma)\}f,$$

$$-\frac{4\pi}{3}\cdot\frac{2}{5}\{\chi(c)-\chi(\gamma)\}g,$$

$$+\frac{4\pi}{3}\cdot\frac{4}{5}\{\chi(c)-\chi(\gamma)\}h.$$

and if these be added to those above, we shall have the whole forces arising from the attraction of the spheroid.

52. Besides these, if the spheroid revolve round its polar axis in the time T, every point will experience a centrifugal force proportional to its distance from the axis. Resolving this in the directions of x and y, we have for the parts in those directions $\frac{4\pi^2}{T^2}.f$ and $\frac{4\pi^2}{T^2}.g$, which must be subtracted from the attractions. Collecting all the terms, the force in the direction of x =

$$\frac{4\pi}{3}\left\{\frac{1}{(f^2+g^2+h^2)^{\frac{3}{2}}}\phi(\gamma)-\frac{12h^2-3f^2-3g^2}{5(f^2+g^2+h^2)^{\frac{7}{2}}}\psi(\gamma)\right.$$
$$\left.-\frac{2}{5}\{\chi(c)-\chi(\gamma)\}-\frac{3\pi}{T^2}\right\}f=F.$$

That in the direction of y =

$$\frac{4\pi}{3}\left\{\frac{1}{(f^2+g^2+h^2)^{\frac{3}{2}}}\phi(\gamma)-\frac{12h^2-3f^2-3g^2}{5(f^2+g^2+h^2)^{\frac{7}{2}}}\psi(\gamma)\right.$$
$$\left.-\frac{2}{5}\{\chi(c)-\chi(\gamma)\}-\frac{3\pi}{T^2}\right\}g=G.$$

That in the direction of z =

$$\frac{4\pi}{3}\left\{\frac{1}{(f^2+g^2+h^2)^{\frac{3}{2}}}\phi(\gamma)-\frac{6h^2-9f^2-9g^2}{5(f^2+g^2+h^2)^{\frac{7}{2}}}\psi(\gamma)\right.$$
$$\left.+\frac{4}{5}\{\chi(c)-\chi(\gamma)\}\right\}h\ldots\ldots=H.$$

53. Prop. 25. To find the conditions of equilibrium of a heterogeneous fluid.

Suppose the fluid to be in equilibrium, and suppose any number of canals terminated in the external surface to be drawn to any point within. It is plain that the separation of the fluid within these canals from the general mass, by the sides of the canals, will not interrupt the equilibrium. But when the fluid within the canals is in equilibrium, if we find the pressure which the fluid in each canal produces upon the place of junction of the canals, all the pressures so found must be equal. For the pressure exerted by the fluid of any one canal upon the fluid at the place of junction will (from the fundamental property of fluids) be transmitted up the other canals, and unless it is counteracted by an equal force produced by the pressure of each of these, it will destroy the equilibrium. It is *necessary* therefore for equilibrium in each imaginary system of canals, and consequently for equilibrium in the whole mass, that the pressure produced by the fluid in a canal terminated at a given point within, be independent of the form of the canal and of the position of the point where it meets the external surface.

This condition is also *sufficient* for equilibrium. For since the fluid has no tendency to run up or down the canals, the equilibrium can be interrupted only by its tendency to burst the canals. But by drawing other canals to any point of those already drawn, it would easily be seen that when the condition above mentioned is satisfied, the external pressure on all sides is equal to the internal pressure: and consequently there is no tendency to burst a canal.

54. Now let P (fig. 17) be any point in one canal, p very near it: let $RP = s$, $Rp = s + \delta s$; let the co-ordinates of P be f, g, h: those of p, $f + \delta f$, $g + \delta g$, $h + \delta h$. Then if a parallelopiped be constructed of which Pp or δs is the diagonal (since it is ultimately a straight line), and whose sides are parallel to f, g, h, respectively, the lengths of those sides will be δf, δg, δh. To find how much the pressure of the fluid at P exceeds that at p, we must resolve each of the forces acting on its particles into two, one in the direction of pP and the other perpendicular to pP; then we

must add together the former, and neglect the latter. The forces are $-F$, $-G$, $-H$, in the directions of f, g and h: the sum of the forces in the direction of Pp is

$$-(F.\cos pPq + G\cos pPr + H\cos pPt)$$

$$= -\left(F\frac{\delta f}{\delta s} + G\frac{\delta g}{\delta s} + H\frac{\delta h}{\delta s}\right).$$

And the matter upon which they act is $\rho\,\delta s$ (the section of the canal being supposed $= 1$): therefore the pressure which they produce ultimately (see the Note to Art 25)

$$= -\rho\,(F\delta f + G\delta g + H\delta h)$$

$$= -\rho\delta f\left(F + G\frac{\delta g}{\delta f} + H\frac{\delta h}{\delta f}\right)$$

$$= -\rho\delta f\left(F + G\frac{dg}{df} + H\frac{dh}{df}\right).$$

If then we can find a quantity V such that

$$\frac{d(V)}{df} = -\rho\left(F + G\frac{dg}{df} + H\frac{dh}{df}\right),$$

V taken between the proper limits will be the pressure.

If now the values of ρ, F, G, H, be such that the expression for $\frac{d(V)}{df}$ cannot be integrated without expressing g and h in terms of f*, we must, in order to find the pressure, substitute for ρ its value in terms of f, g, and h: and for g and h their values given by the equation to the canal. Then $\frac{d(V)}{df}$ will $= -\phi(f)$, and $V = -\psi(f)$, where the form of ϕ and consequently that of ψ is different for every different canal. Let f_0,

---

* This would be the case, if, for instance, $\rho = k$, $F = Cg$, $G = -Cf$: for then $\frac{d(V)}{df} = Ck\left(g - f\frac{dg}{df}\right)$: which cannot be integrated without expressing g in terms of f.

g_0, h_0, be the co-ordinates of the point of junction, and f_1, g_1, h_1, the co-ordinates of the point where one canal meets the surface; then the corrected value of V is $\psi(f_1) - \psi(f_0)$; and this is the pressure produced by the fluid in one canal. If we had another canal of different form terminated in the same points, we should find for the pressure $\chi(f_1) - \chi(f_0)$. It is evident that these cannot generally be equal; and since their equality is necessary for equilibrium, it follows that there cannot be equilibrium on this supposition with regard to the forces.

But if the values of ρ, F, G, H, be such that the expression for $\frac{d(V)}{df}$ can be integrated without expressing g and h in terms of f*, then the expression for the pressure is

$$\omega(f_1, g_1, h_1) - \omega(f_0, g_0, h_0),$$

where the form of the function ω is independent of the form of the canal, and is consequently the same for all canals. If then any number of canals be drawn from the same point at the surface to the same point within it, since f_0, g_0, h_0, f_1, g_1, h_1, will be the same for all, the pressure produced by each at the point of junction will be the same.

It is still necessary that the pressure be the same if different canals from the same point of junction meet the surface at different points. Let f_2, g_2, h_2, be the co-ordinates of the point where another canal meets the surface: then the pressure which the fluid in this canal produces is

$$\omega(f_2, g_2, h_2) - \omega(f_0, g_0, h_0).$$

To make this equal to the last, we must have

$$\omega(f_2, g_2, h_2) = \omega(f_1, g_1, h_1);$$

* This would hold if, for instance, $\rho = k$, $F = Af$, $G = Bg$, $H = Ch$: for then

$$\frac{d(V)}{df} = k\left(Af + Bg\frac{dg}{df} + Ch\frac{dh}{df}\right),$$

and $V = \frac{k}{2}(Af^2 + Bg^2 + Ch^2)$.

or $\omega(f, g, h)$ must be equal for all points at the surface. Its variation therefore from one point of the surface to another must be $= 0$: or at the external surface

$$F + G \cdot \frac{dg}{df} + H \frac{dh}{df} = 0.$$

These two mathematical conditions then {namely, that

$$\rho\left(F + G\frac{dg}{df} + H\frac{dh}{df}\right)$$

be integrable with respect to f without expressing g and h in terms of f, and that at the external surface

$$F + G\frac{dg}{df} + H\frac{dh}{df} = 0\}$$

completely satisfy the physical conditions of equilibrium. The second indicates that the whole force on every point of the external surface must be perpendicular to the surface.

55. Now if the forces F, G, H, be the resolved part of attraction to any particle where the force is any function of the distance, $F + G\frac{dg}{df} + H\frac{dh}{df}$ will be immediately integrable. For let a, b, c, be the co-ordinates of the attracting particle: r or $\sqrt{(f-a)^2 + (g-b)^2 + (h-c)^2}$ its distance: and $\phi(r)$ the force in the direction of r. Then

$$F = -\phi(r) \cdot \frac{f-a}{r}: \quad G = -\phi(r) \cdot \frac{g-b}{r}:$$

$$H = -\phi(r) \cdot \frac{h-c}{r};$$

$$\text{and } F + G\frac{dg}{df} + H\frac{dh}{df}$$

$$= -\frac{\phi(r)}{r}\left\{(f-a) + (g-b)\frac{dg}{df} + (h-c)\frac{dh}{df}\right\}$$

$$= -\frac{\phi(r)}{r} r \frac{dr}{df} = -\phi(r)\frac{dr}{df};$$

the integral of which with respect to f is a function of r, $= -\int_r \phi(r)$. The same thing may be shewn, if the forces arise from attractions to any number of particles; and consequently, if they arise from the attractions to every particle of a solid or fluid mass.

The same thing is true if a part of the forces is produced by centrifugal force. If, (as above) the resolved part of centrifugal force in the direction of $f = \frac{4\pi^2}{T^2}f$; in that of $g = \frac{4\pi^2}{T^2}g$: and in that of $h = 0$: the corresponding part of

$$F + G\frac{dg}{df} + H\frac{dh}{df}$$

$$\text{is } \frac{4\pi^2}{T^2}\left(f + g\frac{dg}{df}\right),$$

the integral of which with respect to f is $\frac{2\pi^2}{T^2}(f^2 + g^2)$.

56. It appears then that the only forces which we have to consider will make $F + G\frac{dg}{df} + H\frac{dh}{df}$ a complete differential without assigning any relation between f, g, and h. Call it then $\frac{dv}{df}$. Then our first condition of equilibrium requires that $\rho\frac{dv}{df}$ be integrable with respect to f. This can happen only if ρ be a function of v. Our first condition then amounts only to this,

$$\rho = \text{a function of } v,$$

$$\text{where } \frac{dv}{df} = F + G\frac{dg}{df} + H\frac{dh}{df}.$$

From this it follows that ρ is constant while v is constant. The equation to a surface of equal density, or *couche de niveau*, is found therefore by making v constant, or $\frac{dv}{df}=0$,

$$\text{or } F+G\frac{dg}{df}+H\frac{dh}{df}=0.$$

The equation to the external surface, found above, is only a particular case of this.

It follows also that, as $\frac{d(V)}{df}=\rho\frac{dv}{df}=\text{function of } v\times\frac{dv}{df}$, V or the pressure will also be a function of v. But ρ is a function of v; consequently V is a function of ρ. That is, the surfaces of equal density will also be surfaces of equal pressure. And the equations of equilibrium will therefore be the same whether the density depend on the pressure or not.

57. Prop. 26. To find the ellipticities of the spheroids of equal density which will satisfy the equation of equilibrium.

For this purpose we shall find the differential equation to a spheroid, and shall make it coincide with the equation found above from the condition of equilibrium.

The polar and equatoreal semi-axes being γ and $\gamma(1+\epsilon)$, the equation to the spheroid is

$$\frac{f^2+g^2}{\gamma^2(1+\epsilon)^2}+\frac{h^2}{\gamma^2}=1,$$

and consequently its differential equation is

$$f+g\frac{dg}{df}+(1+\epsilon)^2\,.\,h\frac{dh}{df}=0.$$

That this may coincide with the equation

$$F+G\frac{dg}{df}+H\frac{dh}{df}=0,$$

we must have

$$f : g : (1+\epsilon)^2.h :: F : G : H;$$

$$\text{or } f : g : (1+2\epsilon).h :: F : G : H.$$

On examining the expressions in (52) it appears that the proportion of F and G is the same as that of f and g, and therefore the only equation resulting from this condition is

$$(1+2\epsilon)\frac{F}{f}=\frac{H}{h}.$$

Substituting in this equation the expressions found in (52), and multiplying 2ϵ into no term but the first, (as all the other terms are small when ϵ is small) we find

$$\frac{2\epsilon.\phi(\gamma)}{(f^2+g^2+h^2)^{\frac{3}{2}}}-\frac{6\psi(\gamma)}{5(f^2+g^2+h^2)^{\frac{5}{2}}}-\frac{6}{5}\{\chi(c)-\chi(\gamma)\}-\frac{3\pi}{T^2}=0.$$

But $f^2+g^2+h^2$ differs from γ^2 only by a quantity which depends upon ϵ: putting therefore γ^2 for $f^2+g^2+h^2$, since all the terms are small,

$$\frac{\epsilon.\phi(\gamma)}{\gamma^3}-\frac{3}{5}.\frac{\psi(\gamma)}{\gamma^5}-\frac{3}{5}\{\chi(c)-\chi(\gamma)\}-\frac{3\pi}{2T^2}=0.$$

Or, since this must be true whatever be the value of γ, we may put for γ the general letter c, and we have

$$\frac{e.\phi(c)}{c^3}-\frac{3}{5}.\frac{\psi(c)}{c^5}-\frac{3}{5}\{\chi(c)-\chi(c)\}-\frac{3\pi}{2T^2}=0.$$

58. If we put for $\phi(c)$, $\psi(c)$, and $\chi(c)$, their values

$$\int_c\rho.\frac{d(c^3.\overline{1+2e})}{dc}\text{ or }\int_c\rho.\frac{d(c^3)}{dc}\text{ nearly, }\int_c\rho.\frac{d(c^5e)}{dc},\text{ and }\int_c\rho.\frac{de}{dc},$$

(49 and 50), and differentiate,

$$\frac{d^2e}{dc^2}+\frac{2\rho c^2}{\int_c\rho c^2}.\frac{de}{dc}+\left(\frac{2\rho c}{\int_c\rho c^2}-\frac{6}{c^2}\right)e=0.$$

This differential equation it is always possible to integrate, at least by series, when ρ is given in terms of c: and the two arbitrary constants will enable us to make the value of e satisfy the equation from which it is derived. Hence, when ρ is given in terms of c, it is always possible to find the ellipticity of every surface of equal density so as to satisfy the condition of equilibrium. And therefore a heterogeneous spheroid constructed in the manner which we have supposed will be a form of equilibrium, if the equation

$$\frac{e \cdot \phi(c)}{c^3} - \frac{3}{5} \cdot \frac{\psi(c)}{c^5} - \frac{3}{5}\{\chi(c) - \chi(c)\} - \frac{3\pi}{2T^2} = 0,$$

be satisfied.

59. Prop. 27. To find the whole force on any point at the external surface.

The forces on any point in the directions of x, y, z, being F, G, and H, the whole force will be $\sqrt{(F^2 + G^2 + H^2)}$. The forces on a point at the surface will be found by putting c for γ, in the expressions of (52). Thus we find for a point at the surface,

$$F = \frac{4\pi}{3}\left\{\frac{1}{(f^2 + g^2 + h^2)^{\frac{3}{2}}}\phi(c) - \frac{12h^2 - 3f^2 - 3g^2}{5(f^2 + g^3 + h^2)^{\frac{7}{2}}}\psi(c) - \frac{3\pi}{T^2}\right\}f,$$

$$G = \frac{4\pi}{3}\left\{\frac{1}{(f^2 + g^2 + h^2)^{\frac{3}{2}}}\phi(c) - \frac{12h^2 - 3f^2 - 3g^2}{5(f^2 + g^2 + h^2)^{\frac{7}{2}}}\psi(c) - \frac{3\pi}{T^2}\right\}g,$$

$$H = \frac{4\pi}{3}\left\{\frac{1}{(f^2 + g^2 + h^2)^{\frac{3}{2}}}\phi(c) - \frac{6h^2 - 9f^2 - 9g^2}{5(f^2 + g^2 + h^2)^{\frac{7}{2}}}\psi(c)\right\}h,$$

and hence $\sqrt{(F^2 + G^2 + H^2)}$, (observing that the first term only of each expression is large,)

$$= \frac{4\pi}{3}\left\{\frac{\phi(c)}{f^2 + g^2 + h^2} - \frac{6h^2 - 3.(f^2 + g^2)}{5(f^2 + g^2 + h^2)^3}\psi(c) - \frac{3\pi}{T^2} \cdot \frac{(f^2 + g^2)}{\sqrt{(f^2 + g^2 + h^2)}}\right\}.$$

If e be the ellipticity of the external surface,

$$f^2 + g^2 + h^2 = c^2 + 2e.(f^2 + g^2),$$

which gives for the force

$$\frac{4\pi}{3}\left\{\frac{\phi(c)}{c^2} - \frac{2e.\phi(c)}{c^4}.\overline{f^2 + g^2} - \frac{6c^2 - 9.\overline{f^2 + g^2}}{5c^6}\psi(c) - \frac{3\pi}{cT^2}.\overline{f^2 + g^2}\right\}.$$

60. Suppose at the equator the centrifugal force $= m \times$ gravity, m being small $\left(\text{in the earth it is } \frac{1}{289}\right)$. The gravity

at the equator, neglecting the small terms, is $\frac{4\pi}{3} \cdot \frac{\phi(c)}{c^2}$: the centrifugal force is $\frac{4\pi^2}{T^2} \cdot a = \frac{4\pi^2}{T^2} c$, nearly. Hence,

$$\frac{4\pi^2}{T^2} c = m \frac{4\pi}{3} \cdot \frac{\phi(c)}{c^2}; \quad \therefore \frac{3\pi}{c\,T^2} = m \frac{\phi(c)}{c^4}.$$

Hence, the whole force at the surface

$$= \frac{4\pi}{3} \left\{ \frac{\phi(c)}{c^2} - \frac{6}{5} \cdot \frac{\psi(c)}{c^4} - \left((2e + m) \cdot \frac{\phi(c)}{c^4} - \frac{9}{5} \cdot \frac{\psi(c)}{c^6} \right) (f^2 + g^2) \right\}.$$

61. The equation of Prop. 26, becomes, at the surface,

$$\frac{e \cdot \phi(c)}{c^3} - \frac{3}{5} \cdot \frac{\psi(c)}{c^5} - \frac{3\pi}{2\,T^2} = 0,$$

or, since $\frac{3\pi}{2\,T^2} = \frac{m}{2} \cdot \frac{\phi(c)}{c^3}$,

$$\left(e - \frac{m}{2} \right) \frac{\phi(c)}{c^3} - \frac{3}{5} \cdot \frac{\psi(c)}{c^5} = 0.$$

Substituting from this the value of $\psi(c)$, we find, for the force at any point of the surface, the following simple expression

$$\frac{4\pi}{3} \cdot \frac{\phi(c)}{c^2} \cdot \left\{ 1 - (2e - m) - \left(\frac{5m}{2} - e \right) \cdot \left(\frac{f^2 + g^2}{c^2} \right) \right\}.$$

62. For the force at the pole we must make $f = 0$, $g = 0$, and the force, therefore,

$$= \frac{4\pi}{3} \cdot \frac{\phi(c)}{c^2} (1 - \overline{2e - m}).$$

For the force at the equator we must make $f^2 + g^2 = a^2 = c^2$, nearly, and, therefore, the equatoreal gravity

$$= \frac{4\pi}{3} \cdot \frac{\phi(c)}{c^2} \cdot \left\{ 1 - (2e - m) - \left(\frac{5m}{2} - e \right) \right\}.$$

The excess of the former above this

$$= \frac{4\pi}{3} \cdot \frac{\phi(c)}{c^2} \cdot \left(\frac{5m}{2} - e\right);$$

and the ratio of this excess to the equatoreal gravity is $\frac{5m}{2} - e$.

Let this $= n$: then, $n + e = \frac{5m}{2}$: a very remarkable proposition, which may be thus stated: "Whatever be the law of the Earth's density, if the ellipticity of the surface be added to the ratio which the excess of the polar above the equatoreal gravity bears to the equatoreal gravity, their sum will be $\frac{5m}{2}$, m being the ratio of the centrifugal force at the equator to the equatoreal gravity." This is called *Clairaut's Theorem.*

63. Prop. 28. To find an expression for gravity at any point at the surface, in terms of the latitude.

Suppose EF, (fig. 16), to represent the Earth's surface, PQ a normal at P. Then, PQN is the latitude of P, and

$$QN = PQ \cdot \cos l.$$

Now, as we shall have to substitute only in the small terms of the equation, $PQ = c$ nearly, and $QN = CN$ nearly $= \sqrt{(f^2 + g^2)}$; hence $\sqrt{(f^2 + g^2)} = c \cdot \cos l$ nearly. Substituting this in the expression of (61), gravity

$$= \frac{4\pi}{3} \frac{\phi(c)}{c^2} \left\{1 - (2e - m) - \left(\frac{5m}{2} - e\right) \cdot \cos^2 l\right\},$$

$$\text{or } = \frac{4\pi}{3} \cdot \frac{\phi(c)}{c^2} \cdot \left\{1 - \left(e + \frac{3m}{2}\right) + \left(\frac{5m}{2} - e\right) \cdot \sin^2 l\right\},$$

$$\text{or } = \frac{4\pi}{3} \cdot \frac{\phi(c)}{c^2} \cdot \left\{1 - \left(e + \frac{3m}{2}\right)\right\} \left\{1 + \left(\frac{5m}{2} - e\right) \cdot \sin^2 l\right\}.$$

Gravity, therefore, may be generally expressed by the formula $E(1 + n \cdot \sin^2 l)$, where E = equatoreal gravity, and

$$n + e = \frac{5m}{2}.$$

64. PROP. 29. To find the ellipticity of the Earth on any assumed law of density of the strata.

Differentiating the equation

$$\frac{e \cdot \phi(c)}{c^3} - \frac{3}{5} \cdot \frac{\psi(c)}{c^5} - \frac{3}{5}\{\chi(c) - \chi(c)\} - \frac{3\pi}{2T^2} = 0 \ldots (\mathrm{I}),$$

we find

$$c^3 \phi(c) \cdot \frac{de}{dc} + 3\psi(c) - 3e \cdot c^2 \phi(c) = 0 \ldots (\mathrm{II}),$$

and differentiating this,

$$\frac{d^2 e}{dc^2} + \frac{2\rho c^2}{\int_c \rho c^2} \cdot \frac{de}{dc} + \left(\frac{2\rho c}{\int_c \rho c^2} - \frac{6}{c^2}\right) e = 0 \ldots (\mathrm{III}).$$

Now, when ρ is given in terms of c, we must substitute it in this last equation, and by integration find e. The expression will contain two arbitrary constants: one of these may, in general, be conveniently determined by substituting in equation (II), and the other by substituting in equation (I). Equation (III) may be transformed into one of a simpler form, thus. Let $\int_c \rho c^2 = p$, and let $pe = v$, or $e = \frac{v}{p}$: then, upon substituting in (III), we get

$$\frac{d^2 v}{dc^2} - \frac{6v}{c^2} - v\frac{c}{p} \cdot \frac{d\rho}{dc} = 0 \ldots\ldots\ldots (\mathrm{IV}).$$

65. *Example.* Suppose $\rho = A \cdot \frac{\sin qc}{c}$, A and q being constant. As this gives a density diminishing from the center to the surface, it is probable that it will pretty nearly represent that of the Earth. On substituting in (IV), we get

$$\frac{d^2 v}{dc^2} - \frac{6v}{c^2} + q^2 v = 0:$$

the complete solution of which is

$$v = C\left\{\sin(qc + C') + \frac{3}{qc}\cos(qc + C') - \frac{3}{q^2 c^2}\sin(qc + C')\right\},$$

C and C' being arbitrary constants. Observing that

$$\phi(c) = 3p,$$

and $$\psi(c) = \int_c \rho \cdot \frac{d}{dc}\left(\frac{c^5 v}{p}\right) = \frac{\rho c^5 v}{p} - \int_c \frac{c^5 v}{p} \cdot \frac{d\rho}{dc},$$

(the integral being made to vanish when $c = 0$), and, after finding these values, substituting then in equation (II), we find that it reduces itself to

$$\frac{45\, C \,.\, \sin C'}{q^2} = 0.$$

C', therefore, must $= 0$; and, therefore,

$$v = C\left(\sin qc + \frac{3}{qc}\cos qc - \frac{3}{q^2c^2}\sin qc\right).$$

Now, making use of this value in equation (I), and observing that

$$\chi(c) = \int_c \rho \frac{d}{dc}\left(\frac{v}{p}\right) = \frac{\rho v}{p} - \int_c \frac{v}{p} \cdot \frac{d\rho}{dc};$$

and putting for $\frac{3\pi}{2T'^2}$ the value $\frac{m}{2} \cdot \frac{\phi(c)}{c^3}$ found in (60), the equation becomes

$$-A \,.\, \frac{3m}{2}\left(\frac{\sin qc}{q^2c^3} - \frac{1}{qc^2}\cos qc\right)$$

$$-\frac{3}{5}C\sin qc \,.\, \frac{\dfrac{\sin qc}{c} + \dfrac{3\cos qc}{qc^2} - \dfrac{3\sin qc}{q^2c^3}}{\dfrac{\sin qc}{q^2} - \dfrac{c\cos qc}{q}}$$

$$+\frac{3}{5}C\left(\frac{q \,.\, \cos qc}{c^2} - \frac{\sin qc}{c^3}\right) = 0.$$

Determining from this equation the value of $\frac{C}{A}$, we find the ellipticity of any stratum, whose polar semi-axis is c

$$= \frac{v}{p} = \frac{5m}{2} \cdot \frac{\left(1 - \dfrac{qc}{\tan qc}\right)^2}{2 - q^2c^2 - \dfrac{qc}{\tan qc} - \dfrac{q^2c^2}{\tan^2 qc}} \cdot \frac{1 - \dfrac{3}{q^2c^2} + \dfrac{3}{qc \,.\, \tan qc}}{1 - \dfrac{qc}{\tan qc}}.$$

For the ellipticity of the surface we must make $c = \mathrm{c}$: thus we obtain

$$\mathrm{e} = \frac{5m}{2} \cdot \frac{\left(1 - \frac{q\mathrm{c}}{\tan q\mathrm{c}}\right) \cdot \left(1 - \frac{3}{q^2\mathrm{c}^2} + \frac{3}{q\mathrm{c} \cdot \tan q\mathrm{c}}\right)}{2 - q^2\mathrm{c}^2 - \frac{q\mathrm{c}}{\tan q\mathrm{c}} - \frac{q^2\mathrm{c}^2}{\tan^2 q\mathrm{c}}}.$$

If $z = 1 - \dfrac{q\mathrm{c}}{\tan q\mathrm{c}}$, $\quad \mathrm{e} = \dfrac{5m}{2} \cdot \dfrac{1 - \dfrac{3z}{q^2\mathrm{c}^2}}{3 - z - \dfrac{q^2\mathrm{c}^2}{z}}$.

66. Suppose $q = \dfrac{5}{6} \cdot \dfrac{\pi}{\mathrm{c}}$. Then $q\mathrm{c} = \dfrac{5\pi}{6}$: $z = 5{,}53452$; and $\mathrm{e} = \dfrac{5m}{2} \cdot 0{,}37703$. And, since n the increase of gravity at the pole $= \dfrac{5m}{2} - \mathrm{e}$, by (62), $n = \dfrac{5m}{2} \cdot {,}62297$. In the Earth $m = \dfrac{1}{289}$: $\dfrac{5m}{2} = \dfrac{1}{115}$; therefore on these assumptions, $\mathrm{e} = \dfrac{1}{305}$, $n = \dfrac{1}{184{,}6}$. These are very nearly the same as the observed values of e and n.

67. Prop. 30. To compare the mean density with the density at the surface.

If the whole spheroid consisted of matter whose density = mean density, its mass would be the same as it actually is. Hence, the mean density $= \dfrac{\text{mass}}{\text{volume}}$. Now, neglecting the ellipticity, the mass may be found by considering it as a series of spherical shells of different density: and since the surface of one of these, whose radius is c, is $4\pi \cdot c^2$, the mass of the shell whose thickness is δc and density ρ, is ultimately $4\pi \cdot \rho c^2 \delta c$, and putting u for the mass,

$$\frac{du}{dc} = 4\pi\rho c^2, \quad u = 4\pi \int_c \rho c^2 = 4\pi \cdot p,$$

and u the whole mass $= 4\pi \text{p}$. And the volume $= \frac{4\pi}{3} \text{c}^3$.

Hence, the mean density $= \frac{3\text{p}}{\text{c}^3}$. And, if the density at the surface $= \rho'$, the ratio of the mean density to that at the surface $= \frac{3\text{p}}{\text{c}^3 . \rho'}$.

68. *Example.* Suppose the same law of density and the same value of q as before.

$$\text{Then } p = A\left(\frac{\sin qc}{q^2} - \frac{c}{q}\cos qc\right) : \rho = \frac{A\sin qc}{c};$$

$$\therefore \frac{3\text{p}}{\text{c}^3\rho'} = \frac{3}{\text{c}^2 . \sin q\text{c}} . \left(\frac{\sin q\text{c}}{q^2} - \frac{\text{c}}{q}\cos q\text{c}\right)$$

$$= \frac{3}{q^2\text{c}^2}\left(1 - \frac{q\text{c}}{\tan q\text{c}}\right) = \frac{3z}{q^2\text{c}^2} = 2,4225.$$

69. PROP. 31. To find the effect produced by the ellipticity of the Earth on the motion of the Moon.

Let the co-ordinates of the Moon, referred to the Earth's center, be f, g, h: h being perpendicular to the plane of the equator, and f in the intersection of the plane of the equator with the ecliptic. Let NP, (fig. 18), be the intersection of the plane of the ecliptic by the plane passing through g and h; and drawing MP perpendicular to NP, let $NP = k$, $PM = l$: then f, k, l, are the three co-ordinates of M, f and k being in the plane of the ecliptic. Let $EP = \frac{1}{u}$, $\tan MEP = s$, $PEN = \theta$, $PNO =$ inclination of ecliptic to equator $= \omega$. Then, if PQ, PR, be drawn parallel to h and g,

$$g = NQ - RP = k\cos\omega - l\sin\omega:$$

$$h = QP + MR = k\sin\omega + l\cos\omega.$$

But $k = \frac{1}{u}\sin\theta$; $l = \frac{1}{u}s$; also $f = \frac{1}{u}\cos\theta$. Thus we get for the original co-ordinates these values,

$$f = \frac{1}{u} \cos \theta,$$

$$g = \frac{1}{u} (\sin \theta \cos \omega - s . \sin \omega),$$

$$h = \frac{1}{u} (\sin \theta \sin \omega + s . \cos \omega).$$

Let F, G, H, be the forces in the directions of f, g, and h; K and L those in the directions of k and l; and P, T, S, the forces parallel to EP, perpendicular to EP in the plane of the ecliptic, and perpendicular to the ecliptic. Then

$$K = G \cos \omega + H \sin \omega,$$

$$L = H \cos \omega - G \sin \omega.$$

Also $P = F \cos \theta + K \sin \theta$;

$$T = F \sin \theta - K \cos \theta;$$

$$S = L.$$

Substituting the values of K and L,

$$P = F \cos \theta + (G \cos \omega + H \sin \omega) \sin \theta,$$

$$T = F \sin \theta - (G \cos \omega + H \sin \omega) \cos \theta,$$

$$S = H \cos \omega - G \sin \omega.$$

70. Now, by Prop. 22.

$$F = \frac{4\pi}{3} \left\{ \frac{\phi(c)}{(f^2 + g^2 + h^2)^{\frac{3}{2}}} - \frac{12h^2 - 3f^2 - 3g^2}{5(f^2 + g^2 + h^2)^{\frac{7}{2}}} \psi(c) \right\} . f,$$

$$G = \frac{4\pi}{3} \left\{ \frac{\phi(c)}{(f^2 + g^2 + h^2)^{\frac{3}{2}}} - \frac{12h^2 - 3f^2 - 3g^2}{5(f^2 + g^2 + h^2)^{\frac{7}{2}}} \psi(c) \right\} . g,$$

$$H = \frac{4\pi}{3} \left\{ \frac{\phi(c)}{(f^2 + g^2 + h^2)^{\frac{3}{2}}} - \frac{6h^2 - 9f^2 - 9g^2}{5(f^2 + g^2 + h^2)^{\frac{7}{2}}} \psi(c) \right\} . h.$$

Upon substituting these values in the expressions for P, T, and S, the quantity multiplied by $\psi(c)$ in each is rather complicated, and it is, therefore, proper to take only those terms which, as remarked in Art. 46. of the *Lunar Theory*, will be much increased by integration. With this restriction, and observing that s is expressed nearly by $k \cdot \sin \overline{g\theta - \gamma}$, where g differs little from 1, we have

$$P = \frac{4\pi}{3}\left\{\frac{u^2 \phi(c)}{(1+s^2)^{\frac{3}{2}}} - \frac{24}{5}\, u^4 . \psi(c) \sin\omega . \cos\omega . \sin\theta . s\right\},$$

$$T = -\frac{4\pi}{3} . \frac{6u^4 \psi(c)}{5} \sin\omega . \cos\omega . \cos\theta . s,$$

$$S = \frac{4\pi}{3} . \left\{\frac{u^2 \phi(c) . s}{(1+s^2)^{\frac{3}{2}}} + \frac{6u^4 \psi(c)}{5} . \sin\omega . \cos\omega . \sin\theta\right\}.$$

Let E be the mass of the Earth. The volume of a spheroid, whose semi-axes are c and $c(1+e)$, is $\frac{4\pi}{3} c^3 (1+2e)$: hence, the volume included between two spheroidal surfaces, is ultimately

$$\frac{4\pi}{3} . \frac{d\{c^3 . (1+2e)\}}{dc} \delta c:$$

and if the density be ρ, the increment of the mass is

$$\frac{4\pi}{3} . \rho . \frac{d\{c^3 . (1+2e)\}}{dc} \delta c;$$

therefore the mass

$$= \frac{4\pi}{3} \int_c \rho d \frac{\{c^3 . (1+2e)\}}{dc} = \frac{4\pi}{3} \phi(c); \quad \therefore\ E = \frac{4\pi}{3} \phi(c).$$

Also, by the equation of (61),

$$\left(e - \frac{m}{2}\right) . \frac{\phi(c)}{c^3} - \frac{3}{5} . \frac{\psi(c)}{c^5} = 0,$$

whence

$$\frac{4\pi}{3} . \frac{6}{5} . \psi(c) = \left(e - \frac{m}{2}\right) . 2c^2 . \phi(c) . \frac{4\pi}{3} = \left(e - \frac{m}{2}\right) 2c^2 . E.$$

Hence, we finally obtain

$$P = \frac{E . u^2}{(1+s^2)^{\frac{3}{2}}} + \left(e - \frac{m}{2}\right) 8c^2 u^4 E . \sin\omega . \cos\omega . \sin\theta . s,$$

$$T = -\left(\mathrm{e} - \frac{m}{2}\right) . 2\mathrm{c}^2 u^4 E . \sin\omega . \cos\omega . \cos\theta . s,$$

$$S = \frac{E u^2 s}{(1+s^2)^{\frac{3}{2}}} + \left(\mathrm{e} - \frac{m}{2}\right) . 2c^2 . u^4 . E . \sin\omega . \cos\omega . \sin\theta.$$

71. The value of $\frac{S - Ps}{h^2 u^3}$ in the equation (l) of Art. 41, *Lunar Theory*, is increased by

$$\left(\mathrm{e} - \frac{m}{2}\right) . 2\mathrm{c}^2 . \frac{u}{h^2} E . \sin\omega . \cos\omega . \sin\theta.$$

$$\text{Now } \frac{E}{h^2} = \frac{E}{\mu} . \frac{\mu}{h^2} = \frac{E}{\mu} a.$$

and $u = a$, nearly:

$$\text{hence, } 2\mathrm{c}^2 . \frac{u}{h^2} E = 2\mathrm{c}^2 . a^2 . \frac{E}{\mu}, \text{ nearly,}$$

and the increase of $\frac{S - Ps}{h^2 u^3}$ is nearly

$$2 . \left(\mathrm{e} - \frac{m}{2}\right) . \mathrm{c}^2 a^2 \frac{E}{\mu} \sin\omega . \cos\omega . \sin\theta.$$

Let the term which s receives in consequence be $A \sin\theta$. Upon substituting this in the equation, since

$$\frac{d^2 (A \sin\theta)}{d\theta^2} + A \sin\theta = 0,$$

the only term, it will be found, which receives an increment of the same form, is

$$\frac{m' u'^3 s}{h^2 u^4} \left\{\frac{3}{2} + \frac{3}{2} \cos 2 . (\theta' - \theta)\right\}, \quad \text{or } n^2 s \left\{\frac{3}{2} + \frac{3}{2} \cos 2 (\theta' - \theta)\right\},$$

nearly, the first part of which will be increased by $\frac{3n^2}{2} A \sin\theta$, n being here the same as m in the *Lunar Theory*. Making then all the additional terms = 0 which depend on $\sin\theta$, we have

$$\frac{3n^2}{2} A + 2 . \left(\mathrm{e} - \frac{m}{2}\right) . \mathrm{c}^2 a^2 \frac{E}{\mu} \sin\omega . \cos\omega = 0$$

$$\text{or } A = \frac{-4}{3n^2} \cdot \left(e - \frac{m}{2}\right) . c^2 a^2 . \frac{E}{\mu} \sin \omega . \cos \omega :$$

and the term in s, or in the tangent of the Moon's latitude,

$$= \frac{-4}{3n^2} \cdot \left(e - \frac{m}{2}\right) . c^2 a^2 . \frac{E}{\mu} \sin \omega . \cos \omega . \sin \theta.$$

This is expressed in parts of the radius; the number of seconds will be found by multiplying it by $\dfrac{180 \times 60 \times 60}{\pi}$.

And $ca = \dfrac{c}{\rho}$ nearly $=$ sin Moon's mean horizontal parallax. The term is, therefore,

$$\frac{-4 . (60)^3}{n^2 . \pi} \cdot \left(e - \frac{m}{2}\right) . \sin^2 \text{paral.} \frac{E}{\mu} . \sin \omega . \cos \omega . \sin \theta.$$

71*. The geometrical interpretation of the term which we have just found is so remarkable as to deserve the reader's attention. The tangent of the Moon's latitude (taking only the principal term of the expression found in the *Lunar Theory*, and combining it with this term), putting C for $\dfrac{4}{3n^2} \cdot \left(e - \dfrac{m}{2}\right) . c^2 a^2 . \dfrac{E}{\mu} \sin \omega . \cos \omega$,

$$\text{is } s = k \sin \overline{g\theta - \gamma} - C \sin \theta.$$

Suppose the ascending node of the Moon's orbit to coincide with the first point of Aries, or to have its longitude $= 0$; that is, suppose $\gamma = 0$; then

$$s = k \sin g\theta - C \sin \theta,$$

or, as g differs very little from 1, the motion for several revolutions will be represented by this formula,

$$s = k \sin \theta - C \sin \theta$$

$$= (k - C) \sin \theta.$$

In this position of the node, therefore, when the Moon's orbit is more inclined to the terrestrial equator than the ecliptic is, the inclination of the orbit to the ecliptic is diminished by the new term, or the plane of the orbit is brought nearer to the plane of the Earth's equator by the new term.

Again, suppose the longitude of the ascending node to be 180°; the expression for the tangent of latitude will be

$$s = k \sin \overline{g\theta - 180^0} - C \sin \theta,$$

which for a few revolutions will be represented with sufficient accuracy by

$$s = k \sin \overline{\theta - 180^0} + C \sin \overline{\theta - 180^0}.$$

$$= (k + C) \sin \overline{\theta - 180^0}.$$

In this position then of the node, when the Moon's orbit is less inclined to the Earth's equator than the ecliptic is, the inclination of the Moon's orbit on the ecliptic is increased by the new term, or the plane of the orbit is brought nearer to the plane of the Earth's equator by the new term.

In all cases, therefore, the plane of the Moon's orbit is brought nearer to the plane of the Earth's equator by the new term. And thus we may represent the effect of that term by supposing the nodes of the Moon's orbit, in their retrograde motion, to regress not upon the ecliptic, but upon a plane lying between the ecliptic and the equator, and supposing the inclination of the Moon's orbit to remain constant, not as respects the ecliptic, but as respects that new plane.

72. In investigating the alteration produced in u, it must be observed that the term

$$-\left(e - \frac{m}{2}\right) . 2c^2 . \frac{u}{h^2} E . \sin \omega . \cos \omega . \cos \theta . s,$$

which is added to $\frac{T}{h^2 u^3}$, upon putting for s its value $k \sin \overline{g\theta - \gamma}$, will contain the terms $\sin \overline{(g-1)\theta - \gamma}$, $\sin \overline{(g+1)\theta - \gamma}$, of which the former will be much increased by integration. The term depending on $\sin \theta . s$, which is added to P, introduces terms (though less important) of the same form. And the term $-a\left(1-\frac{3s^2}{2}\right)$ in $-\frac{P}{h^2u^2}$, will contain $3as\times$ the term added to s, or

$$-\frac{4}{n^2}\,.\,k\sin(g\theta-\gamma)\,.\left(e-\frac{m}{2}\right).\,c^2a^3\,.\,\frac{E}{\mu}\,.\,\sin\omega\,.\,\cos\omega\,.\,\sin\theta,$$

which will also produce terms of the same form. To find the alteration produced in $\frac{dt}{d\theta}$, it is necessary to include many terms depending on $s^3\cos\theta$, &c., and some terms among those added to P.

72*. Since $g\theta-\gamma$ is the Moon's distance in longitude from the mean place of the node, and θ is the Moon's longitude, or her distance in longitude from the first point of Aries, $(g-1)\theta-\gamma$ is the distance of the first point of Aries from the node. Hence the inequalities in longitude depending on the angle $(g-1)\theta-\gamma$ require, in order to go through all their changes, a complete revolution of the Moon's node. For about 9 years therefore, in consequence of these terms, the Moon's true longitude is in advance of her mean longitude, and for the following 9 years her true longitude is behind her mean longitude.

73. The disturbances in the Moon's motion produced by the Earth's ellipticity, though sensible, are very small. The ellipticity of Jupiter produces considerable disturbances in the motions of his satellites, and affects very much the progression of their apsides, and the regression of their nodes.

ON THE METHODS OF ASCERTAINING THE FIGURE OF THE EARTH BY OBSERVATION.

74. THE simplest of all methods, and the least dependent on theory, is that of measuring, by Geodetic operations, the distance between two places which are nearly in the same meridian, and observing the latitude of each. If two measurements of this sort be made in places whose difference of latitude is considerable, the ellipticity of the Earth may be ascertained; as we proceed to shew.

75. Prop. 32. The Earth being supposed a spheroid, to express the length of a small arc of the meridian at any point, in terms of the difference of latitude of its extremities.

Let $APQB$, (fig. 19), be the meridian passing through the extremities P and Q of a small arc of latitude; at P and Q, draw normals meeting in R; then, R is very nearly the center of curvature of p, which bisects PQ. And, if D be the difference of latitude of P and Q, $PRQ = D$;

$$\therefore\ PQ = D\,.\,pR.$$

Now pR (Whewell's *Doctrine of Limits*, p. 63)

$$= \frac{CK^2}{pF} = \frac{AC^2\,.\,BC^2}{pF^3} = \frac{a^2c^2}{pF^3},$$

putting a and c for the semi-axes.

$$\text{But } pF = \frac{c^2}{pL}; \qquad \therefore\ pR = \frac{a^2\,.\,pL^3}{c^4}.$$

And pLN = latitude of $p = l$; $\quad \therefore\ LN = pL\cos L$;

$$\therefore\ CN = \frac{a^2}{c^2}pL\cos l; \text{ and } Np = pL\sin l.$$

Substituting these in the equation $\dfrac{CN^2}{a^2} + \dfrac{Np^2}{c^2} = 1$, we get

$$pL^2\,.\,\frac{a^2\cos^2 l + c^2\sin^2 l}{c^4} = 1;$$

$$\therefore\ pL^3 = \frac{c^6}{(a^2\cos^2 l + c^2\sin^2 l)^{\frac{3}{2}}},$$

$$\text{and } pR = \frac{a^2c^2}{(a^2\cos^2 l + c^2\sin^2 l)^{\frac{3}{2}}}:$$

$$\text{hence } PQ = D\,.\,\frac{a^2c^2}{(a^2\cos^2 l + c^2\sin^2 l)^{\frac{3}{2}}}:$$

If the ellipticity be small, let $a = c\,(1 + e)$;

$$\therefore\ PQ = \frac{D\,.\,c^4(1 + 2e)}{\{c^2(1 + 2e)\cos^2 l + c^2\sin^2 l\}^{\frac{3}{2}}} = Dc(1 + 2e - 3e\cos^2 l),$$

$$\text{or } = D\,.\,c\,(1 - e + 3e\sin^2 l), \text{ nearly}.$$

76. Suppose then two arcs PQ, $P'Q'$, have been measured: suppose the latitudes of the middle points of the arcs to be l, l': the difference of the latitudes of P and Q to be D, that of P' and Q' to be D'. Then

$$\frac{PQ}{D} = c\,(1 - e + 3e\sin^2 l),$$

$$\frac{P'Q'}{D'} = c\,(1 - e + 3e\sin^2 l').$$

Subtracting, $\dfrac{P'Q'}{D'} - \dfrac{PQ}{D} = 3ce\,(\sin^2 l' - \sin^2 l)$;

$$\therefore\ e = \frac{\dfrac{P'Q'}{D'} - \dfrac{PQ}{D}}{3c\,(\sin^2 l' - \sin^2 l)}.$$

As e is small, we may, without sensible error, put for c, $\dfrac{PQ}{D}$, or $\dfrac{P'Q'}{D'}$:

$$\text{hence } e = \frac{1 - \dfrac{PQ}{P'Q'}\cdot\dfrac{D'}{D}}{3(\sin^2 l' - \sin^2 l)}.$$

77. *Example.* By Lambton's measures in India, the arc of the meridian from lat. $8^0.\,9'.\,38'',4$ to lat. $10^0.\,59'.\,48'',9 =$ 1029100,5 feet.

By Svanberg's measures in Sweden, the arc of the meridian from lat. $65^0.\,31'.\,32'',2$ to lat. $67^0.\,8'.\,49'',8 = 593277,5$ feet.

Here $PQ = 1029100,5$; $P'Q' = 593277,5$; $D = 10210'',5$; $D' = 5837'',6$; $l = 9^0.\,34'.\,44''$; $l' = 66^0.\,20'.\,10''$. Make

$$\frac{PQ}{P'Q'}\cdot\frac{D'}{D} = \cos^2\theta,$$

$$\text{or } 2\log\cos\theta = 20 + \log PQ + \log D' - \log P'Q' - \log D$$

$$= \log PQ + \log D' + \text{ar. com.} \log P'Q' + \text{ar. com.} \log D:$$

then, the numerator $= \sin^2\theta$. And the denominator

$$= 3\sin(l + l) \cdot \sin(l' - l).$$

Hence $\log e = 2\log\sin\theta - \log 3 - \log\sin(l' + l) - \log\sin(l' - l)$.

By calculating from the data above, $e = \dfrac{1}{306,3}$.

78. Attempts have also been made to determine the ellipticity of the Earth by measuring the distance between two places on the same parallel, and determining the difference of longitude, either by observation on Jupiter's satellites, or by observing the flash of gunpowder fired on a conspicuous place between them. The difference of longitude may also be determined by mere observation of angles, (see *Phil. Trans.* 1790).

79. Prop. 33. To express the distance of two places on the same parallel, in terms of their difference of longitude.

Let p, q, (fig. 19), be the places; L their difference of longitude. Then, pq (which, when the arc is small, may be measured as a great circle without sensible error,) $= L \cdot CN$.

$$\text{Now } CN = \frac{a^2}{c^2} pL \cdot \cos l = \frac{a^2\cos l}{\sqrt{(a^2\cos^2 l + c^2\sin^2 l)}};$$

$$\therefore\ pq = L \cdot \frac{a^2\cos l}{\sqrt{(a^2\cos^2 l + c^2\sin^2 l)}} = L \cdot c\cos l(1 + 2e - e\cos^2 l)$$

$$= L \cdot c\cos l(1 + e + e\sin^2 l),$$

if the ellipticity be small.

80. If then one arc has been measured in the meridian, and another on a parallel, and if l be the latitude of the middle of the meridional arc, l' that of the parallel, we shall have these equations:

$$\frac{PQ}{D} = c(1 - e + 3e \cdot \sin^2 l),$$

$$\frac{pq}{L} = c\cos l'(1 + e + e\sin^2 l').$$

Eliminating c, e may be found. This method is not considered to be practically accurate: the determination of difference of longitude being attended with great difficulty.

81. The method which on account of its great facility is now very extensively used, is that of observing the intensity of gravity in different latitudes, by means of the pendulum. It is usual to observe the number of vibrations made in a day by the same pendulum, in the different places at which it is proposed to compare the force of gravity; and likewise the number of vibrations made at London or Paris. The observations are commonly made by causing the experimental pendulum to vibrate in front of a clock-pendulum, and by observing the interval between the times at which the two pendulums pass the centers of the arcs of vibration at the same instant and in the same direction: then, since one of the pendulums has in that time gained exactly two vibrations upon the other, the ratio of the actual times of vibration is accurately known: and the time of vibration of the clock-pendulum is found from astronomical observations: and thus the time of vibration of the experimental pendulum, and the number of vibrations which it makes in a day, are found with great accuracy. In some experiments the clock-pendulum itself has been observed. The comparative number of vibrations being found, the comparative force of gravity, or the comparative length of the second's pendulum, can be found: and, as the length of the second's pendulum has been very accurately determined at London and Paris, its length is known at all the places of observation. The French astronomers have used a method more direct, but less convenient, and probably less accurate: it is described at length in the Additions to Biot's *Astronomie Physique*, p. 138.

82. Let p and p' be the lengths of the seconds' pendulum in latitudes l and l', P that at the equator. Since these lengths are proportional to the intensities of gravity, we have, by (63),

$$\left.\begin{array}{l} p = P(1 + n\sin^2 l) \\ p' = P(1 + n\sin^2 l') \end{array}\right\}, \quad \text{where } n = \frac{5m}{2} - e.$$

From these equations,

$$n = \frac{1 - \frac{p}{p'}}{\frac{p'}{p}\sin^2 l' - \sin^2 l} = \frac{1 - \frac{p}{p'}}{\sin^2 l' - \sin^2 l}, \text{ nearly,}$$

which may be calculated as the last example: then $e = \frac{5m}{2} - n$.

83. *Example.* At Madras.

$$l = 13^0.4'.9'', \quad p = 39{,}0234.$$

At Melville Island,

$$l' = 74^0.47'.12'', \quad p' = 39{,}2070.$$

$$\text{Hence, } n = {,}0053214; \text{ and } \frac{5m}{2} = {,}0086505;$$

$$\therefore \; e = {,}0033291 = \frac{1}{300}.$$

84. The ellipticity of the Earth has also been determined from the motion of the Moon. It appears from (71), that in the expression for the tangent of the Moon's latitude, there is this term

$$-\left(e - \frac{m}{2}\right) \cdot \frac{4.60^3}{n^2.\pi} \cdot \sin^2 \text{parallax} \cdot \frac{E}{\mu} \cdot \sin \text{obliquity} \cdot \cos \text{obliquity} \cdot \sin\theta.$$

$$\text{Now } \frac{E}{\mu} = \frac{\text{Earth's mass}}{\text{Earth + Moon}} = \frac{70}{71}, \text{ nearly: } n = \frac{27{,}25}{365{,}25}:$$

mean horizontal parallax = 57′: obliquity = 23°.28′, nearly: hence this term

$$= -\left(e - \frac{m}{2}\right) . 4891''. \sin\theta.$$

It is found by observation that the coefficient = − 8″;

$$\text{hence } \left(e - \frac{m}{2}\right) = \frac{8}{4891} = {,}001635. \quad \text{And } \frac{m}{2} = {,}001730:$$

$$\text{hence } e = {,}003365 = \frac{1}{297}.$$

The ellipticity, found by comparing the observed inequality in longitude with the calculated inequality, differs little from this.

85. The two latter methods, it will be observed, depend entirely upon the theory which we have laid down: the first and second are quite independent of theory. The near agreement of their results is one of the most convincing proofs that the principle of gravitation, and the suppositions upon which our theory is founded, are true.

86. For the calculation of parallax, it is necessary to know the distance of any point on the Earth's surface from the Earth's center, and the angle ACp, (fig. 19), which is called the *corrected* latitude: ALp being the *true* latitude, which = the elevation of the pole. The difference between the true and corrected latitude is called the angle of the vertical.

87. PROP. 34. To find the distance of any point on the Earth's surface from its center, in terms of the latitude of that point.

$$Cp^2 = CN^2 + Np^2 = \frac{a^4}{c^4} LN^2 + Np^2 = Lp^2 . \left(\frac{a^4}{c^4} \cos^2 l + \sin^2 l\right) =$$

(putting for Lp^2 the value found in Prop. 32.)

$$\frac{a^4 \cos^2 l + c^4 \sin^2 l}{a^2 \cos^2 l + c^2 \sin^2 l}; \text{ and } Cp = \sqrt{\frac{a^4 \cos^2 l + c^4 \sin^2 l}{a^2 \cos^2 l + c^2 \sin^2 l}}.$$

If the ellipticity be small,

$$Cp = c\sqrt{\frac{(1 + 4e) \cos^2 l + \sin^2 l}{(1 + 2e) \cos^2 l + \sin^2 l}}$$

$$= c(1 + e \cos^2 l) \text{ nearly} = c(1 + e - e \sin^2 l).$$

88. PROP. 35. To find the angle of the vertical.

Let ACp be the corrected latitude $= l'$.

$$\text{Then } \tan l = \frac{pN}{NL}; \quad \tan l' = \frac{pN}{CN};$$

$$\therefore \frac{\tan l'}{\tan l} = \frac{LN}{CN} = \frac{c^2}{a^2}, \text{ or } \tan l' = \frac{c^2}{a^2} \tan l.$$

$$\text{Hence, } \tan(l - l') = \frac{\tan l - \tan l'}{1 + \tan l \,.\, \tan l'}$$

$$= \frac{\left(1 - \frac{c^2}{a^2}\right) \tan l}{1 + \frac{c^2}{a^2} \tan^2 l} = \frac{(a^2 - c^2) \sin l \cos l}{a^2 \cos^2 l + c^2 \sin^2 l}.$$

When the ellipticity is small, putting $l - l'$ for $\tan(l - l')$,

$$l - l' = 2e \,.\, \sin l \,.\, \cos l \text{ nearly} = e \,.\, \sin 2l.$$

This is in parts of the radius: the number of seconds is

$$\frac{180.60.60}{\pi} e \sin 2l.$$

PRECESSION OF THE EQUINOXES,

AND

NUTATION OF THE EARTH'S AXIS.

ON THE COMPOSITION OF ROTATORY MOTION.

1. PROP. 1. IF a body revolve about an axis AB, (fig. 1) with an angular velocity ω, and if a force be impressed upon it which would make it revolve about the axis AC, with an angular velocity ω'; then the body will not revolve about either of the axes AB, AC, but about an axis AD, in the plane BAC, dividing the angle BAC so that $\sin BAD : \sin CAD :: \omega' : \omega$.

2. It is evident, that the new axis of rotation is that line in the body, which, when the effect of both the original motions is considered, remains at rest. If, then, a line AD in the plane BAC be that axis, the angular motion about AB would tend to raise any point in it, as D, *above* the plane of the paper, as much as the angular motion about AC would tend to depress it *below* the plane of the paper. From D, draw DE, DF, perpendicular to AB, AC. In consequence of the angular motion about AB, the point D would be raised above the paper, with the velocity

$$\omega \times ED = \omega \times AD \, . \sin BAD.$$

And, in consequence of the angular motion about AC, D would be depressed below the paper, with the velocity

$$\omega' \times FD = \omega' \times AD \, . \sin CAD.$$

Making these equal,

$$\omega \, . \, AD \, . \sin BAD = \omega' \, . \, AD \, . \sin CAD;$$

$$\therefore \; \sin BAD : \sin CAD :: \omega' : \omega.$$

3. Now, to shew that AD really is the axis of rotation, take any point P in the body: with center A, suppose a spherical surface described, passing through P, and cutting AB, AC, AD, in B, C, D; let BDC, BP, CP, be arcs of great circles. Let PQ, drawn perpendicular to PB on the surface of the sphere, be the motion of P, produced by the rotation about AB only, in the very small time t; let PR, drawn perpendicular to PC, on the surface of the sphere, be the motion of P, produced by the rotation about AC only, in the same time; then, if the parallelogram QR be completed, PS, the diagonal, is the true motion of P, in that time. Now $CPR = 90^0 = BPQ$; adding RPB to both, $CPB = RPQ$. Also

$$\sin SPQ : \sin SPR :: \sin SPQ : \sin PSQ :: SQ : PQ$$

$$:: PR : PQ.$$

$$\text{But } PR = \omega' t . \sin PC \text{ to radius } AC;$$

$$PQ = \omega t . \sin PB;$$

$$\therefore \sin SPQ : \sin SPR :: \omega' . \sin PC : \omega . \sin PB.$$

$$\text{And } \sin BPD : \sin CPD :: \frac{\sin BD . \sin BDP}{\sin BP}$$

$$: \frac{\sin CD . \sin CDP}{\sin CP} :: \sin PC . \sin BD : \sin PB . \sin CD$$

$$:: \omega' . \sin PC : \omega . \sin PB,$$

$$(\text{since } \sin BD : \sin CD :: \sin BAD : \sin CAD :: \omega' : \omega).$$

$$\text{Hence, } \sin SPQ : \sin SPR :: \sin BPD : \sin CPD.$$

Since, then, the two angles CPB, RPQ, are equal, and are divided into parts whose sines are in the same ratio, it follows that those parts are equal, or $BPD = QPS$. Adding to each BPS,

$$DPS = BPQ = 90^0;$$

and, therefore, PS is perpendicular to the plane APD. In the same manner it may be shewn, that the plane passing through any other point of the body, and through AD, is perpendicular to the motion of that point; and, since the axis of rotation is the line of intersection of all the planes perpendicular to the motion of every point, AD must be the axis of rotation.

4. It is here supposed, that the angular motion about AB tends to raise all the particles between AD and AC, and that the angular motion about AC tends to depress them. If, however, the angular motions about both AB and AC, (fig. 2), tend to raise the particles between AB and AC, produce CA to C': then the angular motion about AC' tends to raise the particles between AC' and AB, and the angular motion about AB tends to depress them. Hence, the new axis of rotation will be the line AD, which makes $\sin BAD : \sin C'AD :: \omega' : \omega$. The same is true, if both angular motions tend to depress the particles between AB and AC.

5. Prop. 2. The angular velocity about the new axis AD, will be to the original angular velocity about AB, as $\sin BAC$ to $\sin DAC$; and the angular velocity about AD to the original angular velocity about AC, as $\sin BAC$ to $\sin BAD$.

6. Let ω'' be the angular velocity about AD;

$$\text{then, } PS = \omega'' t . \sin DP.$$

$$\text{Now } PQ : PS :: \sin PSQ : \sin PQS :: \sin SPR : \sin QPR$$
$$:: \sin DPC : \sin BPC,$$

since BP, DP, CP, are perpendicular to PQ, PS, PR, respectively. But

$$\sin DPC : \sin BPC :: \frac{\sin DC . \sin DCP}{\sin DP} : \frac{\sin BC . \sin BCP}{\sin BP};$$

putting then for PQ and PS their values,

$$\omega t . \sin BP : \omega'' t . \sin DP :: \frac{\sin DC}{\sin DP} : \frac{\sin BC}{\sin BP};$$

whence, $\omega : \omega'' :: \sin DC : \sin BC :: \sin DAC : \sin BAC$.

And, since $\omega' : \omega :: \sin BAD : \sin DAC$ by (2);

$$\therefore \omega' : \omega'' :: \sin BAD : \sin BAC.$$

7. We have supposed, in the enunciations of the Propositions above, that an angular motion about one axis is suddenly impressed upon a body which had previously an angular motion about another axis. It is evident, that the conclusions are the same, if we suppose both angular motions to be impressed at once.

8. From these Propositions, compared with the usual propositions on the composition of forces, it appears, that if two forces in the directions AB, AC, be proportional to ω, ω', their resultant will be in the direction AD, and will be proportional to ω''. And hence, if several angular motions were impressed upon a body at the same time, the new axis of rotation and the angular velocity about that axis would be found, by finding the direction and magnitude of the resultant of forces which are in the directions of the several axes of rotation, and are proportional to the angular velocities.

9. If then a body revolve about an axis, and angular motions about two other axes be impressed upon it, it is indifferent whether we first compound the two impressed motions, and then compound their resultant with the original motion, or compound the original motion with one of the impressed motions, and then compound their resultant with the other motion. For, in compounding the forces proportional to these angular velocities, the order in which they are taken is indifferent.

10. If the angular motion about AC be not produced instantaneously, but by the continued action of a finite force, its effect may be found, by supposing the time divided into a great number of small intervals, and supposing the angular velocity generated in each of those intervals to be impressed at the end of each, and then finding the limit to which we approach, by increasing indefinitely the number of these intervals.

11. Prop. 3. If a uniform force act upon the body, tending to give it a motion of rotation about an axis which is always perpendicular to the axis about which it is at

each instant revolving, and always in the plane BAC, (fig. 1 and 3), the angular velocity will be unaltered.

Let ω be the original angular velocity; and suppose the impressed force such as would generate in 1″ the angular velocity a. Let this 1″ be divided into n parts; then the angular velocity generated in each of these parts, is $\frac{a}{n}$. Compounding the angular velocities ω and $\frac{a}{n}$, of which the axes AB, AC, (fig. 3) are at right angles to each other, we find, by (6) and (8), the new angular velocity about $AD = \sqrt{\left(\omega^2 + \frac{a^2}{n^2}\right)}$. Compounding this angular velocity with the angular velocity $\frac{a}{n}$, generated in the second small interval, and observing that the axis Ac, about which it is produced, is perpendicular to the axis AD, about which the body is now revolving, the angular velocity at the end of the second interval, is

$$\sqrt{\left(\omega^2 + \frac{a^2}{n^2} + \frac{a^2}{n^2}\right)} = \sqrt{\left(\omega^2 + \frac{2a^2}{n^2}\right)}.$$

In the same way the angular velocity at the end of the third interval, is

$$\sqrt{\left(\omega^2 + \frac{2a^2}{n^2} + \frac{a^2}{n^2}\right)} = \sqrt{\left(\omega^2 + \frac{3a^2}{n^2}\right)}, \text{ \&c.};$$

that at the end of the nth interval, or at the end of 1″, is

$$\sqrt{\left(\omega^2 + \frac{na^2}{n^2}\right)} = \sqrt{\left(\omega^2 + \frac{a^2}{n}\right)}.$$

Let n be increased without limit, and this becomes $=\sqrt{(\omega^2)}=\omega$. The angular velocity, therefore, is not altered in the first 1″; and since the same demonstration applies to every succeeding 1″, it follows, that the angular motion is unaltered.

12. Prop. 4. Under the same circumstances, the axis of rotation has a uniform motion in space, from the position AB towards AC; and the angle described in $1'' = \frac{a}{\omega}$.

Suppose 1″ divided into n parts, as in the third Proposition, and suppose AB, AC, (fig. 1), to be at right angles. If we suppose the angular velocity $\frac{a}{n}$, about the axis AC, to be impressed instantaneously, and suppose AD to be the new axis of revolution; then, by Prop. 1,

$$\frac{\sin BAD}{\sin CAD} = \frac{a}{n\omega},$$

or, in the present case,

$$\frac{\sin BAD}{\cos BAD} = \frac{a}{n\omega}, \quad \text{or} \quad \tan BAD = \frac{a}{n\omega}.$$

Suppose n very much increased; then $\tan BAD$ being diminished without limit, we may put the arc for the tangent; hence, $BAD = \frac{a}{n\omega}$. And since, by the last Proposition, the angular velocity remains unaltered, angles equal to BAD will be added to BAD in each succeeding interval; and, therefore, at the end of 1″, the axis of revolution will be inclined to the line which was the axis of revolution at the beginning of that 1″, by the angle $\frac{a}{\omega}$. Since the same is true of every successive 1″, the axis of revolution will move from the position AB, towards the position AC, with the angular velocity $\frac{a}{\omega}$.

13. Prop. 5. Under the same circumstances, if a spherical surface be described in the body about the point A, at which the axes intersect each other; the points at which the successive axes of revolution cut this surface will lie in the circumference of a small circle, whose radius = radius of the sphere $\times \frac{a}{\omega^2}$, nearly.

Let AB, (fig. 3), be the original axis, and, as before, suppose 1″ divided into n parts, and the angular velocity $\frac{a}{n}$, about an axis perpendicular to the axis of rotation, to be impressed at the end of each. At the end of the first interval, the axis will be transported from AB to AD, the angle BAD being $= \frac{a}{n\omega}$; and at the end of the second and succeeding intervals, it will have the positions AD', AD'', &c. in space, each of the angles DAD', $D'AD''$, &c. being $= \frac{a}{n\omega}$.

Now, in $\frac{1''}{n}$, the body revolves through $\frac{\omega}{n}$; hence, if on the surface of the sphere, the angle $D'Dd'$ be made $\frac{\omega}{n}$, and $Dd' = DD'$, d' is the point which, at the end of the second interval, coincides with D'; and Ad' is, therefore, the line in the body, which, at the end of the second interval, is the axis of rotation. In the same manner, if $d'd''$ make with Dd' produced the angle $\frac{\omega}{n}$, and $d'd'' = D'D''$, Ad'' is the line, which, at the end of the third interval, is the axis of rotation, &c. From the construction it is evident, that BD, Dd', $d'd''$, &c., are the sides of a regular polygon. Suppose now, n increased without limit, or the number of the sides of the polygon increased without limit; the limit of the line traced on the spherical surface by its intersections with the successive axes, is a circle.

14. To find the radius of this circle, we observe, that if the circle and polygon be small, the sum of all the angles at D, d', &c. $= 2\pi$; but, since each of them $= \frac{\omega}{n}$, their number, or the number of sides of the polygon $= \frac{2n\pi}{\omega}$.

And the length of each $= AB \cdot \frac{a}{n\omega}$; therefore the circumference of the polygon, or ultimately of the circle $= \frac{2\pi \cdot AB \cdot a}{\omega^2}$; therefore the radius of the circle $= AB \cdot \frac{a}{\omega^2}$.

15. If the force which acts upon the body be nearly, but not exactly, uniform, and if the axes about which it tends to produce motion, be not contained in the plane *BAC*, then the propositions above will be nearly, but not exactly, true. The line traced on the spherical surface in Prop. 5, by the successive poles of rotation, will be a spiral approaching very nearly to a circle: the change of position of the axis in space at every instant, will be in the plane passing through the axis of rotation and the axis of impressed motion at that instant; and the velocity of the change, though not uniform, will be at that instant $\frac{a}{\omega}$.

ON PRECESSION AND NUTATION.

16. Prop. 6. To explain the physical cause of solar precession, and solar nutation.

Let *A*, (fig. 4), be the Earth's center; *AB* the axis of rotation; *S* the Sun; *CHG* the equator of the terrestrial spheroid; *CAG* that diameter of the equator, which is perpendicular to *AS*; and suppose the Earth to be in the position which it has at the summer solstice. In the succeeding investigations, which relate only to the motion of the Earth about its center of gravity, we may suppose the center of gravity to be kept at rest, and the motion of the Earth about this point will be the same as if we supposed it moving freely, (Whewell *On the motion of Points constrained, &c.* Art. 269, Earnshaw's *Dynamics*, Art. 137, Poisson, *Mecanique*, 402). Suppose, then, *A* the Earth's center, to be at rest; and consider the effect which the Sun's attraction would then produce on the Earth. If the Earth were spherical, it is evident that the Sun's attraction would have no tendency to give the Earth any rotation about the center *A*. But the Earth is an oblate spheroid; we must, therefore, consider the effect produced by the Sun's attraction on the parts of the spheroid which are exterior to the sphere that touches the spheroid at its poles. Now the Sun's attraction is inversely as the square of the distance of the matter attracted; and, consequently, the attraction on the spheroidal protuberance at K, is greater than the attraction on that at H. The effect of this, supposing the Earth at rest, would evidently be to give it a motion of rotation about the line CAG, in such a direction as to bring the point K towards B. But the Earth is not at rest, but is revolving about the axis AB, which nearly coincides with the axis of the spheroid, in such a direction as to carry the point C towards K. This, then, is exactly the case considered in Prop. 4. The Earth has a previous motion of rotation; and a force acts on it, which, for a short time at least, is uniform, and which tends to give it a motion of rotation round an axis perpendicular to the axis about which it is revolving. The axis of rotation, therefore, moves from the position AB towards AC, describing in each 1″ the angle $\frac{a}{\omega}$. Let AD be the position of the axis of rotation after a short time; and let AL be perpendicular to the ecliptic. It is evident, that the path of the pole, or the arc joining BD, is a tangent to the circle passing through B, whose center is in AL, and whose plane is perpendicular to AL; and that the motion of the pole in this circle is in a direction opposite to the direction of the Earth's rotation, or is retrograde. This, then, is precession of the equinoxes.

17. If we now consider the situation of the Earth at the winter solstice, (fig. 5) it will be seen, that the Sun's attraction upon H is now greater than the attraction on K; and, therefore, the motion of rotation about CAG, which the Sun's action tends to produce in the Earth, is in the same direction as before. The motion of the axis of rotation is, therefore, in the same direction as at the summer solstice.

18. If the effect of the Sun's action be examined in any other situation of the Earth, it will be found that, as the small line BD is always perpendicular to the plane passing through the Sun and the Earth's axis, it is not always a tangent to the small circle whose center is in AL. The Sun's action, therefore, sometimes increases the incli-

nation of the axis of rotation to the axis of the ecliptic, and sometimes diminishes it; but, (as we shall shew hereafter), does not permanently alter it. This phænomenon is one part of solar nutation. The angular motion of the axis of rotation, about the axis of the ecliptic, is always in the same direction; but as the action of the Sun is different in different positions of the Earth, and is 0 at the equinoxes, this angular motion or precession is irregular. The correction, which it is necessary to apply to a uniform precession, is the other part of solar nutation.

19. Prop. 7. To explain the physical cause of lunar precession, and lunar nutation.

Since the Moon describes, (very nearly), a great circle about the Earth in a month, in the same manner as the Sun in a year, the same explanation which has been given for the precession and nutation produced by the Sun in a year, will apply to those produced by the Moon in a month. But the monthly nutation produced by the Moon is so small, that it is very seldom considered. Since, however, the magnitude and direction of the permanent precession, produced by the Sun, depend upon the inclination of the Earth's axis to the axis of the ecliptic, or the axis of the Sun's apparent orbit, it is easy to see, that the magnitude and direction of the precession, produced by the Moon in one month, depend upon the inclination of the Earth's axis to the axis of the Moon's orbit. Now this is perpetually varying; the axis of the Moon's orbit revolves about the axis of the ecliptic in about 18 years, 7 months, with a motion nearly uniform, and preserving nearly a constant inclination. The velocity and direction of the motion of the Earth's axis, produced by the Moon, are, therefore, irregular. We shall shew, that the precessional motion, though irregular, is permanent; but that the alteration in the inclination to the axis of the ecliptic is periodical; the inclination returning to its former value, in a revolution of the Moon's nodes. This change in the inclination is one part of lunar nutation; the other part is the correction which must be applied to the mean precession, in order to find the true.

20. PROP. 8. The velocity of the Earth's rotation is unaltered by the action of the Sun and Moon.

Since the Sun's action would give the Earth a motion of rotation about an axis in the plane of the equator of the terrestrial spheroid, and the Moon's action would give it a rotation about another axis in the same plane, their combined action would give the Earth a rotation about a third axis in that plane, by (7).

Now to shew that the Earth's angular velocity is unaltered, we must shew, that this axis is always perpendicular to the axis of rotation. Let AC, (fig. 3), be this axis, AB the axis of rotation; by (15), the points of intersection of this axis with a sphere described in the Earth about A, lie nearly in a small circle, whose center is E. The ellipticity of the Earth is produced by its rotation; and since the axis of rotation passes successively through all points of the circle BF in one revolution, the axis of the spheroid will pass through E, the center of that circle. AE, therefore, is perpendicular to AC; or if EC be joined by an arc of a great circle, EC is a quadrant. And EBC is a right angle; hence, BC is also a quadrant, or BAC is always a right angle. Consequently, by Prop. 3, the velocity of rotation is not altered.

21. PROP. 9. To calculate the value of α; the force acting on the Earth being the attraction of a distant body; and the Earth being a homogeneous spheroid.

Let A, (fig. 6), be the Earth's center; AB the axis of the spheroid: S the attracting body; take P, the projection of any point of the Earth, and draw PN perpendicular to SA, and PM perpendicular to the projection of the equator. Let f be the attraction of S upon A; then the attraction of S upon P, is $f.\frac{SA^2}{SP^2}$, or, if $SR = \frac{SA^3}{SP^2}$, it $= f.\frac{SR}{SA}$. If SA, then, be taken to represent the force f, SR will represent the force on P, and RA the difference of forces on P and A; or that difference of forces $= f.\frac{RA}{SA}$, and is in the direction parallel to RA. Now if a force f were applied to every point of the spheroid, in the direction SA, it would produce no effect in giving the Earth a motion about A; without altering the motion, therefore, we may suppose this force applied; that is, we may suppose the only force acting at each point, to be the difference of the force really acting there, and the force at A. In the figure, the point A is evidently the projection of the axis, about which these forces would make the Earth revolve. We must, therefore, find the momentum of the force $f\frac{RA}{SA}$ impressed on a particle δm at P, about the center A. Let $SQ = SA$; then, S being very distant, AQ is nearly perpendicular to SA; and

$$SR = \frac{SQ^3}{(SQ - PQ)^2} = SQ + 2PQ, \text{ nearly};$$

or $RQ = 2PQ$. The force $f.\frac{RA}{SA}\delta m$ may now be resolved into

$$f\frac{RQ}{SA}\delta m \text{ and } f.\frac{QA}{SA}\delta m, \text{ or } f\delta m.\frac{2PQ}{SA} \text{ and } f\delta m.\frac{PN}{SA},$$

acting at P in the directions QP, PN. Their momenta to turn the Earth in the direction KB, are

$$-f\delta m.\frac{2PQ.PN}{SA}, \text{ and } -f\delta m.\frac{PN.PQ}{SA},$$

(considering PN and QA as sensibly equal); the sum of these

$$= -3f.\delta m.\frac{PN.PQ}{SA}.$$

If the absolute force of the attracting body $S = S$, and $SA = r$, then $f = \frac{S}{r^2}$, and the moment of the force on P

$$= -\frac{3S}{r^3}\delta m.PN.PQ.$$

Let $AM = x$; $MP = y$; $BAS = \theta$.

Then $PN = MV - MT = y \sin\theta - x \cos\theta$;

$$PQ = PV + AT = y \cos\theta + x \sin\theta.$$

Substituting these values, the moment of the force on the particle at P

$$= \frac{3S}{r^3}\delta m . \{(x^2 - y^2)\sin\theta\cos\theta + xy(\cos^2\theta - \sin^2\theta)\}.$$

22. We must now find the sums of the expressions $x^2\delta m$, $y^2\delta m$, $xy\delta m$, for every particle of the spheroid. Suppose the spheroid divided into slices by planes parallel to the plane of xy; let two of these be at the distances z and $z + \delta z$ respectively; z being measured perpendicular to the plane of xy, and δz being small; and let the included slice be divided into prisms, by planes parallel to yz; two of these being at the distances x and $x + \delta x$ from the plane of yz; and take a portion of this prism, included between the co-ordinates y and $y + \delta y$. The volume of this portion $= \delta x . \delta y . \delta z$; and if k be the density, the expression $xy\delta m$ becomes, for this portion, $\delta z . \delta x . kxy\delta y$. If p be the sum of $xy\delta m$ for the prism,

$$\frac{dp}{dy} = \delta z . \delta x . kxy,$$

$$\text{or } p = \delta z . \delta x . \frac{kxy^2}{2};$$

taking this between the limits

$$y = -\frac{c}{a}\sqrt{(a^2 - x^2 - z^2)}, \text{ and } y = +\frac{c}{a}\sqrt{(a^2 - x^2 - z^2)},$$

(since the equation to the surface of the spheroid is $\frac{x^2 + z^2}{a^2} + \frac{y^2}{c^2} = 1$),

$$p = 0.$$

Hence, the sum of $xy\delta m$ for the whole spheroid, is $= 0$.

23. For the sum of $x^2\delta m$ it appears in the same manner, that if w be the sum for the prism,

$$\frac{dw}{dy} = \delta z . \delta x . kx^2, \text{ or } w = \delta z . \delta x . kx^2 y;$$

which, taken between the limits $\mp \frac{c}{a}\sqrt{(a^2 - x^2 - z^2)}$, gives

$$w = 2k\delta z \,.\, \delta x \,.\, \frac{c}{a} x^2 \sqrt{(a^2 - x^2 - z^2)}.$$

This is ultimately the increment of the sum of $x^2 \delta m$ for the slice, produced by giving to x the increment δx; calling this sum v,

$$\frac{dv}{dx} = 2k\delta z \,.\, \frac{c}{a} \,.\, x^2 \sqrt{(a^2 - z^2 - x^2)},$$

$$\text{and} \quad v = 2k\delta z \frac{c}{a} \int_x x^2 \sqrt{(a^2 - z^2 - x^2)},$$

z being considered constant in the integration.

The integral is

$$k\delta z \,.\, \frac{c}{a} \,.\, \left\{ -\frac{x}{2}(a^2 - z^2 - x^2)^{\frac{3}{2}} + \frac{a^2 - z^2}{4} x \sqrt{(a^2 - z^2 - x^2)} \right.$$
$$\left. + \frac{(a^2 - z^2)^2}{4} \sin^{-1} . \frac{x}{\sqrt{(a^2 - z^2)}} \right\}.$$

The limits of x are the least and greatest values of x in the slice; that is, the values given by the equation to the surface, upon making $y = 0$; they are, therefore, $\mp \sqrt{(a^2 - z^2)}$; and

$$v = k\delta z \,.\, \frac{c}{a} \,.\, \frac{(a^2 - z^2)^2}{4} \,.\, \pi.$$

Now, if u be the sum of $x^2 \delta m$ for the whole spheroid, v is ultimately the increment of u, arising from giving to z the increment δz; hence,

$$\frac{du}{dz} = k \,.\, \frac{c}{a} \,.\, \frac{\pi}{4} \,.\, (a^2 - z^2)^2 = k \,.\, \frac{c}{a} \,.\, \frac{\pi}{4} \times (a^4 - 2a^2 z^2 + z^4);$$

$$\therefore \; u = k \,.\, \frac{c}{a} \,.\, \frac{\pi}{4} \,.\, \left(a^4 z - \frac{2}{3} a^2 z^3 + \frac{z^5}{5}\right);$$

taking this between the limits $z = \mp a$,

$$u = k \,.\, \frac{c}{a} \,.\, \pi \,.\, \frac{4a^5}{15} = \frac{4\pi}{15} k a^2 c \times a^2.$$

24. In the same manner, it would be found that the sum of the $y^2\delta m$ for the spheroid $= \frac{4\pi}{15} ka^2c \times c^2$. Hence, the moment of all the impressed forces

$$= \frac{3S}{r^3} \cdot \frac{4\pi}{15} ka^2c(a^2 - c^2) \sin\theta \,.\, \cos\theta.$$

25. Now, to find the angular velocity which this would generate in 1″ about the axis whose projection is *A*, we must (Whewell *On the Motion of Points constrained, &c.* Art. 206, Earnshaw's *Dynamics*, Art. 144) divide this by the moment of inertia, or the sum of $(x^2 + y^2)\delta m$ for the whole spheroid. This sum is found in the same manner to be

$$= \frac{4\pi}{15} ka^2c(a^2 + c^2).$$

Hence, the angular velocity generated in 1″, or α,

$$= \frac{3S}{r^3} \cdot \frac{a^2 - c^2}{a^2 + c^2} \cdot \sin\theta \,.\, \cos\theta.$$

26. Prop. 10. To calculate α, supposing the Earth heterogeneous.

Suppose the Earth composed of strata of different densities, bounded by spheroidal surfaces of different ellipticities, as in the Treatise on the Figure of the Earth. Let c be the semi-axis of any one of these spheroids, e its ellipticity; c the semi-axis of the external surface, e its ellipticity; ρ the density at any point; ρ and e being functions of c. Then, (since $a^2 - c^2 = 2c^2e$ nearly, and $a^2 + c^2 = 2c^2$ and $a^2c = c^3$ nearly) as in Prop. 22. of the Figure of the Earth, we shall have for the moment of the impressed forces,

$$\frac{3S}{r^3} \sin\theta \,.\, \cos\theta \,.\, \frac{8\pi}{15} \,.\, \int_c \rho \,.\, \frac{d(c^5e)}{dc}$$

$$= \frac{3S}{r^3} \sin\theta \,.\, \cos\theta \,.\, \frac{8\pi}{15} \psi(\mathrm{c}).$$

And the moment of inertia

$$= \frac{8\pi}{15} \int_c \rho \cdot \frac{d(c^5)}{dc} = \frac{8\pi}{15} \sigma(c), \text{ if } \sigma(c) = \int_c \rho \cdot \frac{d(c^5)}{dc}.$$

Hence, the Earth being heterogeneous,

$$\alpha = \frac{3S}{r^3} \cdot \sin\theta \cdot \cos\theta \cdot \frac{\psi(c)}{\sigma(c)}.$$

27. In the investigation of the value of α, we have supposed that the only force which tends to give the Earth a rotatory motion about AC, (fig. 3 and 4), is the action of a distant body. This, however, is not strictly true; for, since AD, the axis about which the Earth is at any instant revolving, does not coincide with AE the axis of the figure, the centrifugal force will diminish the effect produced by the distant body. With an ellipticity, however, so small as that of the Earth, this diminution is not sensible.

28. Prop. 11. To investigate the quantity of solar precession for any given time.

Suppose EC, (fig. 7), to be the projection of the ecliptic on the surface of a sphere described about the Earth's center; S, P, Q, the projections of the Sun's place, the pole of the Earth, and the pole of the ecliptic; join SP by an arc of a great circle. By Prop. 9 and 10, the value of α is

$$\frac{3B \cdot S}{r^3} \sin\theta \cdot \cos\theta,$$

B being $= \frac{a^2 - c^2}{a^2 + c^2}$ if the Earth be homogeneous, and $= \frac{\psi(c)}{\sigma(c)}$ if the Earth be heterogeneous. Now, θ is the angle made by the Earth's axis with the line joining the centers of the Sun and Earth, and is, therefore, in (fig. 7), represented by SP: hence

$$\alpha = \frac{3B \cdot S}{r^3} \sin SP \cdot \cos SP;$$

and, by Prop. 4, the pole of rotation moves with the velocity

$$\frac{a}{\omega} \quad \text{or} \quad \frac{3B.S}{r^3\omega} \sin SP . \cos SP,$$

in the direction Pp perpendicular to PS. The resolved part of this motion, perpendicular to PQ, or parallel to CE, is

$$\frac{3B.S}{r^3\omega} . \sin SP . \cos SP . \cos SPC.$$

Let ES, the Sun's longitude $= l$; QP the inclination of the equator and ecliptic $= I$. Then

$$\cos SP = \cos CS . \cos CP = \sin l . \sin I;$$

$$\sin SP = \frac{\sin PC}{\sin PSC} = \frac{\cos I}{\sin PSC};$$

$$\cos SPC = \cos SC . \sin PSC = \sin l . \sin PSC;$$

therefore, the velocity of the pole parallel to CS is

$$\frac{3B.S}{r^3\omega} \sin I . \cos I . \sin^2 l,$$

and the motion of the pole in that direction

$$= \frac{3B.S}{\omega} \sin I . \cos I . \int_t \frac{\sin^2 l}{r^3} = \frac{3B.S}{\omega} \sin I . \cos I . \int_l . \frac{\sin^2 l}{r^3} . \frac{dt}{dl}.$$

Now, by Art. 12 of *Lunar Theory*, neglecting the Earth's mass in comparison with the Sun's,

$$\frac{dt}{dl} = \frac{r^2}{\sqrt{\{a(1-e^2).S\}}},$$

where a and e are the semi-axis-major and eccentricity of the Sun's apparent orbit; therefore, the motion of the pole parallel to CS

$$= \frac{3B\sqrt{S}}{\omega\sqrt{\{a(1-e^2)\}}} \sin I . \cos I . \int_l . \frac{\sin^2 l}{r}$$

$$= \frac{3B\sqrt{S}}{\omega\{a.(1-e^2)\}^{\frac{3}{2}}} \sin I . \cos I \int_l \sin^2 l \{1 + e\cos(l-k)\}$$

where k is the longitude of the Sun's perigee. Let $T = 1$ year: by Art. 15 of *Lunar Theory*,

$$T = \frac{2\pi \, . \, a^{\frac{3}{2}}}{\sqrt{S}}:$$

our expression becomes then

$$\frac{6\pi \, . \, B}{T\omega \, . \, (1 - e^2)^{\frac{3}{2}}} \sin I \, . \cos I \int_l \sin^2 l \left\{1 + e \cos(l - k)\right\}$$

$$= \frac{3\pi \, . \, B}{T\omega (1 - e^2)^{\frac{3}{2}}} \sin I \, . \cos I$$

$$\times \left\{C + l - \frac{\sin 2l}{2} + e \, . \sin(l - k) - \frac{e}{2} \sin(l + k) - \frac{e}{6} \sin(3l - k)\right\}.$$

The three last terms of this expression are so small when numerically calculated, that they are rejected; and the motion of the pole parallel to CS, is, therefore,

$$\frac{3\pi \, . \, B}{T\omega (1 - e^2)^{\frac{3}{2}}} \, . \sin I \, . \cos I \left(\overline{C + l} - \frac{\sin 2l}{2}\right),$$

$$\text{or } \frac{3\pi \, . \, B}{T\omega} \sin I \, . \cos I \left(\overline{C + l} - \frac{\sin 2l}{2}\right) \text{ nearly.}$$

29. The first term of this expression never changes sign, but increases as l increases, and is, therefore, nearly proportional to the time from any fixed epoch; it is the precessional motion of the pole. The second term is periodical, going through all its values in half a year: it is the correction which must be applied to the place of the pole found on the supposition of uniform precession, in order to obtain its true place. This is one part of Solar Nutation.

30. To obtain the precession of the equinoxes, or the angular motion of P about Q, we must divide the expression above by $\sin I$: it is, therefore,

$$\frac{3\pi \, . \, B}{T\omega} \cos I \, . \left(\overline{C + l} - \frac{\sin 2l}{2}\right).$$

The first term is the uniform precession; the second is the correction to be applied to it, called the Solar Equation of the Equinoxes in longitude.

31. PROP. 12. To investigate the change in the obliquity of the ecliptic, produced by the Sun's action on the Earth.

In (28) it was found that the velocity of the pole in the direction Pp (fig. 7) is

$$\frac{3B.S}{r^3\omega}\sin SP.\cos SP;$$

therefore, its velocity in the direction PQ

$$=\frac{3B.S}{r^3\omega}\sin SP.\cos SP.\sin SPC.$$

$$\text{But } \sin SP.\sin SPC=\sin SC=\cos l;$$

$$\cos SP=\cos PC.\cos SC=\sin I.\sin l;$$

therefore the motion in the direction PQ, or the diminution of the inclination,

$$=\frac{3B.S}{\omega}\sin I\int_t\frac{\sin l.\cos l}{r^3};$$

which, as in (28), is changed into

$$\frac{6\pi.B}{T\omega}\sin I\int_l\sin l\cos l(1+e\cos\overline{l-k}).$$

Neglecting, as before, the terms depending on e, this is

$$-\frac{3\pi.B}{T\omega}\sin I\frac{\cos 2l}{2}:$$

or, if I be the mean inclination, the true inclination

$$=I+\frac{3\pi.B}{T\omega}\sin I.\frac{\cos 2l}{2}.$$

The term added to I is the second part of Solar Nutation.

32. If we call the two parts of Solar Nutation, mentioned in (28) and (31), x and y, we shall easily perceive that they are connected by this equation,

$$x^2\left(\frac{2T\omega}{3\pi \,.\, B \,.\, \sin I \,.\, \cos I}\right)^2 + y^2\left(\frac{2T\omega}{3\pi \,.\, B \,.\, \sin I}\right)^2 = 1.$$

This is the equation to an ellipse, whose axes are in the ratio of $\cos I : 1$. Thus is explained the construction in Woodhouse's *Astronomy*, new edition, p. 367.

33. PROP. 13. To investigate the motion of the pole produced by the Moon in one sidereal revolution.

By (9), it appears that, instead of considering at once the effect of the Sun and Moon upon the Earth, we may first investigate the effect produced by one, and then add to it the effect produced by the other. For the effect produced by the Moon, the investigation is exactly similar to those of Prop. 11 and 12, and the same figure may be used; observing that, as EC (fig. 7) is the great circle apparently described by the Moon, Q is not now the pole of the ecliptic, but the pole of the Moon's orbit. There is only one difference: putting E for the Earth's absolute force, M for the Moon's, T' for the time of a sidereal revolution of the Moon, I' for the inclination of the Earth's axis to the axis of the Moon's orbit, a' for the mean distance,

$\frac{dt}{dl}$ will be $\frac{r^2}{\sqrt{a'(1-e'^2)(E+M)}}$, and T' will be $\frac{2\pi \,.\, a'^{\frac{3}{2}}}{\sqrt{(E+M)}}$;

and, therefore, instead of $\frac{3\pi B}{T\omega}$, we must put $\frac{3\pi B}{T'\omega} \,.\, \frac{M}{E+M}$.

If the Moon's mass be $\frac{1}{n}$th of the Earth's, this $= \frac{3\pi B}{T'\omega(n+1)}$.

Thus we find for the motion of the pole parallel to the Moon's orbit,

$$\frac{3\pi \,.\, B \sin I' \,.\, \cos I'}{T'\omega(n+1)}\left(\overline{C+l} - \frac{\sin 2l}{2}\right):$$

and for the motion perpendicular to the Moon's orbit,

$$-\frac{3\pi B \,.\, \sin I'}{T' \omega (n+1)} \cdot \frac{\cos 2l}{2},$$

l being now measured from the intersection of the equator with the Moon's orbit.

34. The latter expression, and the second term of the former, are periodical terms, going through all their changes of value twice in a month: their magnitudes, besides, are so small, that they are generally neglected.

Supposing l increased by 2π, we find, for the motion of the pole produced by the Moon's action in a sidereal revolution,

$$\frac{6\pi^2 \,.\, B \,.\, \sin I' \,.\, \cos I'}{T' \omega \,.\, (n+1)};$$

which motion is parallel to the Moon's orbit, or perpendicular to the great circle joining the pole of the Earth with the pole of the Moon's orbit.

35. Prop. 14. To investigate the precessional motion produced by the Moon's action during a long period.

Let Q, (fig. 8), be the pole of the ecliptic; q that of the Moon's orbit; P that of the earth: let them be joined by arcs of great circles; then, by the last article, it appears that by the action of the Moon, the pole is in the time T' carried in the direction Pp, perpendicular to Pq, through the arc

$$\frac{6\pi^2 \,.\, B}{T' \omega (n+1)} \sin qP \,.\, \cos qP.$$

This may be represented by supposing the pole to have the velocity

$$\frac{6\pi^2 \,.\, B}{T'^2 \omega (n+1)} \sin qP \,.\, \cos qP,$$

in the direction Pp. Its velocity then in a direction perpendicular to QP, is

$$\frac{6\pi^2 \,.\, B}{T'^2 \,.\, \omega (n+1)} \sin qP \,.\, \cos qP \,.\, \cos QPq.$$

This we must express in terms of QP, Qq, and the angle PQq.

36. Now,

$$\sin qP \,.\, \cos QPq = \frac{\cos Qq - \cos QP \,.\, \cos qP}{\sin QP};$$

but $\cos qP = \cos QP \,.\, \cos Qq + \sin QP \,.\, \sin Qq \,.\, \cos Q$;
substituting this,

$$\sin qP \,.\, \cos QPq = \sin QP \,.\, \cos Qq - \cos QP \,.\, \sin Qq \,.\, \cos Q.$$

Multiplying by

$$\cos qP = \cos QP \,.\, \cos Qq + \sin QP \,.\, \sin Qq \,.\, \cos Q,$$

we find $\sin qP \,.\, \cos qP \,.\, \cos QPq$

$$\sin QP \,.\, \cos QP \,.\, \cos^2 Qq - (\cos^2 QP - \sin^2 QP) \,.\, \sin Qq \,.\, \cos Qq \,.\, \cos Q$$
$$- \sin QP \,.\, \cos QP \,.\, \sin^2 Qq \,.\, \cos^2 Q.$$

Put I and i for QP and Qq: I and i are nearly constant; and, since the Moon's nodes revolve in a retrograde direction through a great circle in 18,6 years, if $\tau = 18,6$ years,

$$Q = \frac{2\pi t}{\tau}.$$

Hence, the velocity of the pole perpendicular to QP, is

$$\frac{6\pi^2 B}{T'^2 . \omega(n+1)} \,.\, \left\{ \sin I \cos I \,.\, \cos^2 i - \tfrac{1}{2} \cos 2I \,.\, \sin 2i \,.\, \cos \frac{2\pi t}{\tau} \right.$$
$$\left. - \sin I \,.\, \cos I \,.\, \sin^2 i \,.\, \cos^2 \frac{2\pi t}{\tau} \right\}$$
$$= \frac{6\pi^2 B}{T'^2 \,\omega(n+1)} \left\{ \sin I \cos I \,(\cos^2 i - \tfrac{1}{2} \sin^2 i) \right.$$
$$\left. - \tfrac{1}{2} \cos 2I \,.\, \sin 2i \,.\, \cos \frac{2\pi t}{\tau} - \tfrac{1}{2} \sin I \,.\, \cos I \,.\, \sin^2 i \,.\, \cos \frac{4\pi t}{\tau} \right\}.$$

Integrating this with respect to t, we find the precessional motion of the pole

$$= \frac{6\pi^2 B}{T'^2 \,\omega(n+1)} \left\{ \sin I \cos I \,(\cos^2 i - \tfrac{1}{2} \sin^2 i) \,.\, t \right.$$

$$- \frac{\tau}{4\pi} \cos 2I . \sin 2i . \sin \frac{2\pi t}{\tau}$$

$$- \frac{\tau}{8\pi} \sin I . \cos I . \sin^2 i . \sin \frac{4\pi t}{\tau}\Big\} + C.$$

37. The first term of this expression is proportional to the time, and, therefore, increases uniformly. The second term is periodical; it depends upon $\sin \frac{2\pi t}{\tau}$, or $\sin Q$, or sin (180° – longitude of Moon's ascending node), or sin longitude of Moon's ascending node. This is a part of Lunar Nutation. The third term depends upon sin 2 long. Moon's ascending node; it is, therefore, a part of Nutation: but its numerical value is so small, that it is commonly neglected.

38. To obtain the lunar precession of the equinoxes, we must, as before, divide the last expression by $\sin I$. Thus, we get

$$\frac{6\pi^2 B}{T'^2 \omega (n+1)} \Big\{ \cos I (\cos^2 i - \tfrac{1}{2} \sin^2 i) . t$$

$$- \frac{\tau}{4\pi} . \frac{\cos 2I}{\sin I} . \sin 2i . \sin \frac{2\pi t}{\tau} - \frac{\tau}{8\pi} \cos I . \sin^2 i . \frac{\sin 4\pi t}{\tau} \Big\}.$$

The first term, which increases uniformly, is called simply the lunar precession; the second is the lunar equation of the equinoxes in longitude; the third is neglected. The lunar precession for a year is found by putting T for t; it is, therefore,

$$\frac{6\pi^2 B}{T'^2 \omega (n+1)} . \cos I (\cos^2 i - \tfrac{1}{2} \sin^2 i) . T.$$

39. Prop. 15. To investigate the alteration in the obliquity of the ecliptic produced by the Moon's action.

By (35), the velocity of the pole in the direction Pp, is

$$\frac{6\pi^2 B}{T'^2 \omega (n+1)} \sin qP . \cos qP.$$

The velocity, therefore, with which the inclination is increased, is

$$\frac{6\pi^2 . B}{T'^2 \omega (n+1)} . \sin qP . \cos qP . \sin QPq.$$

$$\text{Now, } \sin qP . \sin QPq = \sin i . \sin Q;$$

$$\text{and } \cos qP = \cos I . \cos i + \sin I . \sin i . \cos Q.$$

Their product

$$= \cos I . \sin i . \cos i . \sin Q + \sin I . \sin^2 i . \sin Q . \cos Q;$$

or, the velocity of increase of inclination

$$= \frac{6\pi^2 . B}{T'^2 \omega (n+1)} . \left(\tfrac{1}{2} \cos I . \sin 2i . \sin \frac{2\pi t}{\tau} + \tfrac{1}{2} \sin I . \sin^2 i . \sin \frac{4\pi t}{\tau}\right)$$

Integrating this with respect to t, the inclination

$$= I - \frac{6\pi^2 . B}{T'^2 \omega (n+1)} . \left(\frac{\tau}{4\pi} \cos I . \sin 2i \cos \frac{2\pi t}{\tau} + \frac{\tau}{8\pi} . \sin I . \sin^2 i . \cos \frac{4\pi t}{\tau}\right).$$

The terms subtracted from I are periodical, depending upon $\cos \frac{2\pi t}{\tau}$, or $\cos (180^0 -$ long. Moon's ascending node,) and upon the cosine of twice that angle; the first is a part of Lunar Nutation; the second is usually neglected.

40. If we call x and y the first terms of nutation in (36) and (39), it will easily be seen that they are connected by this equation,

$$x^2 \left\{\frac{2T'^2 \omega (n+1)}{3\pi . B\tau . \cos 2I . \sin 2i}\right\}^2 + y^2 \left\{\frac{2T'^2 \omega (n+1)}{3\pi B . \tau . \cos I . \sin 2i}\right\}^2 = 1.$$

This is the equation to an ellipse, in which the axes have the ratio of $\cos 2I : \cos I$. This explains the construction in Woodhouse's *Astronomy*, page 357.

41. In the two last Propositions we have considered i to be constant, and Q to be proportional to t. It appears, however, from Art. 68, of the *Lunar Theory*, that the inclination is expressed, nearly, by

$$i . \{1 + \frac{3m}{8} \cos 2 . (\text{long. node} - \text{long. Sun})\};$$

and that from the longitude of the node, found on the supposition of its uniform retrogradation, we must subtract

$$\frac{3m}{8} \sin 2 . (\text{long. node} - \text{long. Sun}).$$

Now, the Sun's longitude $= \frac{2\pi t}{T} + C$, nearly: the longitude of the node $= 180^0 - \frac{2\pi t}{\tau}$: hence, for i we ought to put

$$i\left[1 + \frac{3m}{8} \cos\left\{4\pi t\left(\frac{1}{T} + \frac{1}{\tau}\right) + 2C\right\}\right],$$

and for Q we should put

$$\frac{2\pi t}{\tau} - \frac{3m}{8} \sin\left\{4\pi t\left(\frac{1}{T} + \frac{1}{\tau}\right) + 2C\right\}$$

before performing the integrations: and $\sin 2Q$, &c. could be expanded, as in Art. 49, of *Lunar Theory*. But the additional terms thus introduced have small coefficients, and upon integration receive large divisors, so that they become quite insensible. The expressions which we have found are, therefore, subject to no sensible error.

42. If now, in the terms of Nutation, which depend on twice the longitude of the Moon's ascending node, we used not the mean longitude of the node but the true, we should add to the expressions terms which have small coefficients, but which are not integrated, and, therefore, do not receive large divisors. The values thus found for the parts of Nutation at any given time, would, therefore, sen-

sibly differ from those found by the formulæ above. And, since the latter differ from the true ones only by quantities which are insensible, it follows that the values found by using the true longitude of the node, are sensibly erroneous. In calculating Nutation, therefore, the mean longitude of the node must be used, not the true.

43. Prop. 16. Assuming the law of density of the Earth, and the mass of the Moon, to calculate numerically the annual precession and the coefficients of solar and lunar nutation.

It is first necessary to calculate the value of B, or $\frac{\psi(\text{c})}{\sigma(\text{c})}$, which enters into all the expressions. Suppose, then, as in Prop. 29, of the *Figure of the Earth*, $\rho = A \cdot \frac{\sin qc}{c}$. By Art. 61, of the same,

$$\psi(\text{c}) = \frac{5}{3}\left(\text{e} - \frac{m}{2}\right)\text{c}^2 \cdot \phi(\text{c}),$$

$$\therefore\ B = \frac{5}{3}\left(\text{e} - \frac{m}{2}\right)\text{c}^2 \cdot \frac{\phi(\text{c})}{\sigma(\text{c})}.$$

$$\text{Now}\ \ \phi(c) = \textstyle\int_c \rho \dfrac{d \cdot c^3}{dc} = 3\int_c A \cdot c \cdot \sin qc$$

$$= 3A\left(-\frac{c \cdot \cos qc}{q} + \frac{\sin qc}{q^2}\right);$$

which, taken from $c = 0$ to $c = \text{c}$, gives

$$\phi(\text{c}) = 3A\left(-\frac{\text{c} \cdot \cos q\text{c}}{q} + \frac{\sin q\text{c}}{q^2}\right).$$

$$\text{And}\ \ \sigma(c) = \textstyle\int_c \rho \cdot \dfrac{d \cdot c^5}{dc}\ \ \text{by (26)} = 5A\int_c c^3 \sin qc$$

$$= 5A\left(\frac{-c^3 \cdot \cos qc}{q} + \frac{3c^2 \cdot \sin qc}{q^2} + \frac{6c \cdot \cos qc}{q^3} - \frac{6 \sin qc}{q^4}\right);$$

putting c for c, we have the value of σ (c). Hence

$$B = \left(e - \frac{m}{2}\right) \cdot \frac{-q^3 c^3 . \cos qc + q^2 c^2 . \sin qc}{-q^3 c^3 . \cos qc + 3q^2 c^2 \sin qc + 6qc . \cos qc - 6 \sin qc}$$

$$= \left(e - \frac{m}{2}\right) \cdot \frac{z}{2 + z - \frac{6z}{q^2 c^2}}, \text{ where } z = 1 - \frac{qc}{\tan qc}.$$

If we suppose q to have the value used in Art. 66, of the *Figure of the Earth*, namely, $\frac{5}{6} \cdot \frac{\pi}{c}$, and take the value of $e - \frac{m}{2}$, found in Art. 66, we get for the value of B, ,0031677.

44. The solar annual precession, which is found by supposing l to be increased by 2π in the expression of (30), is

$$\frac{6\pi^2 . B . \cos I}{T\omega}.$$

Now $T\omega$ = the angle described by the diurnal revolution of the Earth in 1 year $= 2\pi . 366{,}26$; $I = 23^0 . 28'$: hence, the solar annual precession

$$= \frac{B . 3\pi . \cos 23^0 . 28'}{366{,}26}.$$

This is in parts of the radius: to reduce it to seconds we must multiply it by $\frac{180 . 60 . 60}{\pi}$. Thus the solar annual precession in seconds

$$= \frac{B . 9 . (60)^3 . \cos 23^0 . 28'}{366{,}26}:$$

which, putting for B the value above, $= 15'',42$.

45. From (30) it appears that the coefficient of $\sin 2l$, in the Solar Equation of the equinoxes in longitude, is

to the annual solar precession as $\frac{1}{2} : 2\pi$. This coefficient, therefore,

$$= \frac{15'',42}{4\pi} = 1'',23.$$

And from (31), it appears that the coefficient of $\cos 2l$, in the Solar Nutation in obliquity, is to the annual solar precession as

$$\frac{\sin I}{2} : 2\pi . \cos I;$$

whence this coefficient

$$= \frac{15'',42}{4\pi \cot I} = 0'',53.$$

46. By (38), the lunar annual precession

$$= B . \frac{6\pi^2}{T'\omega} . \frac{T}{T''} . \frac{\cos I}{n+1} . \left(1 - \frac{3}{2}\sin^2 i\right).$$

Now, $T'\omega$ = angle described by the diurnal revolution of the Earth in one revolution of the Moon $= 2\pi . 27,32$:

$$\frac{T}{T'} = \frac{366,26}{27,32} : \; i = 5^0 . 8' . 50'';$$

suppose the Moon's mass $\frac{1}{70}$ that of the Earth, or $n = 70$; hence, the lunar annual precession

$$= B . \frac{3\pi . 366,26 . \cos 23^0 . 28' \times ,98791}{(27,32)^2 . 71}.$$

This reduced to seconds, as the former was

$$= \frac{B . \overline{9.60}|^3 . 366,26 . \cos 23^0 . 28' \times ,98791}{(27,32)^2 . 71};$$

putting the same value for B, this $= 38'',57$.

47. By (38), the coefficient of $\sin \dfrac{2\pi t}{\tau}$, or sin long. Moon's ascending node, in the lunar equation of the equinoxes in longitude, is

$$B \cdot \frac{3\pi \cdot \tau \cdot \cos 2I \cdot \sin 2i}{2\,T'^2 \cdot \omega\,(n+1) \cdot \sin I}.$$

$$\text{Now } \frac{\tau}{T'^2\omega} = \frac{\tau}{T} \cdot \frac{T}{T'} \cdot \frac{1}{T'\omega} = \frac{18{,}6 \cdot 366{,}26}{(27{,}32)^2 \cdot 2\pi};$$

hence, this coefficient

$$= B \cdot \frac{3 \cdot 18{,}6 \cdot 366{,}26 \cdot \cos 2I \cdot \sin 2i}{4\,(27{,}32)^2 \cdot 71 \cdot \sin I},$$

and in seconds

$$= B \cdot \frac{9 \cdot 60^3 \cdot 18{,}6 \cdot 366{,}26 \cdot \cos 46^0 \cdot 56' \cdot \sin 10^0 \cdot 17' \cdot 40''}{4\pi\,(27{,}32)^2 \cdot 71 \cdot \sin 23^0 \cdot 28'}.$$

Giving B the same value as before, the number of seconds is 19,3. And comparing (38) with (39), it is seen that the coefficient just found is to the coefficient of cos long. ascending node in the lunar nutation in obliquity, as

$$\frac{\cos 2I}{\sin I} : \cos I, \text{ or as } 1 : \tfrac{1}{2}\tan 2I:$$

whence this coefficient in seconds

$$= \frac{19{,}3 \cdot \tan 46^0 \cdot 56'}{2} = 10{,}33.$$

The observed values of these coefficients are 18″,036 and 9″,6.

48. Adding together the numbers found in (44) and (46), the whole annual precession = 53″,99. The observed value is 50″,1.

49. Prop. 17. From observations on precession and nutation, to determine the Moon's mass, and the ellipticity of the Earth.

Among the various results of observation we shall select, as those which have been most accurately determined, the whole annual precession in seconds (a), and the coefficient of the lunar nutation in obliquity (b). Comparing the expressions in (38) and (39), the lunar annual precession

$$= b \cdot \frac{4\pi\left(1 - \frac{3}{2}\sin^2 i\right)}{\sin 2i} \cdot \frac{T}{\tau} = b \cdot C, \text{ suppose.}$$

Subtracting this from the whole annual precession, the solar annual precession $= a - bC$, and

$$\frac{\text{solar annual precession}}{\text{lunar annual precession}} = \frac{a - bC}{bC} = \frac{a}{bC} - 1.$$

But, by the expressions in (30) and (38),

$$\frac{\text{solar annual precession}}{\text{lunar annual precession}} = \frac{T'^2(n+1)}{T^2\left(1 - \frac{3}{2}\sin^2 i\right)}:$$

$$\text{hence } \frac{a}{bC} - 1 = \frac{T'^2(n+1)}{T^2\left(1 - \frac{3}{2}\sin^2 i\right)} = D(n+1), \text{ suppose;}$$

$$\text{and } n + 1 = \frac{1}{D}\left(\frac{a}{bC} - 1\right).$$

This determines n, that is, the ratio of the mass of the Earth to the mass of the Moon. If we calculate the values of C and D with the values of i, $\frac{T'}{T}$, and $\frac{\tau}{T}$, already given,

$$n + 1 = \frac{a}{b} \cdot 47{,}52 - 177{,}56.$$

50. The ellipticity of the Earth cannot be determined at once from these data, but an equation can be found depending on the law of density of its strata, such that if a law be assumed with one indeterminate coefficient, that coefficient can be found by approximation, and the ellipticity of the Earth can then be determined. The solar annual precession we have found $= a - b . C$: but the expression for it in seconds, by (44), is

$$B . \frac{9 . 60^3 . \cos 23^0 . 28'}{366{,}26} = B . 4869;$$

making these equal, and putting for C its value,

$$B = a \times {,}0002054 - b \times {,}0007675.$$

This gives the value of

$$\frac{5}{3}\left(\mathrm{e} - \frac{m}{2}\right) . \mathrm{c}^2 . \frac{\phi(\mathrm{c})}{\sigma(\mathrm{c})}.$$

If, as in Art. 65 of the *Figure of the Earth*, we assume for the form of the expression for ρ, $A\,\dfrac{\sin qc}{c}$, we have this equation,

$$B = \left\{\frac{5m}{2} . \frac{\dfrac{3z}{q^2\mathrm{c}^2} - 1}{\dfrac{q^2\mathrm{c}^2}{z} + z - 3} - \frac{m}{2}\right\} . \frac{z}{2 + z - \dfrac{6z}{q^2\mathrm{c}^2}},$$

$$\text{where } z = 1 - \frac{q\mathrm{c}}{\tan q\mathrm{c}}:$$

finding the value of $q\mathrm{c}$ by approximation, the ellipticity of the surface will be found by substituting it in the equation

$$\mathrm{e} = \frac{5m}{2} . \frac{\dfrac{3z}{q^2\mathrm{c}^2} - 1}{\dfrac{q^2\mathrm{c}^2}{z} + z - 3}.$$

51. The values of a and b, according to the observations of most astronomers, are $50'',1$ and $9'',6$. Substituting these in the formulæ above, $n = 68,9$, and $B = ,0029225$. Assuming the law of density mentioned above, $qc = 160^0.53'$, and

$$e = ,0030812 = \frac{1}{324,5}.$$

52. Dr Brinkley, in the *Philosophical Transactions* for 1821, has reduced the value of b to $9'',25$. This gives $n = 78,2$ and $B = ,0031911$; thence $qc = 149^0.10'$, and

$$e = ,0032973 = \frac{1}{303,3}.$$

CALCULUS OF VARIATIONS.

1. In solving problems of maxima and minima by the Differential Calculus, it is necessary to express the quantity, which is to be made a maximum or minimum, in terms of the independent variable, or at least to find an equation between them; in all cases it is necessary to assign a relation between the function (u) and the independent variable (x), so that for any given value of x, the corresponding value of u actually can be found. When this is the case, $\frac{du}{dx}$ can be found; and if we make it $= 0$, an equation is obtained, which, combined if necessary with the original equation, determines the value of x or u, corresponding to the maximum or minimum value of u; and the problem is solved.

2. But it is sometimes necessary to solve problems of maxima and minima, when the relation between u and x cannot be expressed, u depending generally upon an integral $\int_x V$, where V involves y and its differential coefficients, and where it is the *object* of the problem to find the *relation* between y and x. Suppose, for instance, it were required to find the curve of quickest descent from one given point to another: then, if x be measured horizontally from the first point, and y vertically, and if the time $= u$, we have

$$\frac{du}{dx} = \frac{\sqrt{\left\{1 + \left(\frac{dy}{dx}\right)^2\right\}}}{\sqrt{(2gy)}}, \text{ or } u = \int_x \frac{\sqrt{(1+p^2)}}{\sqrt{(2gy)}},$$

$$\text{where } p = \frac{dy}{dx}.$$

This expression plainly cannot be integrated except the relation of y to x be given; but this it is the object of the problem to discover. Here then the methods of the

Differential Calculus entirely fail: and some new process must be devised for the solution of problems similar to the preceding.

3. The method given by Lagrange bears a close analogy to the methods of the Differential Calculus. In order to find what must be the relation between x and y, which will make u a maximum or minimum, we must conceive an expression to be assumed for y, and this expression to be then altered by the quantity δy, δy being some function of x. Now, in order that the assumed value of y may possess the desired property of making u a maximum or minimum, it is necessary that, upon substituting $y+\delta y$ and $y-\delta y$ for y, u may, by both substitutions, be increased, or by both be diminished. From this it follows, by reasoning precisely similar to that employed in the Differential Calculus, that the sum of the terms depending on the first power of δy, in the new values of u, must $= 0$. If then we can by any means express this sum, we shall be able to find the relation between x and y, that will make u a maximum or minimum.

4. This method is sufficient when, supposing u expressed by an integral, the values of x at the limits are known: as, for instance, in the problem of (2), the extreme points of the curve are given. But these may be undetermined, as in the problem "To find the line of quickest descent from one given curve to another given curve." Suppose Aa, Bb, (fig. 1), to be the given curves; APB the curve required. If we suppose δy to have such a form that for the values of x, corresponding to the points A and B, δy is $= 0$, the curve APB will be changed by this variation to such a curve as ApB. But though the consideration of the effect of this variation will assist us in discovering the curve APB, which is the line of shortest descent from the point A to the point B, it plainly will not enable us to determine what points of the given curves must be selected for the extremities of the curve required. If we suppose the form of δy to be such, that δy is not $= 0$ for the values of x corresponding to the points A and B, this variation will change the curve APB into $A'P'B'$, in which the values

of x for A' and B', are the same as those for A and B. Since this new curve is not terminated by the given curves, the variation is not such as the conditions of the problem require. The method which we have given, is, therefore, defective; its deficiency is supplied in the following manner.

5. Instead of supposing the curve to be varied by the variation of only one of its co-ordinates (y), suppose both co-ordinates to be varied; that is, suppose the co-ordinates of the new curve to be

$$x' = x + \delta x, \text{ and } y' = y + \delta y,$$

δx and δy being functions of x. This amounts to supposing, that in fig. 2, x and y being the co-ordinates of P,

$$NN' = \delta x, \text{ and } QP' = \delta y;$$

and that, by taking successively other points of AB, a series of points is determined, through which $A'B'$ is drawn. Now, if we take care to give δx and δy, at the extremities of AB, such a relation, that the point A' will be found in the curve Aa, and the point B' in the curve Bb, the variation will be such, as the conditions of the problem require. And as the form of the function which we suppose to represent δx or δy generally, is absolutely arbitrary, subjected to no other condition, than that δx and δy at the limits, shall have a given ratio; this variation is the most general, which it is possible for us to give.

6. That the value of u may be a maximum or minimum, the sum of the terms depending on the first powers of δx and δy, as is shown by the same kind of reasoning as that used in the Differential Calculus, must $= 0$. In finding the variation of u, therefore, we may confine ourselves to the terms depending on the first power of δx and δy.

7. For this purpose, we will find δp, δq, &c.,

$$\text{where } p = \frac{dy}{dx}, \quad q = \frac{d^2 y}{dx^2}, \text{ \&c.,}$$

as far as the first power of δx and δy.

The original value of p was $\frac{dy}{dx}$; the value, after giving to x and y a variation, is $\frac{dy'}{dx'}$;

$$\text{hence,}\quad \delta p = \frac{dy'}{dx'} - \frac{dy}{dx}.$$

$$\text{Now}\quad \frac{dy'}{dx'} = \frac{\frac{dy'}{dx}}{\frac{dx'}{dx}} = \frac{\frac{dy}{dx} + \frac{d.\delta y}{dx}}{1 + \frac{d.\delta x}{dx}}$$

$$= \frac{dy}{dx} + \frac{d.\delta y}{dx} - \frac{dy}{dx}.\frac{d.\delta x}{dx};$$

$$\therefore\ \delta p = \frac{d.\delta y}{dx} - p.\frac{d.\delta x}{dx}.$$

$$\text{Similarly,}\quad \delta q = \frac{dp'}{dx'} - \frac{dp}{dx} = \frac{\frac{dp'}{dx}}{\frac{dx'}{dx}} - \frac{dp}{dx}$$

$$= \frac{\frac{dp}{dx} + \frac{d.\delta p}{dx}}{1 + \frac{d.\delta x}{dx}} - \frac{dp}{dx} = \frac{d.\delta p}{dx} - q\frac{d.\delta x}{dx}.$$

$$\text{And}\quad \delta r = \frac{d.\delta q}{dx} - r\frac{d.\delta x}{dx},\ \text{\&c.}$$

8. Let $u = \int_x V$; then $\delta u = \int_{x_{,}} V' - \int_x V$,

(V' being the same function of x', y', &c. that V is of x, y, &c.)

$$= \int_x V'\frac{dx'}{dx} - \int_x V = \int_x \left(V'\frac{dx'}{dx} - V\right).$$

Now, suppose V to be a function of x, y, p, q, &c.,

$$\text{and let}\quad \frac{dV}{dx} = M,\quad \frac{dV}{dy} = N,\quad \frac{dV}{dp} = P,\quad \frac{dV}{dq} = Q,\ \text{\&c.};$$

then, (to the first power of δx, &c.)

$$V' = V + M\delta x + N\delta y + P\delta p + Q\delta q + \&c.,$$

$$\text{and } \frac{dx'}{dx} = 1 + \frac{d.\delta x}{dx};$$

$$\therefore \; V'\frac{dx'}{dx} - V = V.\frac{d.\delta x}{dx} + M\delta x + N\delta y + P\delta p + Q\delta q + \&c.$$

Hence, δu, (integrating the first term by parts)

$$= V\delta x - \textstyle\int_x \delta x . \frac{d(V)}{dx} + \int_x (M\delta x + N\delta y + P\delta p + Q\delta q + \&c.).$$

The integrations are performed with respect to x, considering y, p, q, &c. as functions of x; hence, for $\frac{d(V)}{dx}$ we must put

$$\frac{dV}{dx} + \frac{dV}{dy}.\frac{dy}{dx} + \frac{dV}{dp}.\frac{dp}{dx} + \frac{dV}{dq}.\frac{dq}{dx} + \&c.$$

$$= M + Np + Pq + Qr + \&c.;$$

$$\text{then, } \delta u = V\delta x + \textstyle\int_x \{N(\delta y - p\delta x)$$

$$+ P(\delta p - q\delta x) + Q(\delta q - r\delta x) + \&c.\}.$$

9. Now, upon substituting the values found above for δp, δq, &c., we find

$$\delta p - q\delta x = \frac{d.\delta y}{dx} - p\frac{d.\delta x}{dx} - q\delta x = \frac{d(\delta y - p\delta x)}{dx};$$

$$\delta q - r\delta x = \frac{d.\delta p}{dx} - q\frac{d.\delta x}{dx} - r\delta x$$

$$= \frac{d(\delta p - q\delta x)}{dx} = \frac{d^2(\delta y - p\delta x)}{dx^2};$$

and so on. Let $\delta y - p\delta x = \omega$; then $\delta u =$

$$V\delta x + \textstyle\int_x (N\omega + P\frac{d\omega}{dx} + Q\frac{d^2\omega}{dx^2} + \&c.).$$

Integrating by parts,

$$\int_x P\frac{d\omega}{dx} = P\omega - \int_x \omega\frac{d(P)}{dx};$$

$$\int_x Q\frac{d^2\omega}{dx^2} = Q\frac{d\omega}{dx} - \frac{d(Q)}{dx}\omega + \int_x \omega\frac{d^2(Q)}{dx^2};$$

and so for the others: it being observed, that $\frac{d(P)}{dx}$, &c. signify the differential coefficients with respect to x, considering y, p, q, &c. as functions of x. Hence, we have δu

$$= V\delta x + \omega\left\{P - \frac{d(Q)}{dx} + \&c.\right\}$$

$$+ \frac{d\omega}{dx}(Q - \&c.)$$

$$+ \&c.$$

$$+ \int_x \omega\left\{N - \frac{d(P)}{dx} + \frac{d^2(Q)}{dx^2} - \&c.\right\}.$$

And, putting for ω, $\frac{d\omega}{dx}$, &c. their values $\delta y - p\delta x$, $\delta p - q\delta x$, &c.,

$$\delta u = V\delta x + (\delta y - p\delta x).\left\{P - \frac{d(Q)}{dx} + \&c.\right\}$$

$$+ (\delta p - q\delta x).\{Q - \&c.\}$$

$$+ \&c.$$

$$+ \int_x \omega\left\{N - \frac{d(P)}{dx} + \frac{d^2(Q)}{dx^2} - \&c.\right\}.$$

Or, since this quantity results from integration, and must, therefore, be taken between limits, if we put $x_{,}$, $y_{,}$, $p_{,}$, $V_{,}$, &c. for the values of x, y, p, V, &c., at the first limit, and $x_{,,}$, $y_{,,}$, $p_{,,}$, $V_{,,}$, &c. for those at the second limit, we have

$$\delta u = V_{,,}\delta x_{,,} - V_{,}\delta x_{,}$$

$$+ (\delta y_{,,} - p_{,,}\delta x_{,,}).\left\{P_{,,} - \frac{d(Q_{,,})}{dx_{,,}} + \&c.\right\}$$

$$-(\delta y_, - p_,\delta x_,)\left\{P_, - \frac{d(Q_,)}{dx_,} + \&c.\right\}$$

$$+(\delta p_{,,} - q_{,,}\delta x_{,,})(Q_{,,} - \&c.) - (\delta p_, - q_,\delta x_,)(Q_, - \&c.)$$

$$+\ \&c. \qquad\qquad - \&c.$$

$$+\int_x \omega \left\{N - \frac{d(P)}{dx} + \frac{d^2(Q)}{dx^2} - \&c.\right\},$$

the last integral being supposed to be taken between the same limits as the others.

10. When u is a maximum or minimum, the expression above must $= 0$. Now this expression consists of two parts perfectly different; the first involving only the values of δx and δy at the limits, the second being an integral dependent on the general values of δx and δy. Now it would be possible to assign different forms for δx and δy, which should leave their values at the limits unaltered, while the value of the integral

$$\int_x \omega \left\{N - \frac{d(P)}{dx} + \frac{d^2(Q)}{dx^2} - \&c.\right\}$$

should be altered in any way whatever. In order then that δu may $= 0$, whatever be the form of δx and δy, we must make these two parts separately $= 0$, which can be done only by making

$$N - \frac{d(P)}{dx} + \frac{d^2(Q)}{dx^2} - \&c. = 0,$$

$$\text{and } V_{,,}\delta x_{,,} - V_,\delta x_, + (\delta y_{,,} - p_{,,}\delta x_{,,})\left\{P_{,,} - \frac{d(Q_{,,})}{dx_{,,}} + \&c.\right\}$$

$$-(\delta y_, - p_,\delta x_,)\left\{P_, - \frac{d(Q_,)}{dx_,} + \&c.\right\} + \&c. = 0.$$

In the latter of these equations, we must eliminate, by means of given relations, as many as possible of the quantities $\delta x_,$, $\delta x_{,,}$, $\delta y_,$, &c., and make the coefficient of each remaining one separately equal to nothing.

11. The equation

$$N - \frac{d(P)}{dx} + \frac{d^2(Q)}{dx^2} - \&c. = 0,$$

may, in most cases, be rendered more easy of application by integration. Thus, suppose there enter into V only y and p; the equation is reduced to

$$N - \frac{d(P)}{dx} = 0, \qquad \text{or } Np = p \cdot \frac{d(P)}{dx};$$

$$\therefore \frac{d(V)}{dx} = N\frac{dy}{dx} + P\frac{dp}{dx} = p \cdot \frac{d(P)}{dx} + P\frac{dp}{dx};$$

$$\text{integrating, } V = Pp + C.$$

Suppose V involved only p and q; then the equation is reduced to $-\frac{d(P)}{dx} + \frac{d^2(Q)}{dx^2} = 0$,

$$\text{or } \frac{d(P)}{dx} = \frac{d^2(Q)}{dx^2}, \qquad P = \frac{d(Q)}{dx} + C;$$

$$\therefore \frac{d(V)}{dx} = P\frac{dp}{dx} + Q\frac{dq}{dx} = q\frac{d(Q)}{dx} + Cq + Q\frac{dq}{dx};$$

$$\text{integrating, } V = Qq + Cp + C'.$$

Or, suppose V contained only y and q; then

$$N + \frac{d^2(Q)}{dx^2} = 0, \qquad \text{or } N = -\frac{d^2(Q)}{dx^2};$$

$$\therefore \frac{d(V)}{dx} = N\frac{dy}{dx} + Q\frac{dq}{dx} = -p\frac{d^2(Q)}{dx^2} + Q\frac{dq}{dx};$$

$$\text{integrating, } V = Qq - p\frac{d(Q)}{dx} + C.$$

And similarly in other cases.

12. We proceed to illustrate, by examples, the application of these formulæ.

It is required to find the shortest line that can be drawn from one given point to another given point.

$$\text{Here } u = \int_x \sqrt{(1+p^2)}; \quad V = \sqrt{(1+p^2)},$$

into which p only enters; the equation

$$N - \frac{d(P)}{dx}, \text{ \&c.} = 0, \quad \text{becomes } \frac{d(P)}{dx} = 0,$$

$$P = C, \qquad \text{or } \frac{p}{\sqrt{(1+p^2)}} = C; \qquad \therefore p = \frac{C}{\sqrt{(1-C^2)}} = C';$$

and it is a straight line. Since the extremities are invariable, $\delta x_{,}$, $\delta x_{,,}$, $\delta y_{,}$, $\delta y_{,,}$, are all $=0$; and, therefore, the first part of δu vanishes without any relation between V, P, &c.

13. It is required to find the shortest line that can be drawn from one given curve to another given curve.

$$\text{Here } u = \int_x \sqrt{(1+p^2)}, \qquad V = \sqrt{(1+p^2)},$$

and, as before,

$$\frac{d(P)}{dx} = 0, \qquad P \text{ or } \frac{p}{\sqrt{(1+p^2)}} = C, \qquad p = \frac{C}{\sqrt{(1-C^2)}} = C'.$$

The other equation becomes

$$V_{,,}\delta x_{,,} - V_{,}\delta x_{,} + P_{,,}(\delta y_{,,} - p_{,,}\delta x_{,,}) - P_{,}(\delta y_{,} - p_{,}\delta x_{,}) = 0,$$

$$\text{or } P_{,,}\delta y_{,,} + (V_{,,} - P_{,,}p_{,,})\,\delta x_{,,} - P_{,}\delta y_{,} - (V_{,} - P_{,}p_{,})\,\delta x_{,} = 0.$$

Let AB, (fig. 2), be the line required; then, in changing AB to $A'B'$, as the point A' is to be found in the curve Aa, the ratio of δx to δy, or of At to tA', must be the same as the ratio of the increment of x to the increment of y, in the curve Aa. Let $\frac{dy}{dx} = m$ be the differential equation of the curve Aa; then $\delta y_{,}$ must $= m\delta x_{,}$. In the same manner, if $\frac{dy}{dx} = n$ be the differential equation of the curve Bb, $\delta y_{,,} = n\delta x_{,,}$. Also,

$$p_{,} = C'; \qquad p_{,,} = C';$$

$$V_{,} - P_{,}p_{,} = \sqrt{(1+p_{,}^2)} - \frac{p_{,}^2}{\sqrt{(1+p_{,}^2)}} = \frac{1}{\sqrt{(1+p_{,}^2)}} = \frac{1}{\sqrt{(1+C'^2)}};$$

$$V_{,,} - P_{,,}p_{,,} = \frac{1}{\sqrt{(1+C'^2)}};$$

$$\text{and } P_{,} = P_{,,} = \frac{C'}{\sqrt{(1+C'^2)}}.$$

Substituting, the equation becomes

$$\left\{\frac{nC'}{\sqrt{(1+C'^2)}} + \frac{1}{\sqrt{(1+C'^2)}}\right\}\delta x_{,,} - \left\{\frac{mC'}{\sqrt{(1+C'^2)}} + \frac{1}{\sqrt{(1+C'^2)}}\right\}\delta x_{,} = 0,$$

and since we cannot assign any relation which will, in all cases, subsist between $\delta x_{,}$, and $\delta x_{,,}$, we must make each coefficient $= 0$,

$$\text{or } 1 + nC' = 0, \qquad 1 + mC' = 0.$$

These equations shew, that the line required must cut both the given curves at right angles.

14. It is required to find the curve of quickest descent from one given point to another given point.

Let the higher point be made the origin of co-ordinates.

Here $t = \int_x \frac{\sqrt{(1+p^2)}}{\sqrt{(2gy)}}$; and, therefore, $u = \int_x \frac{\sqrt{(1+p^2)}}{\sqrt{(y)}}$ must be a minimum; $V = \frac{\sqrt{(1+p^2)}}{\sqrt{(y)}}$; and as this contains only y and p, we must have, by (11),

$$V = Pp + \frac{1}{C}.$$

$$\text{Now } P = \frac{p}{\sqrt{(y)}\sqrt{(1+p^2)}};$$

$$\therefore \frac{\sqrt{(1+p^2)}}{\sqrt{(y)}} = \frac{p^2}{\sqrt{(y)}\sqrt{(1+p^2)}} + \frac{1}{C},$$

$$\text{or } \sqrt{(y)} \cdot \sqrt{(1+p^2)} = C;$$

$$\therefore \frac{dx}{dy} = \frac{1}{p} = \frac{y}{\sqrt{(C^2 y - y^2)}};$$

$$\text{integrating, } x + C' = \frac{C^2}{2} \text{ ver sin}^{-1} \frac{2y}{C^2} - \sqrt{(C^2 y - y^2)};$$

the equation to a cycloid, whose base is horizontal, whose vertex is downwards, and whose cusp is at the higher of the given points.

15. It is required to find the curve of quickest descent from one given curve to another given curve, the velocity at every point being that acquired by falling from a given horizontal line.

Here, (as in the last), $V = \frac{\sqrt{(1+p^2)}}{\sqrt{(y)}}$; and the equation

$$N - \frac{d(P)}{dx} + \&c. = 0,$$

$$\text{gives } \sqrt{(y)} \sqrt{(1+p^2)} = C.$$

Also, $V - Pp = \frac{1}{C}$; and $P = \frac{p}{\sqrt{(y)}\sqrt{(1+p^2)}} = \frac{p}{C}$.

Hence, the equation for the limits becomes

$$\frac{p_{\prime\prime}}{C} \delta y_{\prime\prime} + \frac{1}{C} \delta x_{\prime\prime} - \frac{p_{\prime}}{C} \delta y_{\prime} - \frac{1}{C} \delta x_{\prime} = 0;$$

or, if $\frac{dy}{dx} = m$, $\frac{dy}{dx} = n$, be the differential equations to the given curves, $\delta y_{\prime}$ must $= m\,\delta x_{\prime}$, and $\delta y_{\prime\prime}$ must $= n\,\delta x_{\prime\prime}$, and

$$\frac{np_{\prime\prime} + 1}{C} \delta x_{\prime\prime} - \frac{mp_{\prime} + 1}{C} \delta x_{\prime} = 0.$$

Since there is no relation between $\delta x_{,}$ and $\delta x_{,,}$, we must have

$$np_{,,} + 1 = 0, \quad mp_{,} + 1 = 0;$$

that is, the cycloid will cut both the curves at right angles. The cusp of the cycloid, as appears from the equation, will not be at the point at which the falling body leaves the first curve, but will be in the given horizontal line.

16. To find the form of a solid of revolution, that the resistance in moving through a fluid in the direction of its axis, on the usual suppositions, may be a minimum.

Let x be measured along the axis of revolution.

The resistance $\propto \int_x y \dfrac{p^3}{1+p^2}$;

$$\therefore \; V = \frac{yp^3}{1+p^2}; \quad P = y\,\frac{3p^2(1+p^2) - 2p^4}{(1+p^2)^2} = y \cdot \frac{3p^2 + p^4}{(1+p^2)^2};$$

and the equation $V = Pp + C$ gives

$$y\,\frac{p^3 + p^5}{(1+p^2)^2} = y\,\frac{3p^3 + p^5}{(1+p^2)^2} + C,$$

$$\text{or} \quad \frac{2yp^3}{(1+p^2)^2} + C = 0,$$

the differential equation to the curve, by whose revolution the solid required is generated.

17. To find the curve, which, of all that can be drawn between two given points, contains between the curve, the evolute, and the radii of curvature at the extremities, the least area.

If h be the increment of x at any point, the corresponding increment of the arc is ultimately $h\sqrt{(1+p^2)}$; the radius of curvature $= \dfrac{(1+p^2)^{\frac{3}{2}}}{-q}$; hence, the increment of the area is ultimately $h\dfrac{(1+p^2)^2}{-2q}$; $\therefore\ u=\int_x \dfrac{(1+p^2)^2}{q}$ must be a minimum.

$$\text{Here, } V=\frac{(1+p^2)^2}{q};$$

and as this involves only p and q,

$$V=Qq+Cp+C', \text{ by (11).}$$

$$\text{And } Q=-\frac{(1+p^2)^2}{q^2}; \quad \therefore \frac{2(1+p^2)^2}{q}=Cp+C'.$$

$$\text{Hence, } \frac{(Cp+C')q}{(1+p^2)^2}=2;$$

$$\text{integrating, } \frac{-C+C'p}{1+p^2}+C'\tan^{-1}p=4(x+a).$$

$$\text{And } \frac{(Cp^2+C'p)q}{(1+p^2)^2}=2p;$$

$$\text{integrating, } \frac{-C'-Cp}{1+p^2}+C.\tan^{-1}p=4(y+b).$$

Eliminating $\tan^{-1}p$,

$$\frac{C'^2-C^2+2CC'.p}{1+p^2}=4\{C(x+a)-C'(y+b)\}.$$

$$\text{Let } C=4f\sin\theta; \quad C'=4f\cos\theta;$$

substituting,

$$f.\frac{\cos 2\theta+\sin 2\theta.p}{1+p^2}=\sin\theta(x+a)-\cos\theta(y+b).$$

To transform this into a more simple equation, we will first change the origin of co-ordinates, preserving their direction; if x' and y' be measured from a point whose co-ordinates are $-a$ and $-b$,

$$\text{then, } x'=x+a, \quad y'=y+b, \quad p'=p,$$

and the equation becomes

$$f.\frac{\cos 2\theta + \sin 2\theta . p'}{1+p'^2} = \sin\theta . x' - \cos\theta . y'.$$

Now, take a new system of co-ordinates x'' and y'', (fig. 3) having the same origin as x' and y', but inclined to them at an angle θ; then

$$y' = x''\sin\theta - y''\cos\theta; \quad x' = x''\cos\theta + y''\sin\theta;$$

$$p' = \frac{dy'}{dx'} = \frac{\frac{dy'}{dx''}}{\frac{dx'}{dx''}} = \frac{\sin\theta - \cos\theta . p''}{\cos\theta + \sin\theta . p''};$$

$$\text{and substituting, } f\,\frac{\cos^2\theta - \sin^2\theta . p''^2}{1+p''^2} = y'';$$

$$\therefore \quad \frac{f}{1+p''^2} = y'' + f\sin^2\theta, \quad \text{or} \quad (y'' + f.\sin^2\theta)(1+p''^2) = f,$$

the equation to a cycloid. If the position of the tangents at the extreme points be given, the constants must be determined, so as to make the cycloid pass through the given points, and touch the given tangents. If the extreme points only be fixed, leaving the directions of the tangents indeterminate, $\delta x_{,}$, $\delta y_{,}$, $\delta x_{,,}$, $\delta y_{,,}$, are $= 0$; and the equation

$$V_{,,}\delta x_{,,} - V_{,}\delta x_{,} + (\delta y_{,,} - p_{,,}\delta x_{,,}) . \left\{P_{,,} - \frac{d(Q_{,,})}{dx_{,,}}\right\}$$

$$- (\delta y_{,} - p_{,}\delta x_{,})\left\{P_{,} - \frac{d(Q_{,})}{dx_{,}}\right\}$$

$$+ (\delta p_{,,} - q_{,,}\delta x_{,,})\,Q_{,,} - (\delta p_{,} - q_{,}\delta x_{,})\,Q_{,} = 0,$$

is reduced to

$$Q_{,,}\delta p_{,,} - Q_{,}\delta p_{,} = 0.$$

Let $\phi_{,}$ and $\phi_{,,}$ be the angles which the directions of the tangents at the extreme points make with the axis of x; then

$$p_{,,} = \tan\phi_{,,}; \quad \delta p_{,,} = (1+p_{,,}^2)\,\delta\phi_{,,}; \quad \delta p_{,} = (1+p_{,}^2)\,\delta\phi_{,};$$

$$\text{also } Q_{,,} = \frac{-(1+p_{,,}^2)^2}{q_{,,}^2}, \quad Q_{,} = \frac{-(1+p_{,}^2)^2}{q_{,}^2};$$

and making equal to nothing the coefficients of $\delta\phi_{,,}$ and $\delta\phi_{,}$, we find

$$\left\{\frac{(1+p_{,,}^2)^{\frac{3}{2}}}{q_{,,}}\right\}^2 = 0, \quad \left\{\frac{(1+p_{,}^2)^{\frac{3}{2}}}{q_{,}}\right\}^2 = 0;$$

that is, at the extremities, the radii of curvature are each = 0; therefore these points are cusps; therefore the curve is a complete cycloid.

18. In all the examples above, we have supposed, as is commonly the case, that V does not involve the limits of the integral. But it may happen, that V will involve the values of x, y, p, &c. at the limits. In that case we must recur to the investigation in (8); instead of giving V' the value which it has there, we must put

$$V' = V + \frac{dV}{dx_{,}}\delta x_{,} + \frac{dV}{dy_{,}}\delta y_{,} + \text{\&c.}$$

$$+ M\delta x + N\delta y + P\delta p + \text{\&c.}$$

By going through the same operation as is there performed, we find δu

$$= V_{,,}\delta x_{,,} - V_{,}\delta x_{,}$$

$$+ (\delta y_{,,} - p_{,,}\delta x_{,,})\left\{P_{,,} - \frac{d(Q_{,,})}{dx_{,,}} + \text{\&c.}\right\}$$

$$- (\delta y_{,} - p_{,}\delta x_{,})\left\{P_{,} - \frac{d(Q_{,})}{dx_{,}} - \text{\&c.}\right\}$$

$$+ \text{\&c.}$$

$$+ \int_x \omega \left\{N - \frac{d(P)}{dx} + \frac{d^2(Q)}{dx^2} - \text{\&c.}\right\}$$

$$+ \int_x \left(\frac{dV}{dx_{,}}\delta x_{,} + \frac{dV}{dy_{,}}\delta y_{,} + \text{\&c.}\right).$$

Now, $\delta x_{,}$, $\delta y_{,}$, &c., are not functions of x, generally, but of the limiting values of x; and, therefore, in the integration they may be considered constant with respect to x, and may be put before the integral sign as multipliers. The last line, therefore, is not to be connected with that immediately preceding, but with the former parts. Thus we have δu

$$= V_{,,}\delta x_{,,} - V_{,}\delta x_{,} + \delta x_{,} \int_x \frac{dV}{dx_{,}} + \delta y_{,} \int_x \frac{dV}{dy_{,}} + \&c.$$

$$+ (\delta y_{,,} - p_{,,}\delta x_{,,}) \cdot (P_{,,} - \&c.) - (\delta y_{,} - p_{,}\delta x_{,}) \cdot (P_{,} - \&c.)$$

$$+ \&c.$$

$$+ \int_x \omega \left\{ N - \frac{d(P)}{dx} + \&c. \right\}.$$

19. Suppose, for instance, it were required to find the curve of quickest descent from one given curve to another given curve, the motion being supposed to commence at the first.

$$\text{Here, } t = \int_x \frac{\sqrt{(1+p^2)}}{\sqrt{\{2g(y-y_{,})\}}};$$

$$\text{let } u = \int_x \frac{\sqrt{(1+p^2)}}{\sqrt{(y-y_{,})}}; \text{ then } V = \frac{\sqrt{(1+p^2)}}{\sqrt{(y-y_{,})}};$$

$$P = \frac{p}{\sqrt{(1+p^2)}\sqrt{(y-y_{,})}};$$

and by (11),

$$V = Pp + \frac{1}{C}, \text{ or } \sqrt{(y-y_{,})}\sqrt{(1+p^2)} = C,$$

and the curve is a cycloid in the same position as before, its cusp being at the point from which the motion begins. Now $x_{,}$, $x_{,,}$, $y_{,,}$, do not enter into V: and to integrate $\frac{dV}{dy_{,}}$, we observe that, from the form of V in this particular case,

$$\frac{dV}{dy_{,}} = -\frac{dV}{dy} = -N = -\frac{d(P)}{dx} \left\{ \text{since } N - \frac{d(P)}{dx} = 0 \right\};$$

$$\therefore \int_x \frac{dV}{dy_{,}} = -P:$$

which, taken between its limits, is $-P_{,,}+P_{,}$. Hence, we have for the first part of δu,

$$V_{,,}\delta x_{,,}+P_{,,}(\delta y_{,,}-p_{,,}\delta x_{,,})-V_{,}\delta x_{,}-P_{,}(\delta y_{,}-p_{,}\delta x_{,})+(P_{,}-P_{,,})\delta y_{,}$$

$$=(V_{,,}-P_{,,}p_{,,})\,\delta x_{,,}+P_{,,}\delta y_{,,}-(V_{,}-P_{,}p_{,})\,\delta x_{,}-P_{,,}\delta y_{,}$$

$$=\frac{1}{C}\,\delta x_{,,}+\frac{p_{,,}}{C}\,\delta y_{,,}-\frac{1}{C}\,\delta x_{,}-\frac{p_{,,}}{C}\,\delta y_{,}.$$

And, if the equations to the limiting curves be

$$\frac{dy}{dx}=m, \quad \frac{dy}{dx}=n,$$

we must, as before, put $\delta y_{,}=m\,\delta x_{,}$, $\delta y_{,,}=n\,\delta x_{,,}$, and we find

$$(1+p_{,,}n)\,\delta x_{,,}-(1+p_{,,}m)\,\delta x_{,}=0:$$

then, since $\delta x_{,}$ and $\delta x_{,,}$ are indeterminate, we shall have

$$1+p_{,,}n=0,$$

which shews that the cycloid cuts the second curve at right angles, and

$$1+p_{,,}m=0,$$

which gives $m=n$, and shews that at the points where the cycloid meets both curves, their tangents are parallel.

20. *There yet remains a very extensive class of problems: those in which the value of one function (v) is given, while another (u) is to be made a maximum or minimum.* For instance; it is required to find the form of a curve, whose length is given, that the area contained by it may be the greatest possible. If we take the variation of u as in (9), we must not as in (10) make the two parts, of which it consists, separately $=0$; for it is not necessary that

δu be $= 0$ for any values whatever of δx and δy, but only for such values as make $\delta v = 0$; a condition which would by that process be entirely neglected. If, however, we make $\delta u + a\delta v = 0$ (a being a constant to be determined), on the supposition that δx and δy have any values whatever, then, the values of δx and δy which make $\delta v = 0$, and no other, will make $\delta u = 0$. And, at the same time, an additional constant is introduced into the equation between x and y, which enables us to give to v the value required in the statement of the problem. Hence, when u is to be made a maximum or minimum, while the value of v is constant, we must make $\delta(u + av) = 0$, proceeding in the same manner as in the simpler cases. And, if it were required that u should be a maximum or minimum, the values of the functions v and w being constant, the same reasoning would shew that we must make $\delta(u + av + bw) = 0$; and so for any number of functions.

21. Taking the instance above, the area $= \int_x y$: the length $= \int_x \sqrt{(1 + p^2)}$;

$$\therefore \quad u + av = \int_x \{y + a\sqrt{(1 + p^2)}\}; \quad \therefore \quad V = y + a\sqrt{(1 + p^2)}:$$

$$\text{and } V = Pp + C, \text{ or } y + a\sqrt{(1 + p^2)} = \frac{ap^2}{\sqrt{(1 + p^2)}} + C;$$

$$\therefore \quad \frac{a}{\sqrt{(1 + p^2)}} = C - y; \quad \frac{dx}{dy} = \frac{1}{p} = \frac{C - y}{\sqrt{\{a^2 - (C - y)^2\}}},$$

$$x = C' - \sqrt{\{a^2 - (C - y)^2\}}; \quad \therefore \quad (x - C')^2 + (y - C)^2 = a^2,$$

the equation to a circular arc. If the limits be fixed, the values of the constant must be determined so as to make the length of the arc equal to the given length, and to make it pass through the two given points. If the limits be not fixed, suppose the first and last ordinates *AM*, *BN*, fig. 4, to be given, their abscissæ being not given; then, since $\delta y_{,} = 0$, $\delta y_{,,} = 0$, the equation for the limits reduces itself to

$$(V_{,,} - P_{,,}p_{,,})\,\delta x_{,,} - (V_{,} - P_{,}p_{,})\,\delta x_{,} = 0;$$

from which

$$V_{,,} - P_{,,}p_{,,} = 0, \quad V_{,} - P_{,}p_{,} = 0.$$

Since $V - Pp = C$, these equations are satisfied by making $C = 0$;

$$\therefore \ (x - C')^2 + y^2 = a^2:$$

that is, when the lengths of the ordinates AM, BN, and the arc AB are given, the area $AMNB$ is a maximum if AB be a circular arc, whose center C is in the line MN. If AM, BN, each $= 0$, MN is the chord of AB: and, it appears that the curve which, with a given length, contains between its chord and its arc the greatest area, is the semi-circle.

22. Given the length of a curve, to find its form, that its center of gravity may be the lowest possible. Let the length $= b$; then, the depth of the center of gravity $= \frac{1}{b} \int_x y \sqrt{(1 + p^2)}$: the length $= \int_x \sqrt{(1 + p^2)}$; hence

$$V = a\sqrt{(1 + p^2)} + \frac{y\sqrt{(1 + p^2)}}{b};$$

and, making $V = Pp + C$, $\dfrac{a + \dfrac{y}{b}}{\sqrt{(1 + p^2)}} = C$,

whence $x = bC \,.\log\{y + ba + \sqrt{(\overline{y + ba}|^2 - b^2C^2)}\} + C'$;

the equation to the catenary. If the extreme points be not fixed, but move on curves, it will be found that the catenary will cut the curves at right angles.

23. Given the surface of a solid of revolution, to find its nature, that the solid content may be a maximum. Let x be measured along the axis of revolution;

the surface $= 2\pi \int_x y \sqrt{(1 + p^2)}$; the solidity $= \pi \int_x y^2$;

$\therefore \ V = y^2 + ay\sqrt{(1 + p^2)}$; $V = Pp + C$ gives $\dfrac{ay}{\sqrt{(1 + p^2)}} = C - y^2$,

$$\text{whence } \frac{dx}{dy} = \frac{C - y^2}{\sqrt{\{a^2y^2 - (C - y^2)^2\}}}.$$

If the first and last ordinates of the generating curve, and the distance between them, be fixed, this equation (supposing it integrated) is sufficient; the three constants which the integral equation will contain must be determined so as to make the first and last ordinates, and the surface, equal to the given quantities. If the distance between the first and last ordinates be not given, then $\delta y_{,} = 0$, $\delta y_{,,} = 0$, and the equation for the limits is

$$(V_{,,} - P_{,,}p_{,,})\,\delta x_{,,} - (V_{,} - P_{,}p_{,})\,\delta x_{,} = 0, \quad \text{or} \quad C\delta x_{,,} - C\delta x_{,} = 0;$$

and, as there is now no relation between $\delta x_{,,}$ and $\delta x_{,}$, $C = 0$.

$$\text{Hence, } \frac{dx}{dy} = \frac{-y^2}{\sqrt{(a^2y^2 - y^4)}} = \frac{-y}{\sqrt{(a^2 - y^2)}};$$

$$\therefore\ x + C' = \sqrt{(a^2 - y^2)},$$

the equation to a circle, whose center is in the line of abscissæ: and the solid is, therefore, a portion of a sphere included between two planes perpendicular to a diameter. If the first and last ordinates be 0, the solid is a whole sphere.

24. Given the whole surface of a solid of revolution (including the circular ends), to find its form, that the solid content may be a maximum. The solidity $= \pi \int_x y^2$: the surface

$$= 2\pi \int_x y \sqrt{(1 + p^2)} + \pi (y_{,}^2 + y_{,,}^2);$$

hence, we must make

$$\delta \left\{ \int_x y^2 + a \int_x y \sqrt{(1 + p^2)} + \frac{a}{2} (y_{,}^2 + y_{,,}^2) \right\} = 0.$$

The part under the sign of integration will be the same as in the last problem, and the equation derived from it will be the same, or

$$\frac{dx}{dy} = \frac{C - y^2}{\sqrt{\{a^2y^2 - (C - y^2)\}^2}}.$$

The part depending on the variation of the limits, besides the usual terms

$$(V_{\prime\prime} - P_{\prime\prime} p_{\prime\prime})\, \delta x_{\prime\prime} - (V_{\prime} - P_{\prime} p_{\prime})\, \delta x_{\prime} + P_{\prime\prime} \delta y_{\prime\prime} - P_{\prime} \delta y_{\prime}$$

must have the terms which express the variation of $\frac{a}{2}(y_{\prime}^2 + y_{\prime\prime}^2)$,

that is, $a(y_{\prime} \delta y_{\prime} + y_{\prime\prime} \delta y_{\prime\prime})$.

Making the whole $= 0$,

$$C\delta x_{\prime\prime} - C\delta x_{\prime} + a y_{\prime\prime} \left\{1 + \frac{p_{\prime\prime}}{\sqrt{(1 + p_{\prime\prime}^2)}}\right\} \delta y_{\prime\prime}$$

$$+ a y_{\prime} \left\{1 - \frac{p_{\prime}}{\sqrt{(1 + p_{\prime}^2)}}\right\} \delta y_{\prime} = 0.$$

The first two coefficients shew that $C = 0$, and that the solid is part of a sphere; the third and fourth require that $y_{\prime\prime} = 0$, $y_{\prime} = 0$; or that $\frac{1}{p_{\prime\prime}} = 0$, $\frac{1}{p_{\prime}} = 0$, which agree with the former; hence, the solid is an entire sphere.

25. Required the curve of quickest descent from one given point to another given point, the length of the curve being given.

$$\text{Here } V = \frac{a\sqrt{(1 + p^2)}}{\sqrt{(y)}} + \sqrt{(1 + p^2)};$$

$$\text{and } V = Pp + C \text{ gives } \frac{a}{\sqrt{(y)}} + 1 = C\sqrt{(1 + p^2)}:$$

the differential equation to the curve. The constants must be determined so as to make the length of the curve equal to the given quantity, and to make the curve pass through both points.

26. Given the mass of a solid of revolution, required its form, that the attraction upon a point in the axis may be a maximum. Let the solid be divided into slices by planes perpendicular to the axis of revolution; then, since the attraction of a circle, whose thickness is h and radius y, upon a point at the distance x from its center, is ultimately

$$2\pi h \left\{1 - \frac{x}{\sqrt{(x^2 + y^2)}}\right\},$$

the attraction of the solid will

$$= 2\pi \int_x \left\{1 - \frac{x}{\sqrt{(x^2 + y^2)}}\right\}.$$

The solidity $= \pi \int_x y^2$. Hence,

$$V = 1 - \frac{x}{\sqrt{(x^2 + y^2)}} + ay^2;$$

the equation $N - \frac{d(P)}{dx} + \&c. = 0$ is reduced to $N = 0$, or

$$\frac{xy}{(x^2 + y^2)^{\frac{3}{2}}} + 2ay = 0, \quad \text{or } x + 2a(x^2 + y^2)^{\frac{3}{2}} = 0,$$

the equation to the curve by the revolution of which the solid is generated. If the first and last values of x be given, a must be determined, so that the included solid = the given solidity. If the first and last values of x are indeterminate, the equation of the limits is

$$V_{\prime\prime}\delta x_{\prime\prime} - V_{\prime}\delta x_{\prime} = 0, \quad \text{or } V_{\prime\prime} = 0, \; V_{\prime} = 0, \quad \therefore \; y_{\prime\prime} = 0, \; y_{\prime} = 0;$$

and the solid must be that generated by the revolution of the whole curve, whose equation is

$$x + 2a(x^2 + y^2)^{\frac{3}{2}} = 0,$$

a being determined so that the whole solidity = the given solidity.

27. The rules and examples above will serve to elucidate all the cases that commonly occur; and the same principles may easily be extended to more difficult problems. For investigations of the cases where u is given by double integration, or by the solution of a differential equation, or where V is a function of two or more independent variables, or where V depends upon another integral, the reader is referred to Lacroix, *Traité du Calcul Différentiel et du Calcul Intégral*, Tome 2eme, or to Woodhouse's *Isoperimetrical Problems*.

ON THE

UNDULATORY THEORY OF OPTICS.

ON UNDULATIONS GENERALLY.

PROP. 1. To explain the nature of an Undulation.

1. The characteristic of an undulation is, the continued transmission in one direction of a *relative state* of particles amongst each other, while the motion of each particle separately considered is a reciprocating motion. The disturbance of the particles from their state of rest, and their motion, may be in any direction whatever.

2. For example: in fig. 1, let the line (a) represent a number of particles in their position of rest: and suppose that in consequence of a disturbance they are at a given time T in the position (β): at the time $T+\frac{\tau}{4}$, in the position (γ): at the time $T+\frac{2\tau}{4}$, in the position (δ): at the time $T+\frac{3\tau}{4}$, in the position (ϵ): and at the time $T+\tau$, in the position (ζ): and in intermediate positions at times intermediate to these. At the time T the particles are in the state of greatest condensation about a, a', and a''. Suppose we fix our attention on one of these condensed groups, as for instance that of which a' is the center. At the time $T+\frac{\tau}{4}$ the center of the condensed group has glided from a' to d', not by the motion of all the particles in that direction, but by such a difference of motions that the particles about a' are not so close together as they were, and the particles

about d' are closer together than they were. At the time $T + \frac{2\tau}{4}$ the point of greatest condensation has advanced to g', precisely the point where at the time T there was the least condensation: at the time $T + \frac{3\tau}{4}$, it has advanced to k': and at the time $T + \tau$, to a''. The particles are now, it may be observed, in just the same state as at the time T, for a'' was then the center of a condensed group. After this, every thing goes on in the same manner, beginning at the time $T + \tau$, as it did beginning at the time T. All that we have said with respect to the condensed mass about a' applies to those about a, a'', and a'''. Now if these motions were really going on before our eyes, we should see several *condensations* (not the condensed particles) passing uniformly and continuously from the left to the right of the line of particles.

But if we fix our attention on any one of these particles, we shall see that it has a reciprocating or oscillating motion. The particle a is advancing from T to $T + \frac{\tau}{4}$ when it has attained its greatest advance: it recedes then to $T + \frac{3\tau}{4}$: it then advances again. The particle d advances from T (when it is at its minimum advance) to $T + \frac{2\tau}{4}$: it then recedes to $T + \tau$. The particle g recedes to $T + \frac{\tau}{4}$, then advances to $T + \frac{3\tau}{4}$, then recedes. And so for the others. The varying state of particles which we have here supposed, satisfies therefore the conditions mentioned in (1), and therefore this is an instance of undulation, the motion of every particle being backwards and forwards in the same line as the direction of transmission of the wave*.

* This is the kind of undulation which in the air produces sound, and is the only kind which, till within a few years, was used for the explanation of the phenomena of Optics.

3. As another example, let (β), (γ), (δ), (ϵ), (ζ), of fig. 2, represent successive states of the particles which when at rest were in the position (a). If we fix our attention on one of the most elevated parts, as for instance k, at T, we find that at $T + \frac{\tau}{4}$ the elevation has passed to a'; at $T + \frac{2\tau}{4}$ to d': &c.: though the particles have had no motion whatever in that direction. And if these motions were actually before us, we should see several elevations passing uniformly and continuously from the left to the right. But if we fixed our attention on any one particle, we should see that it has an oscillating motion above and below the line. The particle a for instance, is at its greatest elevation at $T + \frac{\tau}{4}$, and at its greatest depression at $T + \frac{3\tau}{4}$: d is at its greatest depression at T, at its greatest elevation at $T + \frac{2\tau}{4}$, and at its greatest depression at $T + \tau$: and so for the others. This varying state of particles is therefore another instance of undulation, the motion of every particle being at right angles to the direction of transmission of the wave.*

We might conceive more complicated cases of undulation, as when the motion of the particles is compounded of the two motions supposed in these two cases; or when there is one motion similar to that represented in fig. 2 in the plane of the paper, and another perpendicular to that plane, &c. The last of these suppositions is that to which we shall hereafter refer the phænomena of polarization, and of Optics in general.

Prop. 2. The length of a wave does not depend on the extent of vibration of each particle.

4. It is easily seen that the interval between corresponding points of two waves of condensation in fig. 1 (which is

* This is nearly the kind of undulation which takes place on the surface of deep water in a calm.

the distance from a to a', a' to a'', &c. at T, or the distance from d to d', d' to d'', &c. at $T + \frac{\tau}{4}$, &c.) is wholly independent of the extent of vibration of each particle. For if each particle vibrated only half as far as is now supposed, still at T, a would be a point where the particles are most condensed, and a' would be the next point where they are most condensed, and a'' the next, &c. The interval between similar points of two waves (which we shall call the *length of a wave*, and shall always denote by the letter λ) would be the same as at present: the only difference would be that the particles about a, a', &c. would not be so closely condensed, nor those about g, g', &c., so widely separated as at present. Similarly the length of a wave in fig. 2 would be unaltered if the vibration of the particles were altered in any ratio: the only difference would be that the elevation of the high points and the depression of the low points would be altered in that ratio.

PROP. 3. The length of a wave depends on the velocity of transmission, and on the time of vibration of each particle.

5. In the cases both of fig. 1 and of fig. 2 (and in every other conceivable case of a continued sequence of waves) we see that every particle has returned to the same state at $T + \tau$ as at T, that is, that the vibration of every particle is completed in the time τ. But in this time the wave has appeared to glide over a space equal to the interval between corresponding points of two waves, or λ. Hence we find,

Space described by the wave in the time of vibration of a particle $= \lambda$.

$$\text{Velocity of wave} = \frac{\lambda}{\text{time of vibration of a particle}}.$$

PROP. 4. To express algebraically the transmission of an undulation.

6. The quantity for which we shall seek an expression is, the distance of any point from its point of rest, in a

function of the time and of the distance of that point of rest from some fixed point. Let x be the original distance of any point in the line (a) from some fixed point: to find an expression for its disturbance at the time t, in a function of x and t, consistent with the conditions of an undulation. By the original description of an undulation (1), putting v for the velocity of the wave's transmission, it is easily seen that whatever be the state of disturbance of a particle whose original ordinate is x at the time t, the same state of disturbance must hold at the time $t + t'$ for a particle whose original ordinate is $x + vt'$. Or if ϕ express the form of the function,

$$\phi(x, t) \text{ must} = \phi(\overline{x + vt'}, \overline{t + t'}),$$

whatever be the value of t'. It will be found on trial that $\phi(vt - x)$ satisfies this condition, ϕ being any function whatever. For putting $t + t'$ for t, and $x + vt'$ for x, it becomes

$$\phi(v . \overline{t + t'} - \overline{x + vt'}) = \phi(vt - x),$$

the same as before. But it may be found analytically thus. Expanding the second side we have

$$\phi(x, t) = \phi(x, t) + \frac{d.\phi(x, t)}{dx} vt' + \frac{d.\phi(x, t)}{dt} t' + \&c.,$$

$$\text{or } v\frac{d.\phi(x, t)}{dx} + \frac{d.\phi(x, t)}{dt} = 0,$$

the general solution of which gives $\phi(x, t) = \phi(vt - x)$*.

7. This expression however is too general to be of much use to us, and we will choose a particular form that will be more convenient. Suppose we fix on this condition to determine the form of the function: the vibration of each particle shall follow the same law as the vibration of a cycloidal pendulum. The distance of a cycloidal pendulum from its place of rest is expressed by

---

* This is the expression found, by investigation from *mechanical* principles, for the disturbance of the particles of air when sound passes along a tube of uniform bore, or for the disturbance of an elastic string (as that of a musical instrument) fixed at both ends.

$$a \sin \left\{t\sqrt{\left(\frac{g}{l}\right)} + C\right\} \text{ or } a \sin (nt + C).$$

The required function then for the disturbance of a particle is

$$a \sin \left\{\frac{n}{v} \cdot (vt - x) + A\right\}.$$

For while we consider the motion of a single particle only, x is constant, and the expression is

$$a \sin (nt + C), \text{ where } C = A - \frac{nx}{v}.$$

At the same time it is a function of $vt - x$, and therefore satisfies the condition requisite for an undulation. We will therefore assume as the expression for the disturbance

$$a \sin (nt - \frac{nx}{v} + A).$$

But it is plain that, without any loss of generality, we may get rid of A by altering the origin of time from which t is reckoned, or the origin of linear measure from which x is reckoned. We may therefore take

$$a \cdot \sin \left(nt - \frac{nx}{v}\right)$$

as the expression for the disturbance when one undulation only (consisting of an indefinite number of similar waves) is considered.

8. A form somewhat more convenient may be given thus. The expression

$$a \cdot \sin (nt - \frac{nx}{v} + A)$$

goes through all its periodical values while nt increases by 2π, or while t increases by $\frac{2\pi}{n}$. $\frac{2\pi}{n}$ is therefore the time of vibration of a particle. But by (5)

$$v = \frac{\lambda}{\text{time of vibration}} = \frac{\lambda}{\frac{2\pi}{n}} = \frac{\lambda n}{2\pi}.$$

Consequently $n = \frac{2\pi v}{\lambda}$, and therefore the disturbance

$$= a \sin \left\{ \frac{2\pi}{\lambda} (vt - x) + A \right\}^*,$$

for which we may put

$$a \,.\, \sin \left\{ \frac{2\pi}{\lambda} (vt - x) \right\}$$

when a single undulation only is considered. It is to be observed that a is the maximum vibration of any particle.

PROP. 5. To explain the interference of undulations.

9. By *interference* is meant the co-existence of two undulations in which the length of a wave is the same. The conception of interference is not in any circumstances easy†, and it is more particularly difficult with regard to Physical Optics, from our ignorance of the physical causes to which the undulation is due.

10. ‡If we investigate, from the known properties of air, the motion of the particles (supposed parallel to a fixed line), we find this differential equation for the disturbance of a particle

$$\frac{d^2 \,.\, X}{dt^2} - v^2 \,.\, \frac{d^2 \,.\, X}{dx^2} = 0$$

---

* This is the form of the function tacitly assumed by Newton for the disturbance of particles of air, in his investigation of the velocity of sound. (*Principia*, Lib. II, Prop. 47).

† The simplest illustration is perhaps to be found in the crossing of two waves on the surface of water, each of which affects the surface in the same manner as if the other were not there. If we conceive two series of waves, produced by agitating the surface at two points, to spread in circular forms with equal and uniform velocities, and if one agitation was created a little before the other, so that the wave proceeding from one has proceeded as far (in a given direction) as the hollow between two waves proceeding from the other, then it may be imagined that at every point where this holds, the elevation of one wave may exactly fill up the hollow of the other, and the surface will be, in fact, undisturbed.

‡ The reader who is not familiar with the investigation of the problem of Sound may omit the next three articles.

(v^2 being a constant, $= mgH$ in the common notation, and X being the disturbance from the state of rest), and the solution is

$$X = \phi(vt - x) + \psi(vt + x),$$

where the form of the functions is to be determined by the initial circumstances. Or if we suppose the wave of air to move only in one direction, the expression for the disturbance will be $\phi(vt - x)$. And this may be divided into several different expressions,

$$\psi(vt - x) + \chi(vt - x) + \upsilon(vt - x) + \&c.$$

where the form of each is to be determined by the initial circumstances, or by the cause of the undulation. If there was only a single original cause for the undulations, there would be only a single term $\psi(vt - x)$ to be preserved. But if there were several distinct original causes for the undulations, there would be a single term corresponding to each of these to be preserved, and the whole disturbance would be the sum of all these terms. And it is particularly to be remarked, that the whole disturbance thus found as the effect of all the original causes together, is precisely the sum (with their proper signs) of a number of disturbances, each of which would have been produced by one of the original causes acting separately.

11. Now if we examine to what this property of the solution of the differential equation (namely that it can be broken up into several parts all similar to each other and to the whole) is due, we find it is owing to the circumstance that the differential coefficients of u were raised only to the first power in the equation, or (to express it in other words) that the equation was linear. For the differential coefficient of the sum of a number of functions is the same as the sum of the differential coefficients: but the square of the differential coefficient of a sum of functions is not the same as the sum of their squares, &c. If then the differential coefficients (and the unknown quantity itself if it enters in the equation) be all of the first dimension, the substitution of a sum of functions is the same as the sum of their substitutions separately, and therefore if each of those functions satisfies the equation, their sum will

satisfy the equation. But if they are raised to a higher power, the substitution of the sum is not the same as the sum of the substitutions, and therefore if each function satisfies the equation, their sum will not.

12. If now we retrace the steps of the investigation for air, it will be seen that the *linearity* of the differential equation depends upon this physical fact, that upon altering by a small quantity the relative position of particles, the forces which they exert undergo variations very nearly proportional to that small quantity. And in any other case where this holds, the equations will be linear; and the wave-disturbance of any particle, produced by a number of agitating causes, will be the sum of all the wave-disturbances which these causes would singly have produced. We can hardly conceive any law of constitution of a medium in which undulations are propagated, where this does not hold, and we shall therefore suppose it to be true for light.

13. Taking it then as a fact that the disturbance of every particle produced by two co-existent undulations will be the sum of the disturbances which they would produce separately, we will consider the nature of the disturbance produced by the superposition of two such undulations as those treated of in (7) and (8), each of which is represented geometrically by fig. 1, if the vibrations are in the direction of the wave's transmission, and by fig. 2, if they are perpendicular to that direction. For convenience of figure, we will suppose them of the latter class: but all that we say will apply as well to the former. We will suppose the length of a wave the same in both undulations. In fig. 3, let the Italic letters of (a) represent the state of an undulation, at the time T, where the law of vibration is

$$a \,.\, \sin \frac{2\pi}{\lambda}(vt - x + A),$$

and let (β) represent the state of another undulation at the same time where the law of vibration is

$$b \sin \frac{2\pi}{\lambda}(vt - x + B).$$

If from any point in (α) we measure upwards a distance equal to the elevation of the corresponding point of (β), or measure downwards a distance equal to the depression of the corresponding point of (β), we shall determine the position of the Roman letters. Their distances from the straight line represent the effect of the superposition of the two undulations. This is evidently an undulation of the same kind, and with waves of the same length, as either of the others. But in the instance as we have supposed it, the addition of the undulation (β) to (α) has diminished the maximum vibration of the latter, and has made the maximum to exist at a different part of the line. Thus we see that the magnitude of vibration in an undulation may be diminished by the addition of another undulation transmitted in the same direction. This is a point of great importance, and deserves the reader's attentive consideration.

14. The geometrical figures which we have given are merely illustrations: the conclusion that we have arrived at will be more readily obtained from the algebraic expressions. Adding together the two disturbances

$$a \,.\, \sin\left\{\frac{2\pi}{\lambda}(vt - x) + A\right\} \text{ and } b \,.\, \sin\left\{\frac{2\pi}{\lambda}(vt - x) + B\right\},$$

we have for the whole disturbance

$$(a\cos A + b\cos B)\sin\left\{\frac{2\pi}{\lambda}(vt - x)\right\}$$

$$+ (a\sin A + b\sin B)\cos\left\{\frac{2\pi}{\lambda}(vt - x)\right\}.$$

This may be put under the form

$$c\sin\left\{\frac{2\pi}{\lambda}(vt - x) + C\right\},$$

if $c\cos C = a\cos A + b\cos B$, $c\sin C = a\sin A + b\sin B$.

The sum of the squares of these equations gives

$$c^2 = a^2 + b^2 + 2ab\cos(A - B):$$

and the quotient of the latter by the former gives

$$\tan C = \frac{a \sin A + b \sin B}{a \cos A + b \cos B}.$$

The form of the expression

$$c \sin \left\{\frac{2\pi}{\lambda}(vt - x) + C\right\}$$

shews that the length of a wave is the same as in either of the undulations compounded, but the difference of value of C from A and B shews that the maximum of vibration for a given particle does not generally take place with the same value of t as for either of the undulations compounded. The magnitude of the maximum vibration, which is c or

$$\sqrt{\{a^2 + b^2 + 2ab \cos (A - B)\}}$$

depends on the value of $A - B$: its greatest value is $a + b$, when $A - B = 0$, and its least value is $a \sim b$, when $A \sim B = 180^0$. In these two cases C is equal to one at least of the two angles A and B.

PROP. 6. To examine the effects of interference of two equal and similar undulations: and to shew that when one is $(p + \frac{1}{2}) \times$ length of the wave behind the other (p being a whole number), they will destroy each other.

15. In the case of equal vibration, $a = b$. The value of c is then

$$\sqrt{\{2a^2 + 2a^2 \cos (A - B)\}} = 2a \cos \frac{A - B}{2},$$

$$\text{and } \tan C = \frac{\sin A + \sin B}{\cos A + \cos B} = \tan \frac{A + B}{2}, \text{ or } C = \frac{A + B}{2}.$$

If $A - B = 0$, the value of c is $2a$, and $C = A$: that is, if we add two such undulations as (β) and (ζ) (fig. 2), we shall have an undulation in which the maxima are at the same places, and the maximum vibration is double what it was before. With any other value of $A - B$, c is less than $2a$: and when $B = A \pm 180^0$, c is 0, that is, there is no motion whatever. To understand this clearly, we must consider what is meant by the expression

$$a \cdot \sin\left\{\frac{2\pi}{\lambda}(vt - x) + B\right\},$$

or, (in this case of destruction of the motion)

$$a \sin\left\{\frac{2\pi}{\lambda}(vt - x) + A \pm \pi\right\}.$$

This is the same as

$$a \sin\left\{\frac{2\pi}{\lambda}\left(vt - x \pm \frac{\lambda}{2}\right) + A\right\}.$$

Now this is exactly the same expression as

$$a \cdot \sin\left\{\frac{2\pi}{\lambda}(vt - x) + A\right\},$$

putting $x \mp \frac{\lambda}{2}$ instead of x. That is, the expression for the disturbance in this second undulation, if $B = A \pm 180^0$, is the same as that in the first, provided instead of x we take $x \mp \frac{\lambda}{2}$. That is, one of the undulations may be represented by the same construction as the other, provided we suppose it in advance or in arrear of the other by half the length of a wave. The undulations (β) and (δ), or (γ) and (ϵ) in figures 1 and 2 have this relation to one another. And it will very easily be seen in fig. 2, that if we compound (δ) with (β) by a process similar to that which we used in fig. 3, (13), the elevations of particles in (δ) will correspond to equal depressions in (β), and *vice versâ*, and consequently by their combination the particles will all be brought to their original position. The same will be true after the time $\frac{\tau}{4}$ when (β) has been changed to (γ), and at the same time (δ) has been changed to (ϵ); and at every other time: and therefore there will be continued rest. Thus we arrive at the extraordinary conclusion that one undulation may be absolutely destroyed by another with waves of the same length transmitted in the same direction, provided that the maxima of vibrations are equal, and that one follows the other by half the length of a wave. Since the retardation of a whole length of a wave, or two whole lengths, &c., produces no alteration in an undulation, it is plain that a retardation of $\frac{3\lambda}{2}$, $\frac{5\lambda}{2}$, &c. will produce the same effect as a retardation of $\frac{\lambda}{2}$; and thus two undulations will destroy each other if the maxima of vibration be the same and the waves be of the same length and transmitted in the same direction; and if one follow the other by $\frac{\lambda}{2}$, or $\frac{3\lambda}{2}$, or $\frac{5\lambda}{2}$, &c.

16. The reader is requested particularly to remark this apparently strange conclusion. It is of the greatest importance in Physical Optics, for the following reason. We shall refer hereafter to experiments which shew that the mixture of two pencils of light will produce darkness. This fact seems to defy any attempt at explanation on the supposition that light is occasioned by the emission of material particles. But in consequence of the conclusion at which we have just arrived, it is perfectly explicable on the supposition that light consists of a series of waves of either of the kinds mentioned in (2) and (3), transmitted by some medium which pervades space. It is only necessary for perfect agreement that, in the two pencils which mix, the waves of one precede those of the other by spaces which may be represented by

$$\frac{\lambda}{2}, \text{ or } \frac{3\lambda}{2}, \text{ or } \frac{5\lambda}{2}, \text{ \&c.},$$

which is found experimentally to be true. Any other hypothesis, however, from which the same conclusion could be deduced would be, *primâ facie*, equally entitled to our attention.

PROP. 7. To find the result of the interference of any number of waves.

17. We have shewn in (14) the method of compounding the effects of two waves: the effect of several is found in just the same manner. Suppose for instance, the disturbance produced by one undulation was expressed by

$$a \,.\, \sin\left\{\frac{2\pi}{\lambda}(vt - x) + A\right\}:$$

that of a second, by

$$b \,.\, \sin\left\{\frac{2\pi}{\lambda}(vt - x) + B\right\};$$

that of a third by

$$e \,.\, \sin\left\{\frac{2\pi}{\lambda}(vt - x) + E\right\}, \text{ \&c.}$$

The sum of these is

$$(a \cos A + b \cos B + e \cos E + \text{\&c.}) \sin \frac{2\pi}{\lambda}(vt - x)$$

$$+ (a \sin A + b \sin B + e \sin E + \text{\&c.}) \cos \frac{2\pi}{\lambda}(vt - x)$$

which we will call

$$F \sin \frac{2\pi}{\lambda}(vt - x) + G \cos \frac{2\pi}{\lambda}(vt - x).$$

This as in (14) may be put under the form

$$c \sin\left\{\frac{2\pi}{\lambda}(vt - x) + C\right\}, \text{ if } F = c \cos C, \;\; G = c \sin C;$$

$$\text{whence } c = \sqrt{(F^2 + G^2)}, \;\; \tan C = \frac{G}{F}.$$

In some cases, where the effects of an indefinitely great number of indefinitely small waves are to be combined, F and G may be found by integration: c and C are then determined by the same process as that just given.

18. It remains now only to notice some cases of undulation not included in those already treated of. One is, that a single wave may be transmitted through a medium (as we know to be true with regard to air), and then our theorems about interference are not true. This however will not come

under consideration, as there is reason to think that a single wave in air or in the medium of light would not produce the sensation of sound or colour. We shall generally consider the undulation as a succession of a great (but not infinite) number of waves. Another is, that the magnitude of the maximum vibration of a particle may depend on its situation: for instance, if waves diverge from a center, the vibrations must be more violent in the neighbourhood of that center than at a distance from it. This will be represented by expressing the extent of vibration at any time by

$$\psi(x) \, . \, \phi(vt - x),$$

or by the sum of several such functions. Since two successive waves here would not be equal, our theorem about interferences of waves lagging $\frac{\lambda}{2}$, $\frac{3\lambda}{2}$, &c. would not be strictly true: but it is easily conceivable that, at a distance from the center of divergence, the neighbouring waves would be so nearly equal that our expressions would have no sensible error.

19. Another case is, the interference of undulations whose waves are of different length. We know with respect to air that the velocity of transmission is the same for waves of all lengths: and we might expect the same to hold good with regard to light. We shall afterwards refer to experiments which appear to shew that this is not true. Still whether v be constant or not, it is impossible to unite two such terms as

$$a \, . \sin\left\{\frac{2\pi}{\lambda}(vt - x) + A\right\} \text{ and } b \, . \sin\left\{\frac{2\pi}{\lambda'}(v't - x) + B\right\}$$

so as to destroy the original form. We have seen (17) that when any number of waves is combined, supposing v and λ the same for all, their sum contains no trace of the distinction of waves in which it originated; and fixing upon any one point, or considering x as constant, the vibration is expressed by

$$c \sin\left\{\frac{2\pi v}{\lambda} t + \left(C - \frac{2\pi x}{\lambda}\right)\right\},$$

the same form as that for a cycloidal pendulum. But the two expressions

$$a \sin\left\{\frac{2\pi v}{\lambda}t + \left(A - \frac{2\pi x}{\lambda}\right)\right\} \text{ and } b \sin\left\{\frac{2\pi v'}{\lambda'}t + \left(B - \frac{2\pi x}{\lambda'}\right)\right\}$$

cannot be combined into that form, unless $\frac{v}{\lambda} = \frac{v'}{\lambda'}$, which there is no reason to think true. The consideration therefore of waves of different lengths may be kept perfectly separate, as their ultimate effect will be the same as the sum of all their separate effects, without any possibility of their destroying or modifying one another.

20. The reader is requested to attend to the conventional signification of the following terms.

By a *wave* we mean all the particles included between two which are in similar states of displacement and of motion. For instance, in any one of the cases (β), (γ), (δ), (ϵ), (ζ) of fig. 1 or 2, the particles included between b and b' form a wave: or those between f' and f'' form a wave: &c. It is easily seen that a wave includes particles in every possible state of displacement and of motion consistent with undulatory vibration.

The *length of a wave* we have explained to be the distance between two particles similarly displaced and moving similarly. The interval, in time, of two waves (that is, the interval between the arrival of two successive waves at the same point), it will be recollected, is the same as the time of vibration of any particle, (5).

By the *phase* of a wave, we shall denote the situation of a particle in a wave, considered as affecting its displacement and motion. For instance, b and b' in fig. 1 or 2 are in *similar phases,* because their displacements are equal, and their motions are also equal. But in (β) fig. 2, b and f are not in the same phase: for though their displacements are equal, their motions are in opposite directions. Similarly f and h are not in the same phase, for their displacements are different though their motions are equal. It is readily seen that particles are in the same phase when the distance between them is a multiple of the length of a wave.

We shall say that particles are in *opposite phases* when the displacement and motion of one are equal and opposite to those of the other. For instance, a and g, or e' and l', in (β), (γ), (δ), (ϵ), or (ζ), of fig. 1 or 2 are in opposite phases. It is easy to perceive that particles are in opposite phases when their distance is an odd multiple of half the length of a wave.

In speaking of waves where the displacement of a particle is represented by

$$a \,.\, \sin \left\{ \frac{2\pi}{\lambda}(vt - x) + A \right\},$$

we shall consider the arc

$$\left\{ \frac{2\pi}{\lambda}(vt - x) + A \right\}$$

as the measure of the *phase*.

When we consider a wave as extending over space in a direction different from that in which it is transmitted, we shall use the term *front of a wave* to denote a continuous line passing through all those points which are in the same phase. Thus in fig. 4, suppose that a series of waves of an undulation are passing from the side AB towards CD: and suppose that the line EF passes through a number of points which are simultaneously in the same phase: then EF is a front of a wave. If at the same time GH be another line passing through a number of points which are simultaneously in some other phase, GH is another front of a wave. In considering space of three dimensions it is plain that the front of a wave will generally be a surface.

21. We shall now state a principle of which we shall make extensive use in the calculation of Optical phenomena.

The effect of any wave in disturbing any given point may be found by taking the front of the wave at any given time, dividing it into an indefinite number of small parts, considering the agitation of each of these small parts as the cause of a small wave which will disturb the given point, and finding by summation or integration the aggregate of all the disturbances of the given point produced by the small waves coming from all points of the great wave.

In demonstration of this principle it seems sufficient to say that the agitation of one part at one time is really and truly the cause of the agitation of another part at another time: and that the effect of the great wave, which really is a number of small agitations, will by (12) be the sum of the effects of all the small agitations.

22. A question now arises. What is to limit the waves diverging from each of these small sources of motion? The disturbance spreads generally in a spherical form, so that the front of each little wave is a sphere: are we to suppose the sphere complete, so that each small undulation is propagated backwards as well as forwards?

The following answer appears to be correct, but its application in several cases seems doubtful. The effect of each small wave must be limited by the same considerations which limit the effect of the great wave. Now we know, from the algebraical investigation, that a single wave may be transmitted along a stretched cord, or in air, without being followed by another*. In this case it is plain that the present agitation of one point causes the future agitation of the

* The function X which expresses the disturbance may be *discontinuous*, that is, may be expressed by different algebraical forms for different values of $vt - x$, (which we will call w), provided that $\frac{dX}{dw}$ does not alter *per saltum*. For instance $X = 0$ while w is less than b: $X = a.\text{versin}\frac{2\pi}{\lambda}(w - b)$ from $w = b$ till $w = b + \lambda$: $X = 0$ while w is $> b + \lambda$. This is a discontinuous function expressing a single wave (since the particles are only disturbed when $vt - x$ is $> b < b + \lambda$). And it satisfies the condition just mentioned, since $\frac{dX}{dw}$ is 0 till $w = b$: its value is then $a.\frac{2\pi}{\lambda}\sin\left\{\frac{2\pi}{\lambda}(w - b)\right\}$ from $w = b$ to $w = b + \lambda$, that is, it increases gradually from 0 to $a.\frac{2\pi}{\lambda}$ and diminishes gradually from $a\frac{2\pi}{\lambda}$ to 0: and when w is $> b + \lambda$, $\frac{dX}{dw}$ is 0.

points in the direction in which the wave is going, but of none in the direction from which it came. In figure 4 for instance, if a single wave is going from AB towards CD, and if EF be the front of the wave at any time, then we know that the displacement in EF is the cause of future displacement in GH, because in consequence of the existence of this wave there will hereafter be a wave at GH: but we know that the displacement in EF causes no future displacement between EF and AB, because, though the displacement in EF exists, there will hereafter be no wave between AB and EF. If then we divide EF into a great number of parts, we must consider the displacement in each as causing a hemispherical or nearly hemispherical wave, which diverges only before the front of the great wave and not behind it.

APPLICATION OF THIS THEORY TO THE EXPLANATION OF THE PHÆNOMENA OF LIGHT WHICH DO NOT DEPEND ON POLARIZATION.

23. We shall suppose that light is the undulation of a medium called *ether* which pervades all transparent bodies. Respecting the direction of vibration of each particle we shall make no supposition till we treat of polarized light, as the results of this section are independent of the direction of vibration: to fix the ideas, however, the reader may conceive it to be of the kind represented in fig. 2. We shall suppose that a great number of similar waves follow without interruption, and that the function which expresses the displacement of a particle is

$$a \,.\, \sin\left\{\frac{2\pi}{\lambda}(vt - x) + A\right\}.$$

When in our final results we have found the expression

$$c \sin\left\{\frac{2\pi}{\lambda}vt + C - \frac{2\pi}{\lambda}x\right\}$$

for the displacement of the particles touching a screen or touching the eye, we shall assume the intensity of the light to be represented* by c^2. We shall suppose that the colour of light depends on the value of λ, but that λ is always very small: that for extreme red rays it is 0,0000266 inch: less for yellow: still less for green, blue, and indigo successively, and least for the reddish violet rays, for which $\lambda = 0,0000167$ inch. Common white light we shall suppose to consist of a mixture of waves of all lengths intermediate to these.

---

* We must take some even power of c to represent the intensity, since the undulation where the vibration is expressed by

$$-c \sin \left\{ \frac{2\pi}{\lambda}(vt - x) + C \right\}$$

differs in no respect from that whose vibration is expressed by

$$+c \sin \left\{ \frac{2\pi}{\lambda}(vt - x) + C \right\},$$

except that it is half the length of a wave before or behind it. The propriety of using the second power may be thus shown. If two candles giving the same light be placed near each other and shine on the same screen, we say that there is twice as much light as if one only were shining on it: and this may be regarded as the experimental definition of double the quantity of light. Now if the vibration excited by one of these be

$$c \,.\, \sin \left\{ \frac{2\pi}{\lambda}(vt - x) + C \right\},$$

and that excited by the other be

$$c \,.\, \sin \left\{ \frac{2\pi}{\lambda}(vt - x) + D \right\},$$

the whole vibration will be the sum of these or

$$c\sqrt{2 + 2\cos(C - D)} \,.\, \sin \left\{ \frac{2\pi}{\lambda}(vt - x) + E \right\},$$

$$\text{where } \tan E = \frac{\sin C + \sin D}{\cos C + \cos D}.$$

If the lights be represented by the square of the coefficient we have

light from a single candle $= c^2$

light from two candles $= 2c^2 + 2c^2 \cos(C - D)$.

The value of $C - D$ may be any whatever, and it is probable that in one second of time, from different systems of waves following each other, $\cos(C - D)$ may have gone through all its values many thousand times. To estimate the effect on the eye, we must take the mean of all its values. As the negative values are equal to the positive ones, the mean of all is 0. Thus we have

light from two candles $= 2c^2$.

By using the second power therefore we obtain a result which agrees with our experimental definition: and the same would appear if instead of two candles we had taken three or any other number. But the agreement will not hold if we take any other power of the coefficient as the representative of the illumination.

PROP. 8. A succession of waves, whose fronts are parallel to the screen CD (fig. 5) in which is the comparatively small opening AB, is moving towards the screen: to find the magnitude of vibration on any point G of the semicircle CGD.

24. Let H be the center of AB and of the semicircle, and let $HG=r$, $CHG=\theta$, $HA=b$: divide AB into a great number of small parts, and let the distance of one at Z from H be z, and its breadth δz: then

$$ZG=\sqrt{(r^2-2r\cos\theta\,.\,z+z^2)}.$$

When a wave comes to AB, consider separately the parts corresponding to the small divisions of AB. It seems reasonable to suppose that each of these small parts will cause a diverging wave of equal intensity for all values of θ. For if the medium were air, and a rush of particles took place as in fig. 1, from EF, the only effect in the small part under consideration would be to cause a condensation of air: and this would cause a wave of equal intensity for all values of θ. If the vibrations were like those of fig. 2, and perpendicular to the plane of the paper, the same thing appears evident: if parallel to the plane of the paper, it is not so clear what the proportion of intensities would be. The maximum of vibration when the wave reached G would, if it followed the same laws as in air*, vary nearly as $\frac{1}{ZG}$. Now all the little waves which originate from the different

* When a wave of air diverges symmetrically through any given solid angle, if r be the original distance of any particle from the center, $r+u$ its distance at the time t, $r+h$ the original distance of a second particle; then the distance of the latter at the time t will be

$$r+u+h\left(1+\frac{du}{dr}\right)$$

nearly: and the particles which formerly occupied a volume proportional to r^2h now occupy a volume proportional to

$$(r+u)^2h\left(1+\frac{du}{dr}\right),$$

or nearly proportional to

$$(r^2+2ru)h\left(1+\frac{du}{dr}\right),\quad \text{or to } r^2h\,.\left(1+\frac{2u}{r}+\frac{du}{dr}\right).$$

Consequently,

points of AB are in the same phase: hence we may express the disturbance which the little wave from the part δx produces at G by

$$c \,.\, \delta x \frac{1}{ZG} \sin \frac{2\pi}{\lambda} (vt - ZG).$$

Consequently, if the elasticity be as the mth power of the density, the elasticity $=$ original elasticity $. \dfrac{1}{\left(1+\dfrac{2u}{r}+\dfrac{du}{dr}\right)^m}$

$$= \text{weight of volume } 1 \,.\, gH \,.\, \left(1 - m\frac{2u}{r} - m\frac{du}{dr}\right) \text{ nearly.}$$

If then we take a small column whose base $= 1$, and length $= k$, the excess of the elastic force at one end above that at the other

$$= \text{weight of volume } 1 \,.\, mgH \,.\, k \,.\, \frac{d}{dr}\left(\frac{2u}{r} + \frac{du}{dr}\right) \text{ nearly:}$$

and the mass is weight of volume k: hence the equation for the disturbance of the particles is

$$\frac{d^2u}{dt^2} = mgH \frac{d}{dr}\left(\frac{2u}{r} + \frac{du}{dr}\right),$$

the solution of which is

$$u = \frac{1}{r}\phi'(vt-r) + \frac{1}{r^2}\phi(vt-r) + \frac{1}{r}\psi'(vt+r) - \frac{1}{r^2}\psi(vt+r).$$

If we suppose the wave to travel outwards only,

$$u = \frac{1}{r}\phi'(vt-r) + \frac{1}{r^2}\phi(vt-r).$$

$$\text{If } \phi(vt-r) = -a\cos\left\{\frac{2\pi}{\lambda}(vt-r) + A\right\},$$

$$\phi'(vt-r) \text{ will} = \frac{2\pi a}{\lambda}\sin\left\{\frac{2\pi}{\lambda}(vt-r) + A\right\},$$

and the disturbance

$$= \frac{2\pi a}{\lambda r}\sin\left\{\frac{2\pi}{\lambda}(vt-r) + A\right\} - \frac{a}{r^2}\cos\left\{\frac{2\pi}{\lambda}(vt-r) + A\right\}.$$

This may be put under the form

$$c\sin\left\{\frac{2\pi}{\lambda}(vt-r) + A - C\right\},$$

$$\text{where } c = \sqrt{\left(\frac{4\pi^2a^2}{\lambda^2r^2} + \frac{a^2}{r^4}\right)} = \frac{2\pi a}{\lambda r}\sqrt{\left(1 + \frac{\lambda^2}{4\pi^2r^2}\right)}, \text{ and } \tan C = \frac{\lambda}{2\pi r}.$$

The expression for c shews that at a considerable distance the coefficient will be inversely as r: and as from $r = 0$ to $r = \infty$, C decreases from $\frac{\pi}{2}$ to 0, it appears that the wave is accelerated as it goes on, and ultimately gains a quarter of the length of a wave on the space which it would have described with the uniform velocity v or $\sqrt{(mgH)}$.

Expanding,

$$ZG = r - \cos\theta \, . \, z + \frac{\sin^2\theta}{2r} z^2 + \&c.$$

If z be so small, and r so large, that $\frac{z^2}{2r}$ will never exceed a fraction of λ, (or even if it amounts to several multiples of λ), the terms after the second may be omitted. Then the disturbance produced by one little wave is

$$\frac{c \, . \, \delta z}{ZG} \sin \frac{2\pi}{\lambda} (vt - r + \cos\theta \, . \, z) :$$

and the sum of all the disturbances is

$$c \int_z \frac{1}{ZG} \, . \, \sin \frac{2\pi}{\lambda} (vt - r + \cos\theta \, . \, z).$$

In the integration we shall produce no sensible error if we put r for ZG, and this makes the sum

$$\frac{c}{r} \int_z \sin \frac{2\pi}{\lambda} (vt - r + \cos\theta \, . \, z).$$

Integrating, it is

$$\frac{-c\lambda}{2\pi r \cos\theta} \cos \frac{2\pi}{\lambda} (vt - r + \cos\theta \, . \, z) :$$

and taking this from $z = -b$ to $z = +b$, the disturbance at G is

$$\frac{c\lambda}{2\pi r \cos\theta} \left\{ \cos \frac{2\pi}{\lambda} (vt - r - \cos\theta \, . \, b) - \cos \frac{2\pi}{\lambda} (vt - r + \cos\theta \, . \, b) \right\}$$

$$= \frac{c\lambda}{\pi r \cos\theta} \, . \, \sin \frac{2\pi \, . \, b \, . \cos\theta}{\lambda} \, . \, \sin \frac{2\pi}{\lambda} (vt - r).$$

This represents a vibration whose maximum is

$$\frac{c\lambda}{\pi r \cos\theta} \, . \, \sin \frac{2\pi b \cos\theta}{\lambda} .$$

25. We shall now proceed to compare the values of this expression for different values of θ.

Case 1. Suppose λ much greater than b. (This will generally be the case with sound, as for all audible sounds λ varies from a few inches to several feet.)

Here $\frac{2\pi b \cos\theta}{\lambda}$ will be a small arc, and will not differ much from its sine: putting the arc for the sine, the maximum of vibration becomes

$$\frac{c\lambda}{\pi r \cos\theta} \cdot \frac{2\pi b \cos\theta}{\lambda}, \quad \text{or } \frac{2cb}{r},$$

which is the same for all values of θ.

Case 2. Suppose λ much smaller than b. (This will generally be the case with light.)

For the part nearly opposite to the entering wave, $\cos\theta$ is very small, and

$$\frac{c\lambda}{\pi r \cos\theta} \cdot \sin\frac{2\pi b \cos\theta}{\lambda} = \frac{2cb}{r}.$$

In other parts it is to be observed that the disturbance is 0 when

$$\frac{2\pi b \cos\theta}{\lambda} = \pm\pi, \quad \text{or } = \pm 2\pi, \quad \text{or } = \pm 3\pi, \text{ \&c.},$$

that is, when

$$\cos\theta = \pm\frac{\lambda}{2b}, \quad \text{or } = \pm\frac{2\lambda}{2b}, \quad \text{or } = \pm\frac{3\lambda}{2b}, \text{ \&c.}$$

Hence there is a succession of points in which there is absolute darkness. Of the intermediate parts, the brightest will be found (nearly) by making

$$\sin\frac{2\pi b \cos\theta}{\lambda} = \pm 1:$$

then the maximum of vibration is $\frac{c\lambda}{\pi r \cos\theta}$. Consequently the intensity of light at the brightest part of one of the bright portions is to that of the part nearly opposite to the entering wave, as $\frac{c^2\lambda^2}{\pi^2 r^2 \cos^2\theta}$ to $\frac{4c^2b^2}{r^2}$, or as λ^2 to $4\pi^2 . b^2 . \cos^2\theta$, or as $\frac{\lambda^2}{4\pi^2 b^2 \cos^2\theta}$ to 1.

If $\frac{\lambda^2}{b^2}$ be so small as for light (for instance if $\lambda = 0{,}00002$ and $b = 0{,}1$ inch, $\frac{\lambda^2}{b^2} = \frac{1}{25000000}$), it is plain that the value of this ratio will be extremely small when $\cos\theta$ has any sensible value, and we may say without perceptible error that, except nearly opposite to the entering wave, there will be complete darkness all round.

26. If in the investigation we had included the terms depending on z^2, we should merely have had a very small addition to

$$\frac{2\pi}{\lambda}(vt - r + \cos\theta . z),$$

and this addition would not sensibly have altered in value while $\frac{2\pi}{\lambda}(vt - r + \cos\theta . z)$ increased by 2π. It will easily be seen that in Case 1 this would have produced no effect, and in Case 2 the maxima of brightness and the absolute darkness would only have been shifted a little way; their number, relative position, and the intensity of light, remaining sensibly the same*. And if for ZG we had put its more accurate value $r - \cos\theta . z$, the terms added to the expression would not have been sensible.

27. The conclusion in Case 2 may also be obtained thus. In fig. 6 divide AB into a number of equal parts Aa, ab, bc, &c. such that the distance of A from G is less than that of a by $\frac{\lambda}{2}$: that of a less than that of b by $\frac{\lambda}{2}$; and so on. The waves from corresponding parts of Aa and ab are, at starting, in the same phase. Consequently when they reach G, the wave from a part of Aa is in advance of that

* We shall hereafter consider cases in which these terms are sensible.

from the corresponding part of *ab* by $\frac{\lambda}{2}$, or they are in opposite phases, and therefore, by (15), they destroy each other. Thus every part of *Aa* destroys a corresponding part of *ab*, and therefore the whole effect of *Ab* is 0. Similarly the whole effect of *bd* is 0; &c. Thus if the number of parts be even, there is no vibration produced at *G*: if it be odd, there is only the vibration produced by the last of the small parts. But, for the position nearly in front of the wave, all the parts are nearly at the same distance, and the vibration is produced by the added effects of all the small waves coming from every part of *AB*. If Case 1 be considered in the same way, it appears that the difference of the paths of the waves from different parts of the opening is so small in proportion to the length of a wave, that all when they fall on *G* may be considered to be in the same phase, and therefore every part of the semicircle is in the same state of vibration.

28. The conclusions at which we have arrived are very important as removing the original objection to the undulatory theory of light. It was objected that if light were produced by an undulation similar to that producing sound, it ought to spread in the same manner as sound: that if light coming from a bright point entered a room by a small hole, it ought (instead of going on straight to illuminate a spot on the opposite side) to spread through the room in the same manner as a sound coming in the same direction and through the same hole. The answer appears in the results of the last investigation: the length of the waves of air is much greater than the aperture, that of the waves of light much less: and the same investigation which shews that in the former case the sound ought to spread equally in all directions, shews that in the latter the light ought to be insensible except nearly in front of the hole. We have reason to think that when sound passes through a very large aperture, or when it is reflected from a large surface (which amounts nearly to the same thing) it is hardly sensible except in front of the opening, or in the direction of reflection.

29. Our conclusion with regard to light is also important as removing one source of doubt in several succeeding investigations. In our ignorance of the law of intensity of the vibrations propagated from a center in different directions, we have supposed the intensity equal in all directions: and yet with this supposition we have found that when the aperture is much larger than λ, there is no sensible illumination except nearly in the direction in which the wave was going before it reached the aperture. The same would be true if the intensity diminished according to some function of the angle made with the original direction of the wave. Since then the illumination is (as far as the senses will be able to judge) nothing, except the obliquity is small, whatever be the function, we may assume that function of any form most convenient, provided that it does not alter rapidly in the neighbourhood of the original direction, and does not increase considerably as the angle of obliquity increases.

30. From the result of this investigation it appears also that the motion of every small part of the wave is perpendicular to the front of the wave. For in fig. 5 that part of the wave which passes through the orifice AB illuminates only that part of the semi-circle which is defined by drawing a straight line perpendicular to the front: and in the same manner if we had covered AB and opened another orifice, we should have found that the only illumination was on the part determined by drawing a straight line through the new orifice perpendicular to the front. In this we see the origin of the idea of rays of light. The reader is particularly requested to observe that this theorem is proved only by the demonstration of the proposition above, and depends entirely on this assumption, that the waves of light move with the same velocity in all directions. We shall hereafter speak of cases in which the motion of the wave is not perpendicular to its front.

It will readily be seen that the whole of this applies as well to the motion of the small parts of a wave whose front is not plane.

PROP. 9. To explain the reflection of light on the undulatory theory.

31. We shall again refer to the motion of sound for an analogical illustration of this point. In fig. 7 let $ABCD$ be the front of a wave (which for simplicity we suppose plane, every part moving in parallel directions) advancing in the direction BB' or CC'' and meeting the smooth wall $C''B'$. Then it appears from the investigation of sound*

---

* Let x, y, z, be the original co-ordinates of any particle of air: and at the time t let them be $x+X$, $y+Y$, $z+Z$. Then the particle which originally had for co-ordinate $x+\delta x$ will at the time t have

$$x+\delta x+X+\frac{dX}{dx}\,\delta x \text{ nearly;}$$

or the distance between two particles in the direction of x, which was originally δx is, at the time t, $\delta x\left(1+\frac{dX}{dx}\right)$ nearly. Similarly the distances in the directions of y and z which were originally δy and δz are, at the time t,

$$\delta y\left(1+\frac{dY}{dy}\right) \text{ and } \delta z\left(1+\frac{dZ}{dz}\right).$$

Consequently, the air which occupied the rectangular parallelopiped whose sides were δx, δy, δz, now occupies the parallelopiped, nearly rectangular, whose sides are

$$\delta x\left(1+\frac{dX}{dx}\right),\quad \delta y\left(1+\frac{dY}{dy}\right),\quad \delta z\left(1+\frac{dZ}{dz}\right).$$

And if the elasticity (represented by the pressure upon a unit of surface) was originally P, and varied as $(\text{density})^m$, $\left(m \text{ being nearly } \frac{4}{3}\right)$, the elasticity of the air in this parallelopiped is nearly

$$P\left(1-m\frac{dX}{dx}-m\frac{dY}{dy}-m\frac{dZ}{dz}\right).$$

This then is the expression for the elasticity of the air about that point whose co-ordinates were originally x, y, z: the alteration of elasticity being supposed small.

Consequently, at the time t, the elasticity about that point whose co-ordinates were originally $x+h$, y, z, is

$$P\left(1-m\frac{dX}{dx}-m\frac{dY}{dy}-m\frac{dZ}{dz}\right)+P\frac{d}{dx}\left(1-m\frac{dX}{dx}-m\frac{dY}{dy}-m\frac{dZ}{dz}\right)h.$$

And therefore if there is a small parallelopiped whose sides were h, k, l, the excess of pressure which urges it on in the direction of x is

$$-P\frac{d}{dx}\left(1-m\frac{dX}{dx}-m\frac{dY}{dy}-m\frac{dZ}{dz}\right)h.kl$$

$$=m.P.hkl\frac{d}{dx}\left(\frac{dX}{dx}+\frac{dY}{dy}+\frac{dZ}{dz}\right).$$

that after any part of this wave, as *BCD*, has come in contact with the wall, it will proceed in the direction $C''E$,

---

And if W were the original weight of the air in volume 1, the weight of this parallelopiped is $W.hkl$. Consequently the acceleration in direction of x is

$$\frac{mPhkl}{Whkl}\cdot\frac{d}{dx}\left(\frac{dX}{dx}+\frac{dY}{dy}+\frac{dZ}{dz}\right)\times \text{acceleration produced by gravity}$$

$$=\frac{mP}{W}g\cdot\frac{d}{dx}\left(\frac{dX}{dx}+\frac{dY}{dy}+\frac{dZ}{dz}\right);$$

$$\text{or } \frac{d^2(x+X)}{dt^2}=\frac{mP}{W}g\frac{d}{dx}\left(\frac{dX}{dx}+\frac{dY}{dy}+\frac{dZ}{dz}\right).$$

Let H be the height of a homogeneous atmosphere: by which we mean that the pressure P would support a column whose height is H and base 1, weighing W for every unit of volume: that is $P=HW$. Then

$$\frac{d^2(x+X)}{dt^2},\ \text{or}\ \frac{d^2X}{dt^2}=mgH\frac{d}{dx}\left(\frac{dX}{dx}+\frac{dY}{dy}+\frac{dZ}{dz}\right):$$

and similar equations hold for y and z. These are the general equations for the small disturbances of air.

These equations cannot be integrated generally: but a number of different integrals can be found, adapted to particular purposes. For instance, putting $mgH=v^2$,

$$X=\frac{x-a}{r^2}\phi'(vt-r)+\frac{x-a}{r^3}\phi(vt-r)$$

$$+\frac{x-a}{r^2}\psi'(vt+r)-\frac{x-a}{r^3}\psi(vt+r)$$

$$\text{where } r=\sqrt{(x-a)^2+(y-b)^2+(z-c)^2};$$

with similar expressions, *mutatis mutandis*, for Y and Z. This is the general expression for spherical waves going to and from the center whose co-ordinates are a, b, c.

$$\text{Again } X=\cos\alpha\,.\,\phi(vt-x\cos\alpha-y\cos\beta-z\cos\gamma),$$

$$Y=\cos\beta\,.\,\phi(vt-x\cos\alpha-y\cos\beta-z\cos\gamma),$$

$$Z=\cos\gamma\,.\,\phi(vt-x\cos\alpha-y\cos\beta-z\cos\gamma),$$

where $\cos^2\alpha+\cos^2\beta+\cos^2\gamma=1$; $\cos\alpha$, $\cos\beta$, $\cos\gamma$, being positive or negative.

This is the equation for plane waves going in the direction of a line which makes with x, y, and z, the angles α, β, γ.

It is particularly to be remarked that the sum of any number of these solutions may be taken for one solution: and also that the functions may be discontinuous, with the limitation mentioned in the note to (22).

Now suppose the plane wave to be interrupted by a wall, whose equation is $x=c$. Since the particles of air constantly remain in contact with the wall, we must have $X=0$ when $x=c$, whatever be the values of y and z. It is plain that the form above given for X will not satisfy this condition, and that no single function would do so. But with the addition of another function we may satisfy it. Thus, if

$$X=\cos\alpha\,.\,\phi(vt-x\cos\alpha-y\cos\beta-z\cos\gamma)$$

$$-\cos\alpha\,.\,\phi\{vt+(x-2c)\cos\alpha-y\cos\beta-z\cos\gamma\},$$

making with the wall the same angle as CC'', but on the opposite side of the perpendicular: and the front of the wave will be $B'C'D'$, making with the wall the same angle as BCD, but on the opposite side: the extent of vibration &c. remaining as before. And this appears to justify us sufficiently in the assertion that the waves of light may be reflected in the same manner. With regard to the smoothness of the reflecting surface, all that is necessary is that the elevations or depressions do not exceed a fraction of λ.

32. The following is a more independent method of arriving at the same result, and is perhaps satisfactory. In fig. 8 let ABC be the front of a wave going in the direction of AA'. As soon as each successive small portion of this has reached the surface, we will consider it as causing an agitation in the ether next in contact with the surface, and will suppose that agitation to be the center of a spherical wave, diverging with the same velocity as the plane wave {see the note to (24)}. Let us now consider the state of things when A has reached A'. B has reached B' some time before: and would at this time have arrived at D if not interrupted. Consequently it has diverged into a sphere ab whose radius $= B'D$. C reached the surface at a still longer time previous, and would at this time have reached E: it has therefore diverged into a sphere cd whose radius is CE. The same holds for every intermediate point. If now we examine the nature of the front of the grand wave formed by all these little waves, we see that it must be

---

$$Y = \cos\beta \, . \, \phi\,(vt - x\cos\alpha - y\cos\beta - z\cos\gamma)$$
$$+ \cos\beta \, . \, \phi\,\{vt + (x - 2c)\cos\alpha - y\cos\beta - z\cos\gamma\},$$
$$Z = \cos\gamma \, . \, \phi\,(vt - x\cos\alpha - y\cos\beta - z\cos\gamma)$$
$$+ \cos\gamma \, . \, \phi\,\{vt + (x - 2c)\cos\alpha - y\cos\beta - z\cos\gamma\}.$$

The differential equations are satisfied, and the condition $X = 0$ when $x = c$ is also satisfied. The first terms of X, Y, Z, taken together, express the original wave, whose direction makes angles α, β, γ, with x, y, z; the second express the new or reflected wave, whose direction makes angles $180^\circ - \alpha$, β, and γ with x, y, and z. This shews that the direction of reflection follows the law commonly enounced as the law of reflection. The intensity of the reflected wave is the same as that of the incident wave. This is the theory of oblique echos.

the plane which touches all the spheres, and which evidently makes the same angle with CA' that $A'E$ or AC makes with it, but inclined the opposite way. The motion of the wave, which is perpendicular to this front, makes therefore the same angle after reflection as before.

It must be remarked that this demonstration is in no wise affected if we suppose *all* the spherical waves to be accelerated or retarded by the same quantity. If then we should find occasion hereafter to assert that in some cases the direction of vibration is changed at reflection, or (which amounts to the same) that half the length of a wave must be added to or subtracted from the path after reflection, the demonstration of the law of reflection will not be invalidated.

PROP. 10. To explain the refraction of light on the undulatory theory.

33. We must assume that the waves are transmitted with smaller velocity in glass, water, &c., and in all substances commonly called refracting media, than in what we call vacuum. This assumption appears in the highest degree probable, whether we suppose the vibrations in the refracting media to be vibrations of the same ether, incumbered by its connexion with the particles of the refracting body, or we suppose the vibrations to be vibrations of the particles of the refracting body.

34. Now in fig. 9, let ABC be the front of a wave going in the direction of AA'. As soon as each successive small portion of this wave has reached the surface CA' of the refracting medium, suppose it to cause an agitation in the ether or the particles of the medium at that surface, and consider that agitation to be the center of a wave, which diverges in a spherical form in the medium with a less velocity than the velocity of the plane wave. Consider the state of the particles when A has arrived at A'. B has reached B' some time before: and would have arrived at D if not interrupted by the refracting medium. Consequently it has diverged into a sphere ab whose radius is less than

$B'D$ in the proportion in which the velocity is diminished. Let μ be the number expressing this proportion: then

$$B'b = \frac{1}{\mu} B'D.$$

Similarly C has reached the surface still longer before, and has therefore diverged into a sphere whose radius

$$Cc = \frac{1}{\mu} CE.$$

The same holds for every intermediate point. Now the front of the grand wave formed by all these little waves is evidently the plane which touches all the spheres; and therefore makes with the refracting surface an angle whose sine is $\frac{Cc}{CA'}$. This angle is equal to the angle which the direction of the wave (or the perpendicular to its front) makes with the perpendicular to the refracting surface: it is therefore the angle of refraction. Consequently

$$\text{sin. refraction} = \frac{Cc}{CA'}.$$

$$\text{Similarly sin. incidence} = \frac{CE}{CA'}.$$

$$\text{Therefore } \frac{\text{sin refraction}}{\text{sin incidence}} = \frac{Cc}{CE} = \frac{1}{\mu}.$$

35. It is easily seen that a similar demonstration applies when the waves are transmitted in the second medium with greater velocity than in the first: which we suppose to represent the circumstances of light coming out of a refracting medium into vacuum. If in this case, fig. 10, the angle of incidence is so great that Cc (the radius of the wave diverging from C) is greater than CA', that is, if $AA' \cdot \mu$ is greater than CA' or $\frac{AA'}{\sin ACA'}$, or $\sin ACA'$ is greater than $\frac{1}{\mu}$, or sin. incidence greater than $\frac{1}{\mu}$, no plane

can be found which touches all the spheres. There will be no grand wave therefore: and the little waves causing displacements in different directions will very soon destroy each other. Thus there will be no refracted ray. This is a well known law of optics.

It must be remarked that the demonstrations of (32) and (34) are not free from obscurity, for the reason mentioned in (22).

36. There is another phænomenon attendant on refraction which we can explain but vaguely, though it is easily seen that the explanation is not without foundation. The particles of ether next in contact with the glass, (if we suppose glass to be the refracting medium) communicating motion to the denser ether within the glass, may be considered as small bodies striking large ones. Now if they followed the same law as elastic bodies*, a certain motion would be communicated to the large bodies, and the small bodies would lose their original motion and would receive a motion in the *opposite* direction. The motion of the struck bodies causes the refracted wave of which we have just spoken; the motion remaining to the striking bodies will cause a reflected wave in the ether. The magnitude of the reflection will plainly be diminished as the difference between the particles is diminished. Thus refraction is always accompanied by reflection: and the reflection is more feeble as the vibrating media on both sides of the surface approach more nearly to the same state: that is, as the refractive index approaches to 1. This is experimentally true.

37. If, however, the rays are passing from glass to air, we must represent the state of the particles by large bodies striking small ones. The small bodies receive a motion, which causes the refracted wave: the large bodies will preserve a part of their motion in the *same* direction, and this

* The motions would follow this law, if the particles acted on one another like those of air by condensation or rarefaction of the fluid between them: or in the manner which Fresnel supposes (to be alluded to hereafter): or in any way which makes the force equal at equal distances of the particles. We are not therefore making the forced supposition of particles impinging on each other.

will cause a reflected wave. Thus when light passes through glass there will be reflection at both surfaces. But there is this difference between the two reflections: one is caused by a vibration in the same direction as that of the incident ray, and the other by a vibration in the direction opposite to that of the incident ray. We shall find this distinction important in explaining a fundamental experiment (65).

The same thing may be thus shewn. If we suppose a mass of glass to be cracked and the separated parts to be again pressed close together, there will be no more reflection than from the interior of a mass of glass: that is, there will be none at all. Still as there are really two surfaces in contact, each of which separately reflects, we must suppose the reflections to be of such a kind that they destroy each other. Consequently if the vibration from one reflection be in one direction, that from the other reflection must be in the opposite direction.

38. We shall now state a difficulty in the undulatory theory of refraction which has not yet been entirely surmounted, but which does not appear by any means insurmountable. The index of refraction we have found to be the proportion of the velocities of the waves in vacuum and in the refracting medium. Now it is well known that, experimentally, the refractive index is different for rays of different colours, that is, for waves whose lengths are different. It is evident then that waves whose lengths are different are transmitted with different velocities either in vacuum*, or in the refracting medium, or in both. The difference does not depend on the extent of vibration of each particle, for the refractive index is the same for a bright light as for a feeble one, but merely on the length of the wave, or on the time of vibration†. We are unable to explain this; and the

* If the velocities for different rays were different in vacuum, the aberration of stars (which is inversely as the velocity) would be different for different colours, and every star would appear as a spectrum whose length would be parallel to the direction of the Earth's motion. We know of no reason to think that this is true.

† The difficulty might perhaps be explained thus. We have every reason to think that a part of the velocity of sound depends on the circumstance that the

analogy which has guided us in so many instances fails here entirely. There is no such difference between the velocities of transmission of long and short waves of air. The sounds of the deepest and the highest bells of a peal are heard at any distance in the same order. Happily another analogy comes in to our aid, derived from undulations in which the vibrations have a much greater similarity than those of air to the vibrations which appear to constitute light. The velocities of waves of water vary with the length of the wave, the long waves always travelling with the greatest speed. In this respect there is a fair analogy between the relations of the velocity to the length of the wave, in glass or other refracting substances (as regards luminiferous waves), and in water (as regards waves of water). But the proportion of the velocity of a long wave of water to a short one is not invariable, but depends upon the depth of the water. In this may be found an analogy, of a less satisfactory kind, to the change in the relation of the velocity to the length of the wave, on comparing one refractive medium with another. We may remark, however, that if we calculate theoretically the velocity of a wave of light, on the supposition that the distance between the particles of ether is a sensible quantity in comparison with the length of a wave, we find that the velocity is different for waves of different lengths. (See the note to Art. 103). The very close approximation to observed proportions of velocity which has already been attained by calculations of this kind seems to offer good reason for supposing that the explanation will be made complete. On the whole, we regard the slight imperfection of the theory as

---

law of elasticity of the air is altered by the *instantaneous* development of latent heat on compression, or the contrary effect on expansion. Now if this heat required *time* for its development, the quantity of heat developed would depend on the time during which the particles remained in nearly the same relative state, that is on the time of vibration. Consequently the law of elasticity would be different for different times of vibration, or for different lengths of waves: and therefore the velocity of transmission would be different for waves of different lengths. If we suppose some cause which is put in action by the vibration of the particles to affect in a similar manner the elasticity of the medium of light, and if we conceive the degree of development of that cause to depend on time, we shall have a sufficient explanation of the unequal refrangibility of differently coloured rays.

deserving attention, but not as offering the most trifling difficulty to the general adoption of the theory.

PROP. 11. To explain the course of waves after reflection or refraction.

39. First we must give a construction for determining the front of a wave at any time from knowing its front at any previous time. In fig. 11 let AB be a small part of the curved front of a wave: and draw Aa, Bb, normals to the front, and let $Aa = Bb$. Then by (30), the motion of the part A of the wave is in the direction Aa, and the motion of the part B is in the direction Bb: and their velocities are equal. Consequently after a short time the front of the wave will pass through a, b, and through every other point as c, d, &c., determined by this condition that Cc, Dd, &c. are perpendicular to the original front and are equal to Aa. Now by (30) the motion of the part of the wave at a will be perpendicular to the new front $abcd$. But as $Bb = Aa$, it is easily seen that ab is parallel to AB and is therefore perpendicular to Aa: and therefore the motion of the new front will be in the direction of Aa produced: that is in the same straight line as before. In the same manner it would appear that all its succeeding motion will be in the same straight line. And the velocity will be equal for all points of the wave. Thus we arrive at the following construction. Draw AP, BQ, CR, DS, &c. normals to the original front, and make AP, BQ, CR, DS, &c. equal to the space through which a wave will travel in the time for which the form is required. The front of the wave at that time will be the locus of the points P, Q, R, S. The relation between the old and the new front is therefore this: every line which is normal to one is also normal to the other. If we confine ourselves to space of two dimensions, this is comprehended in saying that they have the same evolute. This applies equally to the motion of a wave in vacuum or in glass, or in any other refracting medium, provided that the velocity of a wave's motion is equal in all directions.

40. Suppose now (to fix upon a particular case) a plane wave is received on a reflecting paraboloid whose axis is

parallel to the direction of the wave's motion, or perpendicular to its front. In fig. 12 let AD be one position of the wave, $A'D'$ a succeeding position, and so on. From Prop. 9 it appears that the front of each small part of the wave makes the same angle with the surface after reflection as before, but on the opposite side of the normal: and that consequently the line representing the direction of the wave's motion, and which is perpendicular to the front, makes the same angle with the normal before and after reflection. As all the lines representing the direction of motion of different points of the wave are parallel to the axis of the paraboloid, those which represent the direction of motion after reflection (by a well known theorem) converge to F the focus. Consequently the form of the wave, which by (39) is the surface to which all these lines are normals, is a spherical surface whose center is F. Thus then at one time $A'D'$ will be the front of the wave: at a later time BC will be the form of that part which is not reflected, and $A''B$, $D''C$, the form of those parts which are reflected, the part incident at A' having been reflected to A'': at a still later time, bc will be the form of the part not reflected, and $A'''B'b$, $D'''C'c$ the form of the reflected parts, the part incident at A' having been reflected to A''', and that incident at B having been reflected to B', &c: and when the whole has been reflected, all trace of the original form of the wave will be lost, and the existing form will be only a spherical surface of which F is the center. The concave spherical wave goes on towards F, contracting till it passes through that point, when all the different small parts cross, and then they form a diverging spherical wave of which F is the center.

It is easily seen that an explanation of exactly the same kind applies to the effects of refraction, the velocity of the wave being supposed to be altered in a given ratio as in Prop. 10, and the direction of the motion of each part of the wave being always supposed perpendicular to that part of the front.

41. We have explained the motion of the wave after reflection or refraction as if the terminating edges of the front of the wave did not cause any disturbance beyond the

line perpendicular to the front: as if for instance there were a certain disturbance all along the line $A''B$ which afterwards arrived at $A'''B'$ without causing the least disturbance in the ether beyond A'''. This however is not true, and we shall hereafter take into account the effects of the lateral spread of the waves.

42. From the nature of the demonstration it appears that whenever all the small parts of a wave meet each other after reflection or refraction, they have described paths corresponding to equal times. In the case of reflection, this is the same as saying that the whole paths (consisting of the sum of those before and after reflection) must be the same for every point: but in the case of refraction a different statement is necessary. For if the waves move in glass (or other refracting media) with a velocity which is $\frac{1}{\mu} \times$ that in vacuum, then the path in glass, as compared with that in air, is not to be estimated by its length, but by $\mu \times$ its length. And therefore when all the small parts of a wave meet each other after refraction, the sum of the path in vacuum and $\mu \times$ the path in glass is the same for all.

43. This principle may be advantageously applied in the solution of some problems. Suppose for instance it is required to find the form of a refracting surface BP, fig. 13, which shall cause the wave diverging from A to converge to C. The principle above mentioned gives us at once this equation,

$$AP + \mu \, . \, PC = \text{constant}:$$

which is the same as Newton's in the 97th Proposition of the Principia.

44. A *focus* therefore may be defined as the point to which a spherical wave converges, or from which it diverges. It may also be defined as the point at which little waves from all parts of a great wave arrive at the same time. It will readily be seen that our demonstration and our definition include equally real and imaginary foci, in the same manner as the theorems and definitions of common optics.

45. It appears from (40) that the wave, after converging to a point, diverges from it in the same manner as if that point were a center of excitation, or a source of light. In all experiments therefore in which it is wished to produce a series of waves diverging from a point, the image of the Sun's disk, produced by a lens of short focal length, may be used instead of a luminous body.

Prop. 12. A series of waves diverges from a point A, fig. 14, and falls upon two plane mirrors BC, CD, inclined at a very small angle α, and touching each other in the line whose projection on the paper is C: to find the intensity of illumination on different parts of the screen EF where the streams of light reflected from the two mirrors are mixed.

46. Let G be the virtual image (determined by the rules of common optics) of A, produced by reflection at BC: H that produced by reflection at CD. Then instead of supposing the light to have come originally from A, we may without error in our results suppose it to originate in two sources at G and H. For the course of any part of a wave after reflection from BC is just the same as if it had come from G; and the length of its path measured in a straight line from G is the same as the sum of the paths of the incident and reflected ray, since the distance of A and of G from the point of reflection is the same (thus the part of the wave which is incident at N is reflected in the direction NM which is the same as GN produced, by (32) and (39), and the length of its path $AN + MN$ is the same as GM, since $AN = GN$). But that the circumstances may be exactly represented, we must suppose the fictitious wave to start from G at the same time at which the real wave starts from A, and to have the same intensity. Similarly we must suppose the other wave to start from H at the same time and with the same intensity as that which starts from A; and therefore at the same time and with the same intensity as that which starts from G. The problem is therefore reduced to this: to find the intensity of illumination on the screen when waves start at the same instant and with the same intensity from G and H.

47. Join GH, bisect it in L, and produce LC to meet the screen in O: let M be any point at a small distance from O; $AC = a$: $CO = b$. Since the angle between the mirrors is α, it is easily seen that $GCH = 2\alpha$. And since $GC = AC = HC$, CL is perpendicular to GH and bisects the angle GCH. Consequently

$$GL = LH = a \cdot \sin\alpha: \text{ and } LO = a\cos\alpha + b.$$

Then for the disturbance produced at M by the wave coming from G {taking the same expression as in (8) and (24)} we have

$$\frac{c}{GM}\sin\frac{2\pi}{\lambda}(vt - GM + A).$$

The variation in the value of GM is so small that without sensible error we may in the coefficient put GO or LO instead of GM; and thus the disturbance produced at M by the wave coming from G is

$$\frac{c}{LO}\sin\frac{2\pi}{\lambda}(vt - GM + A).$$

Similarly the disturbance produced by the wave coming from H is

$$\frac{c}{LO}\sin\frac{2\pi}{\lambda}(vt - HM + B).$$

B however must be equal to A, because the waves on leaving G and H respectively are in the same phase at the same time (which is represented by putting 0 for GM and HM, and requires B to be the same as A). Hence the whole disturbance of the ether at M is

$$\frac{c}{LO}\left\{\sin\frac{2\pi}{\lambda}(vt - GM + A) + \sin\frac{2\pi}{\lambda}(vt - HM + A)\right\}$$

$$\text{or } \frac{2c}{LO}\cdot\cos\left\{\frac{\pi}{\lambda}(GM - HM)\right\}\cdot\sin\frac{2\pi}{\lambda}\left(vt - \frac{GM + HM}{2} + A\right),$$

and consequently, by (23), the intensity of the light at M is represented by

$$\frac{4c^2}{LO^2} \cdot \cos^2 \left\{ \frac{\pi}{\lambda} (GM - HM) \right\}.$$

48. Now

$$GM^2 = LO^2 + (GL + OM)^2 = (a \cos \alpha + b)^2 + (a \sin \alpha + OM)^2,$$

and consequently

$$GM = a \cos \alpha + b + \tfrac{1}{2} \cdot \frac{(a \sin \alpha + OM)^2}{a \cos \alpha + b} \text{ nearly.}$$

Similarly

$$HM = a \cos \alpha + b + \tfrac{1}{2} \cdot \frac{(a \sin \alpha - OM)^2}{a \cos \alpha + b} \text{ nearly:}$$

therefore

$$GM - HM = \frac{2a \,.\, \sin \alpha \,.\, OM}{a \cos \alpha + b} = \frac{2a \sin \alpha}{a + b} \,.\, OM \text{ nearly:}$$

and therefore the brightness at M is represented by

$$\frac{4c^2}{(a+b)^2} \cdot \cos^2 \left(\frac{2\pi}{\lambda} \cdot \frac{a \sin \alpha}{a + b} \,.\, OM \right).$$

This varies according to the position of M.

(1) Suppose M to coincide with O: $OM = 0$: and the brightness is $\dfrac{4c^2}{(a+b)^2}$. This is its greatest value.

(2) Suppose $OM = \pm \dfrac{a+b}{a \sin \alpha} \cdot \dfrac{\lambda}{4}$.

Then $\dfrac{2\pi}{\lambda} \cdot \dfrac{a \sin \alpha}{a+b} \,.\, OM = \pm \dfrac{\pi}{2}$,

and the brightness

$$= \frac{4c^2}{(a+b)^2} \cos^2 \frac{\pi}{2} = 0,$$

or there is absolute blackness.

(3) Suppose $OM = \pm \frac{a+b}{a \sin \alpha} \cdot \frac{2\lambda}{4}$.

$$\text{Then } \frac{2\pi}{\lambda} \cdot \frac{a \sin \alpha}{a+b} \cdot OM = \pm \pi,$$

and the brightness

$$= \frac{4c^2}{(a+b)^2} \cos^2 \pi = \frac{4c^2}{(a+b)^2},$$

the same as when $OM = 0$.

(4) Suppose $OM = \pm \frac{a+b}{a \sin \alpha} \cdot \frac{3\lambda}{4}$:

$$\text{then } \frac{2\pi}{\lambda} \cdot \frac{a \sin \alpha}{a+b} \cdot OM = \pm \frac{3\pi}{2},$$

and the brightness

$$= \frac{4c^2}{(a+b)^2} \cos^2 \frac{3\pi}{2},$$

or there is darkness.

(5) Generally, if

$OM = \pm 2n \cdot \frac{a+b}{a \sin \alpha} \cdot \frac{\lambda}{4}$, n being a whole number,

the brightness has its maximum value $\frac{4c^2}{(a+b)^2}$:

and if $OM = \pm (2n+1) \cdot \frac{a+b}{a \sin \alpha} \cdot \frac{\lambda}{4}$,

there is blackness.

Between these values of OM, the brightness has intermediate values.

Thus it appears that there will be a series of points at equal distances along the line IK, at which the illumination is alternately maximum and minimum, beginning at O where it is maximum; and that at the points of minimum illumination the light is actually evanescent. Now considering the screen as extended in the direction perpendicular to the paper, and observing that the investigation which applies to M applies to every point in the line perpendicular to the paper at M, it is easily seen that the appearance on the screen is a series of bars alternately bright and black.

49. We have supposed the plane of reflection to be perpendicular to the edge where the two mirrors touch. If however it had been inclined in any manner, the result would have been precisely the same.

50. We have not yet considered the effect of a mixture of light of different colours in the same pencil (such as exists in white sun-light, and in most kinds of artificial light). According to the suppositions made in (23) this is represented by supposing the light to consist of different series of waves which may or may not be intermixed, the value of λ being different for each different series: and by (19) these different series cannot affect one another, and therefore the effect of each in producing illumination of its peculiar colour is to be considered separately, and the sum of the effects of all the illuminations to be taken afterwards.

51. Now if we examine the expression for the illumination, it will appear that at O the intensity is $\dfrac{4c^2}{(a+b)^2}$ whatever be the value of λ. Consequently, that point is illuminated by each of the differently coloured lights, with four times the quantity of illumination which there would have been if the light from one mirror only had fallen upon it. But there is no other point in similar circumstances. For if we put λ', λ'', λ''', &c. for the lengths of waves of differently coloured lights, λ' being the smallest and corresponding to the blue light, and c', c'', c''', &c. for the coefficient of displacement, and if we consider the point where $OM = \dfrac{a+b}{a \sin \alpha} \cdot \dfrac{\lambda'}{2}$, we find

for the intensity of the blue light, $\dfrac{4c'^2}{(a+b)^2}$, for the intensity of the next kind of light, $\frac{4c''^2}{(a+b)^2}\cos^2\frac{\pi\lambda'}{\lambda''}$,

for the intensity of the third kind of light, $\frac{4c'''^2}{(a+b)^2}\cos^2\frac{\pi\lambda'}{\lambda'''}$,

and so on. If they had been in the same proportion as in the light reflected from a single mirror, the intensities would have been

$$\frac{4c'^2}{(a+b)^2},\quad \frac{4c''^2}{(a+b)^2},\quad \frac{4c'''^2}{(a+b)^2},\quad \&c.$$

The different colours therefore are not mixed in the same proportion as in the original light. The same may be shewn of any other point: and thus if the original light be white, no point of the screen will be illuminated with white light except the middle of the central bright bar.

52. The same thing may be thus shewn. The breadth of the bright and dark bars for each colour is proportional to the value of λ for that colour (48). Consequently the bars are narrower for the blue rays than for the green: narrower for the green than for the yellow; &c. by (23). The line passing through the paper at O is however to be the center of a bright bar of each colour. In that line therefore there will be a perfect mixture of all the colours: at a line on each side there will be nearly a total absence of all: but beyond this the red bars will sensibly overshoot the yellow and green and blue bars, and the more as we recede farther from the center. Consequently the bars will be coloured, the bright bars being red on the outside and blue on the inside. And after two or three bars, the outside of the red bars will mingle with the inside of the next blue bars, and there will be no such thing as a black bar. This will continue as we recede from O till the colours become mixed in such a way that it is impossible to distinguish the bars, and the whole is a mass of tolerably uniform white light. This indistinctness of bars, and ultimately their disappearance, always take place when one of the mixed streams of light has described a path longer than that of the other

by several lengths of waves. In general, when white light is used, no bars can be seen where the length of one path exceeds that of the other by ten or twelve times the mean value of λ.

53. The quantity λ, as we have mentioned in (23), is so small that it could not be made sensible to the eye. But $\sin a$ may be made as small as we please, and consequently $\frac{a+b}{a \sin a} \lambda$ may be made large enough to be easily visible to the eye. It is by this and similar means that the lengths of waves for differently coloured light have been measured.

54. The agreement of the facts of experiment with these conclusions from the theory is most complete. And this may be considered as the fundamental experiment on which the undulatory theory is established. It is perfectly certain in this experiment that the mixture of two streams of light whether white or coloured does produce black. The bars next the central white are remarkably black: and the dark bars beyond the next bright bars are also very black: and upon intercepting either stream of light all these dark bars become bright. It appears plain that no theory of emission of particles can explain this fact: and it seems difficult to conceive that any theory except that of undulations can explain it.

55. We shall occasionally have to mention the system of bright and dark bars described in (52). We shall generally call them the *fringes of interference*.

56. The reader is particularly requested to observe that when $a+b$ or the distance of G and H from the screen is given, the breadth of the bars for any given colour is inversely as $a \cdot \sin a$ (48), or inversely as GH. And generally, the nearer together are the two sources of waves which interfere, the broader are the fringes of interference.

Prop. 13. A series of waves diverging from a point A fig. 15, falls upon the prism BCD, each of whose sides BC,

CD, makes the small angle a with the third side: to find the intensity of illumination on different parts of the screen EF where the two streams are mixed.

57. The investigation is exactly similar to that of the last proposition, with this difference only. By common Optics,

$$AG = AH = CA . (\mu - 1) . \sin \alpha \text{ nearly}, = (\mu - 1) . a \sin \alpha \text{ nearly}.$$

In the former investigation we had $GL = LH = a \sin \alpha$. Consequently where we find $a \sin \alpha$ in the former investigation we may put $(\mu - 1) . a . \sin \alpha$, and we shall have the correct expression for this case. Thus the intensity of illumination

$$= \frac{4c^2}{(a+b)^2} . \cos^2 \left\{ \frac{2\pi}{\lambda} . \frac{(\mu - 1) . a \sin \alpha}{a + b} OM \right\}$$

and the interval at which the centers of the bright and black bars succeed each other is

$$\frac{a + b}{(\mu - 1) a \sin \alpha} . \frac{\lambda}{4} .$$

58. The results are exactly similar to those obtained in (50), (51), (52), (the absolute breadth of the bars being different) with the following exception. The breadth of the bars for different colours does not (as before) depend simply on λ, but on $\frac{\lambda}{\mu - 1}$. Now μ varies with λ: it is greatest for the blue rays or those for which λ is least, and less for those for which λ is greater, through all the different colours. Consequently the breadths of the bars formed by the different colours are not in the same proportion as before, but are more unequal. The mixture of colours therefore at the edges of those bars which are a little removed from the central bar is not the same as before; and after a smaller number from the center, the colours of the different bars are mixed with each other (52).

Prop. 14. Suppose that in the experiment of Prop. 12 or 13 a thin piece of glass PQ is placed in the path of one of the pencils of light: to find the alteration produced in the fringes of interference.

59. Let T be the thickness of the glass: and consider the case of Prop. 12. It is plain that, as in (42), the length of that portion of the path of one pencil which traverses the glass is not to be estimated by its linear measure, but by $\mu \times$ that measure, inasmuch as the motion of the wave is slower in glass than in air by that proportion. We must consider therefore that instead of describing the space T in air, the wave describes a space equivalent to μT in air; and therefore the effect of the glass is the same as that of lengthening the path by $(\mu - 1)T$. Instead of HM in the expression of (47) we must put $HM + (\mu - 1)T$: and the intensity of light at M is now

$$\frac{4c^2}{LO^2} \cdot \cos^2 \frac{\pi}{\lambda} \{GM - HM - (\mu - 1)T\},$$

which as in (48) is changed to

$$\frac{4c^2}{(a+b)^2} \cdot \cos^2 \frac{2\pi}{\lambda} \left(\frac{a \sin a}{a+b} OM - \frac{\mu - 1}{2} T\right):$$

and the places of maximum brightness are now determined by making

$$\frac{a \sin a}{a+b} OM - \frac{\mu - 1}{2} T = 0, \quad \text{or} = \pm \frac{\lambda}{2}, \quad \text{or} = \pm \lambda, \text{ \&c.}:$$

or by making

$$OM = \frac{a+b}{2a \sin a} \cdot (\mu - 1)T, \quad \text{or} = \frac{a+b}{2a \sin a} \cdot \{(\mu - 1)T \pm \lambda\},$$

$$\text{or} = \frac{a+b}{2a \sin a} \{(\mu - 1)T \pm 2\lambda\}, \text{ \&c.}$$

60. Now if $\mu - 1$ were the same for rays of all colours, it is evident that these expressions would be precisely the same as those found for the bright points in (48), increased by a constant

$$\frac{a+b}{2a \sin a} (\mu - 1)T.$$

That is, the whole system of fringes would be shifted towards K, without any other alteration. As $\mu - 1$ is not constant this is not strictly true: the fringes are shifted, but there is also an alteration of colour arising from the difference of spaces, (not even proportional to the breadth of the bars), through which the differently coloured bars are shifted.

61. It will readily be imagined that if a piece of common glass were interposed, the lengths

$$GM \text{ and } HM + (\mu - 1)\,T$$

would, on account of the exceeding smallness of λ, differ in every point between I and K by many multiples of λ, and therefore (52) no fringes would be visible. The experiment may however be successfully performed by taking two pieces of glass from the same plate, whose difference of thickness will be very small, and placing one in the path of one pencil, and the other in the path of the other. But it may be better performed by taking a pretty uniform piece of glass, cutting it across the middle, and holding one half perpendicular to the path of one pencil, and the other half inclined to the path of the other. It is evident that the obliquity of passage produces the same effect as the use of a thicker piece of glass: and by gentle inclination the difference of paths may be made as small as we please.

62. The difference of paths is to be calculated thus. In fig. 16 let $WXYZ$ be the path of a portion of the wave perpendicular to one half, and $RSTV$ the path of another portion (which for simplicity we suppose moving in a parallel direction) through the other half whose angle of inclination is β. Let T' be the thickness. From T draw Ts perpendicular to RS produced. Since the front of the wave in air, when the portion in question is incident at S, is perpendicular to RS at the point S; and since, when the portion has reached T, the front of the wave in air is perpendicular to TV at the point T; it is plain that the wave has advanced in the direction perpendicular to its front only through the space Ss. But the time which has been occupied in this progress is the time of describing ST in glass

or $\mu . ST$ in air. Consequently the retardation (measured by the space which the wave would have advanced further, if it had moved in air) is

$$\mu . ST - Ss.$$

And the retardation of the portion incident at X is

$$(\mu - 1)\, XY \text{ or } (\mu - 1)\, T'.$$

Therefore the upper pencil is more retarded than the lower by

$$\mu . ST - Ss - (\mu - 1)\, T'.$$

The angle of incidence is β: and if γ be the angle of refraction,

$$ST = \frac{T'}{\cos\gamma}, \text{ and } Ss = \frac{T' . \cos(\beta - \gamma)}{\cos\gamma}:$$

$$\text{also } \mu = \frac{\sin\beta}{\sin\gamma};$$

$\therefore$ the retardation =

$$T'\left\{\frac{\sin\beta}{\sin\gamma\cos\gamma} - \frac{\cos(\beta-\gamma)}{\cos\gamma} - \frac{\sin\beta}{\sin\gamma} + 1\right\}$$

$$= 2\, T'\left(\sin^2\frac{\beta}{2} - \mu . \sin^2\frac{\gamma}{2}\right).$$

If β be small,

$$\sin\frac{\gamma}{2} = \frac{1}{\mu} . \sin\frac{\beta}{2} \text{ very nearly,}$$

and the retardation

$$= 2\, T'\left(1 - \frac{1}{\mu}\right)\sin^2\frac{\beta}{2} \text{ nearly.}$$

This is to be substituted for $(\mu - 1)\, T$ in the expressions of (59) and (60).

63. These conclusions are fully supported by experiment: and this is important as establishing one of the fundamental points of the undulatory theory of Optics, namely

that light moves more slowly in glass than in air. The whole of the investigation of the last proposition depends on this assumption.

PROP. 15. A series of waves is incident upon two plates of glass separated by a very small interval (fig. 17); part of the light is reflected at the lower surface of the first glass and part at the upper surface of the second glass: and these portions interfere: to find the intensity of the mixture.

64. Let AB be the path of one portion which is refracted in the direction BC, and of which one part is reflected in the direction CD, while another part is refracted at C and falls on the second plate at E, is partially reflected to F, and partially refracted in the direction FG parallel to CD. Draw FD perpendicular to CD. Then the path which one wave has described in going from C to D, measured by the equivalent path in vacuum, is $\mu . CD$: while that which the other has described in going from C to F (where its front has the same position as the front of that which has reached D) is $CE + EF$. The excess of the latter above the former is

$$CE + EF - \mu . CD.$$

Let D be the distance of the plates, γ the angle of incidence at C, β the angle of refraction. Then

$$CE + EF = \frac{2D}{\cos\beta}:$$

$$\text{and } CD = FC . \sin\gamma = 2D . \tan\beta . \sin\gamma;$$

$$\text{also } \mu = \frac{\sin\beta}{\sin\gamma}:$$

therefore the excess

$$= \frac{2D}{\cos\beta} - \frac{2D . \sin^2\beta}{\cos\beta} = 2D\cos\beta.$$

If then the extent of vibration in the light reflected from C be $A \sin \frac{2\pi}{\lambda}(vt - x)$, where the distance x is measured by the equivalent path in air; then the extent of vibration in the wave reflected from E will be represented by

$$B \sin \frac{2\pi}{\lambda}(vt - x - 2D\cos\beta);$$

and the whole intensity will be the intensity of the light in which the displacement of a particle is represented by the sum of these quantities. It must be recollected that by the reasoning in (37) we are entitled to suppose that the signs of A and B are different.

65. We have here however omitted the consideration of that part of the light which is reflected from F to H, again partially reflected at H and partially refracted at K: and the other parts successively reflected. It is plain that (putting V for $2D\cos\beta$) the part refracted at K will be retarded $2V$: that at the next point $3V$: and so on. Now suppose that when light goes from glass to air, the incident vibration being

$$a \sin \frac{2\pi}{\lambda}(vt - x),$$

the reflected vibration is

$$b \, . \, a \, . \sin \frac{2\pi}{\lambda}(vt - x),$$

and the refracted vibration

$$c \, . \, a \, . \sin \frac{2\pi}{\lambda}(vt - x):$$

and suppose that when light goes from air to glass, the coefficient is multiplied by e for the reflected vibration, and by f for the refracted vibration. Then if the coefficient for the light passing in the direction BC is a, that for the vibration reflected at C will be ab: that for the vibration refracted at F, $acef$: that for the vibration refracted at K, ace^3f: and so on. Thus the whole vibration is

$$ab\sin\frac{2\pi}{\lambda}(vt-x)+acef\{\sin\frac{2\pi}{\lambda}(vt-x-V)$$

$$+e^2\sin\frac{2\pi}{\lambda}(vt-x-2V)+e^4\sin\frac{2\pi}{\lambda}(vt-x-3V)+\&c.\}$$

$$=a\left\{b\sin\frac{2\pi}{\lambda}(vt-x)+cef.\frac{\sin\frac{2\pi}{\lambda}(vt-x-V)-e^2\sin\frac{2\pi}{\lambda}(vt-x)}{1-2e^2\cos\frac{2\pi}{\lambda}V+e^4}\right\}.$$

We shall anticipate so much of succeeding investigations (see Art. 125 and 126) as to state that, whether the vibrations are in or perpendicular to the plane of incidence*, there is reason to think that

$$\frac{e}{b}=-1,\ \text{and}\ cf=1-e^2.$$

Using these equations to simplify the expression; resolving it into the form

$$F\sin\frac{2\pi}{\lambda}(vt-x)+G\cos\frac{2\pi}{\lambda}(vt-x)$$

as in (17); and, as in (17) and (23), taking F^2+G^2 to represent the intensity, we find for the brightness of the reflected light

$$\frac{4a^2e^2\sin^2\frac{\pi}{\lambda}V}{1-2e^2\cos\frac{2\pi}{\lambda}V+e^4},\ \text{or}\ \frac{4a^2e^2\sin^2\frac{2\pi}{\lambda}D\cos\beta}{(1-e^2)^2+4e^2\sin^2\frac{2\pi}{\lambda}D\cos\beta}.$$

66. The supposition that we have made is that of a thin plate of air or vacuum inclosed between plates of glass, or mica, &c. But it is plain that the investigations

---

* When there are vibrations of both these kinds, it is necessary to calculate the illumination from each, and to take their sum.

apply in every respect to a thin plate of fluid with air on both sides: as for instance a soap-bubble. To examine particular cases,

(1) If $D = 0$, the intensity $= 0$ whatever be the value of λ. Thus it is found that where plates of glass &c., are absolutely in contact or very nearly so, there is no reflection: and when a soap-bubble has arrived at its thinnest state, just before bursting, the upper part appears perfectly black.

(2) The intensity is also 0 if

$$D \cos \beta = \frac{\lambda}{2}, \ \lambda, \ \frac{3\lambda}{2}, \ \&c.$$

But when light of different colours is mixed, it will be impossible to make the light of all the different colours vanish with the same value of D, and thus no value of D will produce perfect blackness.

(3) If $D \cos \beta = \frac{\lambda}{4}$, and if we take the value of λ corresponding to the mean rays (as the green-yellow), the intensity of light in the different colours will be nearly in the same proportion as in the incident light, or the reflected light will be nearly white. But this will not take place on increasing the value of D, or the reflected light will be coloured till D is become so large that for a great number of different kinds of light, corresponding to very small differences of λ, $\frac{4D \cos \beta}{\lambda}$ has the values of successive odd numbers.

Prop. 16. In the circumstances of the last proposition, to find the intensity of the light refracted into the second plate.

67. It is readily seen that the coefficient of the vibration refracted at E is $a.cf$: that of the vibration refracted at H

is $a.ce^2f$: and so on. Also the wave entering at H is behind that which entered at E by the same quantity V as before. Hence the sum of the vibrations will be

$$a.cf\left\{\sin\frac{2\pi}{\lambda}(vt-x)+e^2\sin\frac{2\pi}{\lambda}(vt-x-V)+\&c.\right\}$$

$$=a.cf.\frac{\sin\frac{2\pi}{\lambda}(vt-x)-e^2\sin\frac{2\pi}{\lambda}(vt-x+V)}{1-2e^2\cos\frac{2\pi}{\lambda}V+e^4}.$$

Treating this in the same manner as in (65), the intensity of light is found to be

$$\frac{a^2(1-e^2)^2}{(1-e^2)^2+4e^2\sin^2\frac{\pi V}{\lambda}}.$$

68. The proportional variations of this expression are much smaller than those of the expression of (65); its greatest value being a^2, and its least $\frac{a^2(1-e^2)^2}{(1+e^2)^2}$. The absolute variations are however exactly the same: and in fact the sum of the two expressions is always $=a^2$. This is expressed by saying that one of the intensities is *complementary* to the other. This relation spares us the necessity of examining every particular case of the value of D. If for any particular value of D the expression of (65) is maximum for any particular colour, that of (67) is minimum for the same colour: and so on. Thus if for some value of D the expression of (65) gave maximum intensity of red light, less of yellow, the mean intensity of green, less of blue, and nothing of violet (the mixture of which would produce a rich yellow): then the expression of (67) would give the minimum intensity of red light, more of yellow, the mean intensity of green, more of blue, and the maximum of violet (the mixture of which would produce a greenish blue diluted with much white). It is to be remarked that in the case of transmitted light the colours can never be so vivid as in reflected light, because none of the colours ever

wholly disappears, as no values of D and λ will make the expression of (67) = 0.

PROP. 17. Two glass prisms, right-angled or nearly so, (fig. 18) are placed with their hypotenusal sides nearly in contact: light is incident in such a manner that the angle of internal incidence at the hypotenusal side is nearly equal to the angle of total reflection: part of the light is reflected through the first prism, and part is refracted through the second: to find the expressions for the intensities.

69. The investigations and results are exactly the same as those of Prop. 15 and 16: but this case deserves a particular consideration for the following reason. In this case there is no difficulty (which there is in the others) in making the angle of incidence approach as near as we please to the angle of total reflection, and consequently no difficulty in making β (which is the angle of refraction from the first prism into air) as nearly $= 90^0$ as we please, or $\cos\beta$ as small as we please. Consequently $D\cos\beta$ may be made extremely small without making D very small. Now if $D\cos\beta$ in Prop. 15 and 16 were *moderately* small (as for instance $\frac{1}{1000}$ inch), we might find about twenty different colours of light dividing the colours from red to violet by tolerably equal shades, each of which, in consequence of the difference of their values of λ, would make

$$\sin^2\left(\frac{2\pi}{\lambda}D\cos\beta\right) = 1:$$

and between these colours, the expression would have all its changes of value. The mixture of light would therefore be produced by taking parcels from all the various shades of colour, and mixing them in the same proportion as in common light; and therefore would be nearly white. But when $D\cos\beta$ in this proposition is *extremely* small (for instance less than any value of λ, or not many times greater), not one colour perhaps, or not more than one or two, can be found, for which

$$\sin^2\left(\frac{2\pi}{\lambda}D\cos\beta\right) = 1:$$

and thus there will be an excess of some colours, and the light will be strongly coloured.

70. There is also another reason. By a small change of the angle of incidence we produce a small change in γ; and

$$\text{as } \frac{d\beta}{d\gamma} = \frac{\tan\beta}{\tan\gamma},$$

this produces a great change in β (β being nearly $= 90^0$). Consequently the change in $D\cos\beta$ is considerable: and the expression for the intensity of light will be varied much. If then the light of the clouds fall in different directions on this combination of prisms, or if the sun light be made (by a lens) to fall on it in different directions, the light both reflected and transmitted will form on a screen bands of light. As the position and breadth of these bands are different for every different colour, the mixture forms a very splendid series of coloured bands, in which the succession of colours differs from that produced by almost every other phenomenon of interferences. The same effect may be seen as well if the combination of prisms be held to the eye: when the light coming in different directions to the eye will exhibit the bands in great perfection.

PROP. 18. Two convex lenses of small curvature, or a convex lens and plane glass, (fig. 19) are placed in contact: to find the intensity of the light reflected and transmitted at any point *M*.

71. The two surfaces at *M* will be so nearly parallel that we may without sensible error consider them as parallel:* and therefore the investigations of Prop. 15 and 16 apply. It is only necessary to find an expression for *D*

* As we shall suppose in the investigation that the separation of the two surfaces at *M* is but a small multiple of λ, it is evident that for the points immediately about *M* the defect from parallelism will produce an error amounting only to a very small fraction of λ: and therefore the small waves in (32) will have their effects added together in the direction in which light is reflected from one of the surfaces, nearly in the same degree as in the direction in which it is reflected from the other surface.

in terms of OM, O being the point where the lenses are in contact. Let r be the radius of the lower surface of the upper lens: r' that of the upper surface of the lower lens. Then D or the separation at M is the sum of the versed sines of two circles whose radii are r and r', to the arcs whose chord is OM: and therefore

$$D = \frac{OM^2}{2r} + \frac{OM^2}{2r'} \text{ nearly } = OM^2\left(\frac{1}{2r} + \frac{1}{2r'}\right).$$

The intensity of reflected light (65) will therefore be

$$\frac{4a^2 e^2 \sin^2 \frac{\pi}{\lambda} V}{(1 - e^2)^2 + 4e^2 \sin^2 \frac{\pi}{\lambda} V},$$

and that of the transmitted light (67) will be

$$\frac{a^2 (1 - e^2)^2}{(1 - e^2)^2 + 4e^2 . \sin^2 \frac{\pi}{\lambda} V},$$

$$\text{where } V = OM^2 . \cos\beta \left(\frac{1}{r} + \frac{1}{r'}\right).$$

(1) The reflected light vanishes when

$$V = 0, \text{ or } = \lambda, \text{ or } = 2\lambda, \text{ \&c.},$$

or when

$$OM^2 = 0, \text{ or } = \frac{\lambda \sec\beta}{\frac{1}{r} + \frac{1}{r'}}, \text{ or } = \frac{2\lambda \sec\beta}{\frac{1}{r} + \frac{1}{r'}}, \text{ or } = \frac{3\lambda \sec\beta}{\frac{1}{r} + \frac{1}{r'}}, \text{ \&c.}$$

Consequently for any particular colour there will be a dark spot at O and a series of dark rings of which O is the center, and the squares of whose diameters are in the proportion of 1, 2, 3, &c.

(2) The most brilliant light is reflected when

$$OM^2 = \frac{\lambda \sec \beta}{2\left(\frac{1}{r} + \frac{1}{r'}\right)}, \text{ or } = \frac{3\lambda \sec \beta}{2\left(\frac{1}{r} + \frac{1}{r'}\right)}, \text{ or } = \frac{5\lambda \sec \beta}{2\left(\frac{1}{r} + \frac{1}{r'}\right)}, \text{ \&c.}$$

Consequently between the dark rings there are bright rings, the squares of the diameters of whose most brilliant parts are in the proportion of

$$\frac{1}{2}, \frac{3}{2}, \frac{5}{2}, \text{ \&c.}$$

(3) The diameters of these rings are *cæteris paribus* as $\sqrt{(\sec \beta)}$. Consequently on inclining the incident ray, or on depressing the position of the eye by which they are viewed, the diameters of the rings increase.

(4) For differently coloured rays, the diameters of the rings vary as $\sqrt{(\lambda)}$. The different colours which are mixed in white light produce therefore a series of rings whose diameters are different, and which overlapping each other produce a series of colours analogous to those mentioned in (52). The colours at last become mixed to such a degree that no traces of rings are visible.

(5) The diameters of the rings vary, *cæteris paribus*, as

$$\frac{1}{\sqrt{\left(\frac{1}{r} + \frac{1}{r'}\right)}}.$$

To make the rings large, therefore, $\frac{1}{r} + \frac{1}{r'}$ must be small, or r and r' must be large. If the lower glass be plane, or $\frac{1}{r'} = 0$, the diameters of the rings vary as $\sqrt{(r)}$.

(6) The transmitted light, just as in (68), produces rings complementary to those produced by the reflected light. The center therefore is bright, and is surrounded by a dark ring, which is followed by bright and dark rings alternately, which soon become coloured and finally cease to be visible. The diameter of each ring is the same as that of the ring of opposite character produced by reflected light.

72. These are commonly called *Newton's rings*, from the circumstance that the measures which suggested the most important part of Sir Isaac Newton's theory of light, and which have served in a great degree for the foundation of all theories, were made by him on those rings. The colours of the successive rings, arising from the different mixtures of all the colours producing white light, are commonly called *Newton's scale of colours.* In describing them it is usual to describe them by the number of the ring (including the central spot in the reckoning) in which they occur. For instance, the blue of the second order is not the blue which surrounds the central black spot, but the blue which surrounds the first black ring. This scale is valuable, as giving us an invariable series of colours which can at any time be produced without difficulty. The colours described in (52) as resulting from the experiment of Prop. 12 would do as well, but the adjustment of the apparatus for that experiment is much more troublesome.

Prop. 19. Light diverging from a center A, (fig. 20) is allowed to pass through a small aperture BC: to find the illumination on different points of the screen DE.

73. Suppose the wave diverging from A to proceed in a spherical form till it reaches CB: there suppose every part of it (within the limits of the aperture) to be the origin of a little wave proportional in intensity to the superficial extent of that part. By the principle of (21), the sum of the disturbances which each of these produces at M is to be taken for the whole disturbance there: and this being found, the intensity of light will be found as in

the preceding problems. Draw AO perpendicular to the screen, and consider it as the axis of z, A being the origin: let x be in the plane of the paper perpendicular to AO, and y perpendicular to the paper. Let $AB = a$, $AO = a + b$: (then b is very nearly the distance of the screen from the aperture): let x, y, and z be co-ordinates of any point P of the wave, and p and q co-ordinates of any point M of the screen in the directions of x and y (that in the direction of z being $a + b$). Suppose the front of the wave divided by lines perpendicular to the paper into narrow parallelograms, x and $x + \delta x$ being the co-ordinates of two of these lines: and suppose the parallelogram whose breadth is δx to be divided into small parallelograms by lines parallel to the paper, the co-ordinates of two lines being y and $y + \delta y$, or the surface of the small parallelogram formed by the intersection of these with the others being $\delta x . \delta y$. Then the displacement which the little wave originating at this surface would cause at M will be represented by

$$\delta x . \delta y . \sin \frac{2\pi}{\lambda}(vt - PM).$$

Now $PM^2 = (p - x)^2 + (q - y)^2 + (a + b - z)^2$

which, since $x^2 + y^2 + z^2 = a^2$,

$$\text{is } = (a + b)^2 + a^2 - 2px - 2qy - 2(a + b) . z.$$

$$\text{But } z = \sqrt{(a^2 - x^2 - y^2)} = a - \frac{x^2}{2a} - \frac{y^2}{2a} \text{ nearly,}$$

as x and y are supposed to be small even at their greatest values: therefore

$$-2(a + b)z = -2a^2 - 2ab + \frac{a+b}{a}x^2 + \frac{a+b}{a}y^2:$$

$$\text{and } PM^2 = b^2 + \frac{a+b}{a}x^2 - 2px + \frac{a+b}{a}y^2 - 2qy,$$

whence

$$PM = b + \frac{a+b}{2ab}x^2 - \frac{p}{b}x + \frac{a+b}{2ab}y^2 - \frac{q}{b}y \text{ nearly,}$$

$$\text{or } = b - \frac{ap^2}{2b(a+b)} - \frac{aq^2}{2b(a+b)} + \frac{a+b}{2ab}\left(x - \frac{ap}{a+b}\right)^2$$

$$+ \frac{a+b}{2ab}\left(y - \frac{aq}{a+b}\right)^2.$$

Put B for $b - \dfrac{ap^2}{2b(a+b)} - \dfrac{aq^2}{2b(a+b)}$:

then the displacement caused by the little wave is

$$\delta x \,.\, \delta y \,.\, \sin\frac{2\pi}{\lambda}\left\{vt - B - \frac{a+b}{2ab}\left(x - \frac{ap}{a+b}\right)^2 - \frac{a+b}{2ab}\left(y - \frac{aq}{a+b}\right)^2\right\}$$

$$= \delta x \,.\, \delta y \,.\, \sin\frac{2\pi}{\lambda}(vt - B) \,.\, \cos\frac{\pi}{\lambda} \,.\, \frac{a+b}{ab}\left\{\left(x - \frac{ap}{a+b}\right)^2 + \left(y - \frac{aq}{a+b}\right)^2\right\}$$

$$- \delta x \,.\, \delta y \,.\, \cos\frac{2\pi}{\lambda}(vt - B) \,.\, \sin\frac{\pi}{\lambda} \,.\, \frac{a+b}{ab}\left\{\left(x - \frac{ap}{a+b}\right)^2 + \left(y - \frac{aq}{a+b}\right)^2\right\}.$$

Call this $\delta x \,.\, \delta y \,.\, W$: and let v be the sum of all the disturbances for the slice between the lines whose ordinates are x and $x + \delta x$: w that for the whole surface. Since v increases by $\delta x \,.\, W \,.\, \delta y$ upon increasing y by δy,

$$\frac{dv}{dy} = \text{ultimate value of } \frac{\delta v}{\delta y} = \delta x \,.\, W,$$

$$\text{whence } v = \delta x \textstyle\int_y W.$$

This integration is to be performed, and the limits of the values of y for the integral will be expressed in terms of x from a knowledge of the shape of the aperture. The effect of the narrow parallelogram being $\delta x \int_y W$, it is found in the same manner that the effect of the whole aperture is $\int_x \int_y W$. As $\sin\dfrac{2\pi}{\lambda}(vt - B)$ and $\cos\dfrac{2\pi}{\lambda}(vt - B)$ are not concerned in the integration, the whole displacement may be expressed thus:

$$\sin\frac{2\pi}{\lambda}(vt-B)\int_x\int_y\cos\frac{\pi}{\lambda}\cdot\frac{a+b}{ab}\left\{\left(x-\frac{ap}{a+b}\right)^2+\left(y-\frac{aq}{a+b}\right)^2\right\}$$

$$-\cos\frac{2\pi}{\lambda}(vt-B)\int_x\int_y\sin\frac{\pi}{\lambda}\cdot\frac{a+b}{ab}\left\{\left(x-\frac{ap}{a+b}\right)^2+\left(y-\frac{aq}{a+b}\right)^2\right\}.$$

Let the integrals be E and F: then the whole displacement

$$=E\,.\sin\frac{2\pi}{\lambda}(vt-B)-F\,.\cos\frac{2\pi}{\lambda}(vt-B),$$

and hence {as in (17) and (23)} the illumination is represented by E^2+F^2.

74. Now

$$E=\int_x\int_y\left\{\cos\frac{\pi}{\lambda}\cdot\frac{a+b}{ab}\left(x-\frac{ap}{a+b}\right)^2\times\cos\frac{\pi}{\lambda}\cdot\frac{a+b}{ab}\cdot\left(y-\frac{aq}{a+b}\right)^2\right.$$

$$\left.-\sin\frac{\pi}{\lambda}\cdot\frac{a+b}{ab}\left(x-\frac{ap}{a+b}\right)^2\times\sin\frac{\pi}{\lambda}\cdot\frac{a+b}{ab}\cdot\left(y-\frac{aq}{a+b}\right)^2\right\}$$

proceeding therefore according to the precepts of (73), the first thing to be done is to find

$$\int_y\cos\left\{\frac{\pi}{\lambda}\cdot\frac{a+b}{ab}\cdot\left(y-\frac{aq}{a+b}\right)^2\right\}\text{ and }\int_y\sin\left\{\frac{\pi}{\lambda}\cdot\frac{a+b}{ab}\cdot\left(y-\frac{aq}{a+b}\right)^2\right\}.$$

These integrals cannot generally be found in terms of y.

Tables however have been formed* expressing the numerical value of the integrals

---

* Suppose U to be the function of unknown form which is the integral of S a given function of s. Then the values of U corresponding to $s-h$ and $s+h$

$$\text{are } U-\frac{dU}{ds}h+\frac{d^2U}{ds^2}\cdot\frac{h^2}{2}-\frac{d^3U}{ds^3}\cdot\frac{h^3}{2.3}+\&c.$$

$$\text{and } U+\frac{dU}{ds}h+\frac{d^2U}{ds^2}\cdot\frac{h^2}{2}+\frac{d^3U}{ds^3}\cdot\frac{h^3}{2.3}+\&c.$$

$$\int_s \cos\frac{\pi s^2}{2} \text{ and } \int_s \sin\frac{\pi s^2}{2}$$

for different values of s, and there is no difficulty in applying them to the numerical expression of the *first* integral in our problem. But as the integral thus found is given in numbers, and not in an expression involving y and which can be expressed in terms of x, it is seldom possible to perform the second integration.

75. In certain cases however this may be effected. For example, suppose the aperture to be a parallelogram whose sides are $2e$ and $2f$ in the direction of x and y, the line AO passing through its center. Let

$$\frac{a+b}{\lambda ab}\left(y - \frac{aq}{a+b}\right)^2 = \frac{s^2}{2};$$

$$\therefore y - \frac{aq}{a+b} = s\sqrt{\frac{\lambda ab}{2(a+b)}}, \text{ and } \frac{dy}{ds} = \sqrt{\frac{\lambda ab}{2(a+b)}};$$

$$\therefore \int_y \cos\left\{\frac{\pi}{\lambda}\cdot\frac{a+b}{ab}\cdot\left(y - \frac{aq}{a+b}\right)^2\right\} = \sqrt{\frac{\lambda ab}{2(a+b)}}\cdot\int_s \cos\frac{\pi s^2}{2};$$

and this is to be taken from

$$y = -f \text{ to } y = +f,$$

---

and therefore the value of the integral from $s-h$ to $s+h$ is

$$2\frac{dU}{ds}h + 2\frac{d^3U}{ds^3}\cdot\frac{h^3}{2.3} + \&c.$$

$$\text{or } 2Sh + 2\frac{d^2S}{ds^2}\cdot\frac{h^3}{2.3} + \&c.$$

which can be easily calculated: and by taking h small enough, one or two terms will be sufficient. In this way the values of the integral can be computed for successive limits, and the sum will be the integral up to any given value of s. A table of $\int_s \cos\frac{\pi s^2}{2}$ and $\int_s \sin\frac{\pi s^2}{2}$ will be given at the end of this Tract.

or from

$$s = -\sqrt{\left\{\frac{2(a+b)}{\lambda ab}\right\}}\left(f + \frac{aq}{a+b}\right)$$

$$\text{to } s = \sqrt{\left\{\frac{2(a+b)}{\lambda ab}\right\}}\left(f - \frac{aq}{a+b}\right).$$

This will be the sum of the two numbers, in the table of $\int_s \cos\frac{\pi s^2}{2}$, corresponding to

$$s = \sqrt{\left\{\frac{2(a+b)}{\lambda ab}\right\}}\left(f + \frac{aq}{a+b}\right)$$

$$\text{and } s = \sqrt{\left\{\frac{2(a+b)}{\lambda ab}\right\}}\left(f - \frac{aq}{a+b}\right).$$

Let these be A_1, A_2: and after proceeding in a similar way for the sine, let the numbers in the table of $\int_s \sin\frac{\pi s^2}{2}$ corresponding to

$$s = \sqrt{\left\{\frac{2(a+b)}{\lambda ab}\right\}}\left(f + \frac{aq}{a+b}\right)$$

$$\text{and } s = \sqrt{\left\{\frac{2(a+b)}{\lambda ab}\right\}}\left(f - \frac{aq}{a+b}\right)$$

be B_1, B_2,

Then $E =$

$$\int_x \left\{\sqrt{\left\{\frac{\lambda ab}{2(a+b)}\right\}}(A_1 + A_2)\cos\frac{\pi}{\lambda}\cdot\frac{a+b}{ab}\left(x - \frac{ap}{a+b}\right)^2 - \sqrt{\left\{\frac{\lambda ab}{2(a+b)}\right\}}(B_1 + B_2)\sin\frac{\pi}{\lambda}\cdot\frac{a+b}{ab}\left(x - \frac{ap}{a+b}\right)^2\right\}.$$

In the same manner $F =$

$$\int_x \left\{\sqrt{\left\{\frac{\lambda ab}{2(a+b)}\right\}}(A_1 + A_2)\sin\frac{\pi}{\lambda}\cdot\frac{a+b}{ab}\left(x - \frac{ap}{a+b}\right)^2 + \sqrt{\left\{\frac{\lambda ab}{2(a+b)}\right\}}(B_1 + B_2)\cos\frac{\pi}{\lambda}\cdot\frac{a+b}{ab}\left(x - \frac{ap}{a+b}\right)^2\right\}.$$

Integrating with respect to x in exactly the same manner between $x = -e$ and $x = +e$, and putting A_3, A_4, for the numbers in the table of $\int_s \cos \frac{\pi s^2}{2}$ corresponding to

$$s = \sqrt{\left\{\frac{2(a+b)}{\lambda a b}\right\}} \left(e + \frac{ap}{a+b}\right)$$

$$\text{and } s = \sqrt{\left\{\frac{2(a+b)}{\lambda a b}\right\}} \left(e - \frac{ap}{a+b}\right),$$

and B_3, B_4 for those in the table of $\int_s \sin \frac{\pi s^2}{2}$ corresponding to the same,

$$E = \frac{\lambda a b}{2(a+b)} \{(A_1 + A_2)(A_3 + A_4) - (B_1 + B_2)(B_3 + B_4)\}$$

$$F = \frac{\lambda a b}{2(a+b)} \{(A_1 + A_2)(B_3 + B_4) + (B_1 + B_2)(A_3 + A_4)\}.$$

The intensity or $E^2 + F^2$

$$= \frac{\lambda^2 a^2 b^2}{4(a+b)^2} \{(A_1 + A_2)^2 + (B_1 + B_2)^2\} . \{(A_3 + A_4)^2 + (B_3 + B_4)^2\}.$$

This expression (omitting the first factor) is the product of two factors, of which one depends entirely on q and the other depends entirely on p. If a certain value of p makes its factor small, every part of the screen for which p has that value will have a small intensity of light. A similar remark applies to the values of q which make the other factor small. Thus the screen will be crossed by bars of light of different intensity at right angles to each other.

76. Our limits will not allow us to examine in detail this or other cases. The discussion of the values for particular values of p and q depends entirely upon examination of the numerical results: and this must be done for a great number of values of p and q before any conjecture can be formed as to the fringes &c. about the edges. One of the

simplest cases is, to find the intensity of light produced by the shadow of a plate bounded by a straight line. If y is parallel to the edge, and x for the edge $= 0$, then the limits of the first integration are

$$\text{from } y = -\infty \text{ to } y = +\infty,$$

and those of the second

$$\text{from } x = 0 \text{ to } x = \infty.$$

This case has been fully considered by M. Fresnel, and he has arrived at this conclusion. If a plane be drawn through the bright point and the edge of the plate, and if the intersection of this with the screen be called the geometrical shadow: then on the dark side of the geometrical shadow the intensity of the light diminishes rapidly without increasing at all, and soon becomes insensible: but on the bright side the light increases and diminishes by several alternations before it acquires sensibly its full brightness, forming a succession of several bands on the bright side of the geometrical shadow. And as, for the same point, the limits for which the tabular numbers are taken are different for different values of λ, and as the factor of the whole varies with λ, the intensity of the differently coloured lights will be differently proportioned at different points, and thus the bands will be coloured, nearly as in (52). This phænomenon, known by the name of *Grimaldi's coloured fringes*, had long been observed, and an imperfect explanation was given by Newton. In Fresnel's *Mémoire sur la Diffraction* it was shewn, from accurate measures with various values of a and b, to be a consequence of the theory of undulations, (*Memoires de l' Institut*, 1821).

77. Another instance is, if the form of the plate be a square *corner*, then besides the bands on the outside of the geometrical shadow there are seen within it hyperbolical curves as in fig. 21. The accurate investigation* may be

* In this and the preceding case, it is necessary to consider the effect produced by small waves diverging from distances sensibly different. In the investigation we suppose that the absolute effect of each of these is the same as the effect of a wave diverging from a surface of equal extent at a smaller distance. This is manifestly incorrect: but it produces no sensible error in the result, for the reason mentioned in (29).

performed as above: but a general explanation may be given thus. Let P and Q be points similarly situated on the two sides: the small waves diverging from their neighbourhood would, as in (46), produce bands by their interferences; and the breadth of these, by (56), would be inversely as the distance of P and Q. Consequently the nearer P and Q are taken to the solid angle, the broader will the bands be, and their form will therefore resemble the hyperbola. In this we have omitted the effects of interference of other portions of the light nearer to and further from the angle, but as the omitted parts would at different points produce effects nearly similar, it is probable that the general form of the curves will be similar to hyperbolas.

78. Another instance is, if the light fall on a very narrow slit, coloured bands of much greater breadth are thrown on the screen. The second case of (25) sufficiently explains their origin. If the slit be triangular, it was observed by Newton that the bands are rectangular hyperbolas, the asymptotes being parallel and perpendicular to the axis of the triangle. This appears from the same investigation, as the intervals between the bands are inversely as b the breadth of the aperture, or inversely as the distance from the geometrical shadow of the triangle's vertex.

79. In the following instance we may find the intensity at one point without much trouble. Suppose the aperture in (73) to be a round hole: to find the intensity at that point of the screen which is the projection of its center. Divide the circle into rings, the inner and outer radii of one being r and $r + \delta r$, or its surface being $2\pi r \delta r$. The distance of every point of this ring from the point of the screen is

$$b + \frac{a+b}{2ab} r^2 \text{ nearly,}$$

and hence the whole displacement at the central point of the screen is

$$\int_r 2\pi r \sin \frac{2\pi}{\lambda} \left(vt - b - \frac{a+b}{2ab} r^2\right),$$

$$\text{or} \quad \frac{\lambda ab}{a+b} \cos \frac{2\pi}{\lambda}\left(vt - b - \frac{a+b}{2ab} r^2\right).$$

If c be the radius of the hole, this is to be taken from $r = 0$ to $r = c$, and its value is

$$\frac{2\lambda ab}{a+b} . \sin \frac{2\pi}{\lambda}\left(vt - b - \frac{a+b}{4ab} c^2\right) . \sin \frac{2\pi}{\lambda} . \frac{a+b}{4ab} c^2.$$

The intensity of illumination is consequently

$$\frac{4\lambda^2 a^2 b^2}{(a+b)^2} . \sin^2 \left(\frac{2\pi}{\lambda} . \frac{a+b}{4ab} c^2\right).$$

On referring to (71) it will be seen that this expression is nearly similar to the expression for the intensity of the reflected light in Newton's rings, considering the denominator in (71) as constant, and making

$$V = \frac{c^2 (a+b)}{2ab}:$$

and consequently the colours are nearly the same for the same values of V. To obtain the colours corresponding to those of the inner rings, we must have V as small as possible, or $\frac{1}{a} + \frac{1}{b}$ must be as small as possible, and therefore b must be as great as possible. On diminishing b, V increases. Consequently if we first place the screen at a very great distance and then bring it nearer to the aperture, the series of colours at the center will be the same as those found on tracing Newton's rings outwards: but as we cannot make $\frac{1}{a} + \frac{1}{b} = 0$, we cannot have all the orders beginning from the central black. This agrees with observation. For any other point of the screen, the intensity can be found only by the general method of (73)*.

* The reader will find little trouble in applying the same principles to the demonstration of the following phænomena:

The shadow of a long and narrow parallelogram consists of several bands, the central band being white, and the others coloured, and separated by darker bands.

The central point of the shadow of a very small circle is white, and its brightness is sensibly the same as if the circle did not obstruct the light.

PROP. 20. Every thing remaining as in the last problem, except that, close to the hole, a lens is placed of such focal length that light diverging from A will be made to converge to O: to find the intensity of light on the screen.

80. From (44) it appears that the form of the front of the wave after refraction by the lens will be a sphere of which O is the center. Let O, fig. 22, be the origin of co-ordinates: p and q the co-ordinates of a point M on the screen: x, y, z, those of P, z being parallel to OA. Then

$$PM^2 = (p - x)^2 + (q - y)^2 + z^2.$$

But by the equation to the surface of a sphere,

$$x^2 + y^2 + z^2 = b^2:$$

hence

$$PM^2 = b^2 + p^2 + q^2 - 2px - 2qy,$$

$$\text{and } PM = b + \frac{p^2 + q^2}{2b} - \frac{px}{b} - \frac{qy}{b} \text{ nearly.}$$

The terms depending on x^2 and y^2 will be insensible, as they will be multiplied by the very small quantities p^2 and q^2. Put B for $b + \frac{p^2 + q^2}{2b}$: then, as in (73), the whole displacement at M is

$$\int_x \int_y \sin \frac{2\pi}{\lambda}\left(vt - B + \frac{px}{b} + \frac{qy}{b}\right).$$

This expression is much simpler than that of (73), as there are no terms involving x^2 and y^2. The first integration can always be performed: it gives

$$-\frac{b\lambda}{2\pi q} \cos \frac{2\pi}{\lambda}\left(vt - B + \frac{px}{b} + \frac{qy}{b}\right):$$

and if y' and y'' are the least and greatest values of y for a given value of x (given by the equation to the aperture in terms of x), the first integral is, between these limits,

$$\frac{b\lambda}{2\pi q}\left\{\cos \frac{2\pi}{\lambda}\left(vt - B + \frac{px}{b} + \frac{qy'}{b}\right) - \cos \frac{2\pi}{\lambda}\left(vt - B + \frac{px}{b} + \frac{qy''}{b}\right)\right\}$$

$$= \cos \frac{2\pi}{\lambda}(vt - B) \cdot \frac{b\lambda}{2\pi q} \cdot \left\{\cos \frac{2\pi}{b\lambda}(px + qy') - \cos \frac{2\pi}{b\lambda}(px + qy'')\right\}$$

$$- \sin \frac{2\pi}{\lambda}(vt - B) \cdot \frac{b\lambda}{2\pi q} \cdot \left\{\sin \frac{2\pi}{b\lambda}(px + qy') - \sin \frac{2\pi}{b\lambda}(px + qy'')\right\}.$$

Let the integrals of the terms within the brackets (with respect to x and between the proper limits) be P and Q: then the coefficients of

$$\cos \frac{2\pi}{\lambda}(vt - B) \text{ and } \sin \frac{2\pi}{\lambda}(vt - B)$$

are respectively $\frac{b\lambda}{2\pi q} P$ and $-\frac{b\lambda}{2\pi q} Q$,

and the intensity of the light is $\frac{b^2\lambda^2}{4\pi^2 q^2}(P^2 + Q^2)$.

81. Ex. Let the aperture be a parallelogram whose sides are $2e$ and $2f$ in the direction of x and y. Here

$$y' = -f, \quad y'' = +f:$$

$$\cos \frac{2\pi}{b\lambda}(px + qy') - \cos \frac{2\pi}{b\lambda}(px + qy'')$$

$$= \cos \frac{2\pi}{b\lambda}(px - qf) - \cos \frac{2\pi}{b\lambda}(px + qf)$$

$$= 2 \sin \frac{2\pi}{b\lambda} px \cdot \sin \frac{2\pi}{b\lambda} qf,$$

the integral of which is

$$-\frac{b\lambda}{\pi p} \cdot \sin \frac{2\pi}{b\lambda} qf \cdot \cos \frac{2\pi}{b\lambda} px;$$

which from $x = -e$ to $x = +e$ gives $P = 0$. Also

$$\sin \frac{2\pi}{b\lambda}(px + qy') - \sin \frac{2\pi}{b\lambda}(px + qy'')$$

$$= \sin \frac{2\pi}{b\lambda}(px - qf) - \sin \frac{2\pi}{b\lambda}(px + qf)$$

$$= -2 \cos \frac{2\pi px}{b\lambda} \cdot \sin \frac{2\pi qf}{b\lambda}$$

the integral of which is

$$-\frac{b\lambda}{\pi p} \cdot \sin \frac{2\pi qf}{b\lambda} \cdot \sin \frac{2\pi px}{b\lambda},$$

which from $x = -e$ to $x = +e$ gives

$$Q = -\frac{2b\lambda}{\pi p} \cdot \sin \frac{2\pi qf}{b\lambda} \cdot \sin \frac{2\pi pe}{b\lambda}.$$

Hence the intensity is

$$\frac{b^4\lambda^4}{\pi^4 p^2 q^2} \cdot \sin^2 \frac{2\pi qf}{b\lambda} \cdot \sin^2 \frac{2\pi pe}{b\lambda},$$

$$\text{or } 16e^2f^2 \cdot \left(\frac{b\lambda}{2\pi qf} \sin \frac{2\pi qf}{b\lambda}\right)^2 \cdot \left(\frac{b\lambda}{2\pi pe} \sin \frac{2\pi pe}{b\lambda}\right)^2.$$

This expression is maximum when $p = 0$, $q = 0$: so that there is a bright point in the place of the image determined by common Optics. It is 0 when p is any multiple of $\frac{b\lambda}{2e}$, or when q is any multiple of $\frac{b\lambda}{2f}$. This shews that the screen is traversed by rectangular dark bars at equal intervals, the intervals in the direction of the length of the parallelogram being shorter than the others. For a given value of p, the brightness is greatest when $q = 0$, or when q has one of the values which makes $\frac{b\lambda}{2\pi qf} \cdot \sin \frac{2\pi qf}{b\lambda}$ maximum. Thus it appears that there will be a brilliant point at the center; a four-rayed cross through the center, the rays being interrupted at intervals; and a series of less bright patches in square arrangement in the angles of the cross: also the distances from the center are greater for the red rays than for the blue. When the parallelogram is narrow, the bright parts in the direction of one side form one of the kinds of spectra described by Fraunhofer.

82. Let the aperture be an equilateral triangle. Take x in the direction of the perpendicular to one side, and let the angle opposite this side be the origin of co-ordinates; let e be the whole length of the perpendicular. Then

$$y' = -x.\tan 30^0: \quad y'' = +x.\tan 30^0.$$

Hence

$$P = \int_x \left\{\cos\frac{2\pi x}{b\lambda}(p - q.\tan 30^0) - \cos\frac{2\pi x}{b\lambda}(p + q\tan 30^0)\right\}$$

$$= \frac{b\lambda}{2\pi(p - q\tan 30^0)}\sin\frac{2\pi x}{b\lambda}(p - q\tan 30^0)$$

$$-\frac{b\lambda}{2\pi(p + q\tan 30^0)}\sin\frac{2\pi x}{b\lambda}(p + q\tan 30^0),$$

the value of which from $x = 0$ to $x = e$ is found by putting e for x. And

$$Q = \int_x \left\{\sin\frac{2\pi x}{b\lambda}(p - q\tan 30^0) - \sin\frac{2\pi x}{b\lambda}(p + q\tan 30^0)\right\},$$

the value of which from $x = 0$ to $x = e$ is

$$\frac{b\lambda}{2\pi(p - q\tan 30^0)}\left\{1 - \cos\frac{2\pi e}{b\lambda}(p - q\tan 30^0)\right\}$$

$$-\frac{b\lambda}{2\pi(p + q\tan 30^0)}\left\{1 - \cos\frac{2\pi e}{b\lambda}(p + q\tan 30^0)\right\}.$$

The sum of the squares is $\left(\text{omitting the factor } \frac{b^2\lambda^2}{4\pi^2}\right)$

$$\frac{1}{(p - q\tan 30^0)^2}.\left\{2 - 2\cos\frac{2\pi e}{b\lambda}(p - q\tan 30^0)\right\}$$

$$+\frac{1}{(p+q\tan 30^0)^2}\left\{2-2\cos\frac{2\pi e}{b\lambda}(p+q\tan 30^0)\right\}$$

$$-\frac{2}{p^2-q^2\tan^2 30^0}\times$$

$$\left\{1+\cos\frac{4\pi eq\tan 30^0}{b\lambda}-\cos\frac{2\pi e}{b\lambda}(p-q\tan 30^0)-\cos\frac{2\pi e}{b\lambda}(p+q\tan 30^0)\right\}$$

$$=\frac{2p^2+6q^2\tan^2 30^0}{(p^2-q^2\tan^2 30^0)^2}-\frac{4pq\tan 30^0+4q^2\tan^2 30^0}{(p^2-q^2\tan^2 30^0)^2}\cos\frac{2\pi e}{b\lambda}(p-q\tan 30^0)$$

$$+\frac{4pq\tan 30^0-4q^2\tan^2 30^0}{(p^2-q^2\tan^2 30^0)^2}\cos\frac{2\pi e}{b\lambda}(p+q\tan 30^0).$$

$$-\frac{2p^2-2q^2\tan^2 30^0}{(p^2-q^2\tan^2 30^0)^2}\cos\frac{4\pi eq\tan 30^0}{b\lambda}.$$

Let $p=r\cos\theta$, $q=r\sin\theta$: which is the same as referring M to the central point of the screen by polar co-ordinates. Then observing that $\tan^2 30^0=\frac{1}{3}$, and restoring the factors

$$\frac{b^2\lambda^2}{4\pi^2}\text{ and }\frac{b^2\lambda^2}{4\pi^2q^2},$$

this may be put in the form

$$\frac{3.b^4\lambda^4}{32\pi^4.r^4}\cdot\frac{1}{\sin^2\theta.\sin^2(\theta-60^0).\sin^2(\theta-120^0)}$$

$$\times\left\{\frac{3}{4}-\sin(\theta-60^0).\sin(\theta-120^0).\cos\left(\frac{4\pi re}{b\lambda\sqrt{(3)}}\sin\theta\right)\right.$$

$$-\sin(\theta-120^0).\sin(\theta-180^0).\cos\left(\frac{4\pi re}{b\lambda\sqrt{(3)}}\sin(\theta-60^0)\right)$$

$$\left.-\sin(\theta-180^0).\sin(\theta-240^0).\cos\left(\frac{4\pi re}{b\lambda\sqrt{(3)}}.\sin(\theta-120^0)\right)\right\}.$$

The maximum value it will be found is when $r=0$, and is

$= \frac{e^4}{3}$. The value is also considerable when $\theta = 0$, or $= 60^0$, or $= 120^0$, or $= 180^0$, or $= 240^0$, or $= 300^0$, when it is

$$\frac{\lambda^2 e^2 b^2}{3\pi^2 r^2}\left(1 - \frac{\lambda b}{\pi r e}\sin\frac{2\pi r e}{\lambda b}\right) + \frac{\lambda^4 b^4}{6\pi^4 r^4}\left(1 - \cos\frac{2\pi r e}{\lambda b}\right).$$

This points out exactly the star-like form observed by Sir J. Herschel (*Encycl. Metrop.* Light, Art. 772).

83. Let the aperture be a great number m of equal parallelograms of length $2f$ and breadth e at equal distances g. Here $y' = -f$, $y'' = +f$: and the expression to be integrated is

$$\cos\frac{2\pi}{\lambda}\left(vt - B + \frac{p}{b}x - \frac{qf}{b}\right) - \cos\frac{2\pi}{\lambda}\left(vt - B + \frac{p}{b}x + \frac{qf}{b}\right)$$

$$= 2\sin\frac{2\pi qf}{\lambda b}\,.\,\sin\frac{2\pi}{\lambda}\left(vt - B + \frac{p}{b}x\right).$$

The general integral is

$$-\frac{\lambda b}{\pi p}\sin\frac{2\pi qf}{\lambda b}\,.\cos\frac{2\pi}{\lambda}\left(vt - B + \frac{p}{b}x\right).$$

If k be the value of x corresponding to the first side of the first parallelogram, that corresponding to the first side of the $(n+1)^{\text{th}}$ parallelogram will be $k + n(e+g)$, and that corresponding to its last side $k + n(e+g) + e$. The integral therefore for the $(n+1)^{\text{th}}$ parallelogram is

$$\frac{\lambda b}{\pi p}\,.\sin\frac{2\pi qf}{\lambda b}\,.\left\{\cos\frac{2\pi}{\lambda}\left(vt - B + \frac{pk}{b} + \frac{pn(e+g)}{b}\right)\right.$$

$$\left. - \cos\frac{2\pi}{\lambda}\left(vt - B + \frac{pk}{b} + \frac{pn(e+g)}{b} + \frac{pe}{b}\right)\right\}$$

$$= \frac{2\lambda b}{\pi p}\,.\sin\frac{2\pi qf}{\lambda b}\,.\sin\frac{\pi pe}{\lambda b}\,.\sin\frac{2\pi}{\lambda}\left\{vt - B + \frac{pk}{b} + \frac{pe}{2b} + \frac{p(e+g)}{b}n\right\}.$$

Let $B - \frac{pk}{b} - \frac{pe}{2b} = C$: then the whole displacement at M $\{$restoring the factor $\frac{b\lambda}{2\pi q}$ from (80)$\}$ is

$$\frac{\lambda^2 b^2}{\pi^2 pq} \sin\frac{2\pi qf}{\lambda b} . \sin\frac{\pi pe}{\lambda b} \Sigma \sin\frac{2\pi}{\lambda}\left\{vt - C + \frac{p(e+g)}{b} n\right\}.$$

The finite integral of the last term with respect to n is

$$\frac{-1}{2\sin\frac{p(e+g)\pi}{b\lambda}} . \cos\frac{2\pi}{\lambda}\left\{vt - C + \frac{p(e+g)}{b}(n - \tfrac{1}{2})\right\},$$

which from $n = 0$ to $n = m$ is

$$\frac{\sin\frac{mp(e+g)\pi}{b\lambda}}{\sin\frac{p(e+g)\pi}{b\lambda}} \sin\frac{2\pi}{\lambda}\left\{vt - C + \frac{p(e+g)}{b} . \frac{m-1}{2}\right\}.$$

Thus, putting

$$C - \frac{p(e+g)}{b} . \frac{m-1}{2} = D,$$

we have for the whole displacement

$$\frac{\lambda^2 b^2}{\pi^2 pq} . \sin\frac{2\pi qf}{\lambda b} . \sin\frac{\pi pe}{\lambda b} . \frac{\sin\frac{mp(e+g)\pi}{b\lambda}}{\sin\frac{p(e+g)\pi}{b\lambda}} . \sin\frac{2\pi}{\lambda}(vt - D):$$

and consequently the intensity of the light is

$$4e^2 f^2 . \left(\frac{\lambda b}{2\pi qf} . \sin\frac{2\pi qf}{\lambda b}\right)^2 . \left(\frac{\lambda b}{\pi pe} . \sin\frac{\pi pe}{\lambda b}\right)^2 . \left\{\frac{\sin . \frac{p(e+g)\pi}{b\lambda} m}{\sin\frac{p(e+g)\pi}{b\lambda}}\right\}^2 .$$

84. The most remarkable variation of this depends on the last term. This may be represented by $\left(\frac{\sin m\theta}{\sin\theta}\right)^2$, where

m is a large whole number. This has a great number of maxima corresponding nearly to the values of θ which make $m\theta$ an odd multiple of $\frac{\pi}{2}$; but the greatest maximum is that found by making $\sin\theta = 0$. Its value is then m^2, which is much greater than any of the others. For, the next maximum, when $m\theta = \frac{\pi}{2}$ nearly, is $\frac{1}{\left(\sin\frac{\pi}{2m}\right)^2}$ nearly $= \frac{4m^2}{\pi^2}$: the next is nearly $= \frac{4m^2}{9\pi^2}$ &c.: and when $\sin\theta$ is nearly $= 1$, the maximum is nearly 1. As we approach to the value $\theta = \pi$, one or two values are sensible, and then we reach the large maximum, of the same value as before. Suppose now we have placed on the object glass of a telescope a grating consisting of 100 parallel wires. There will be a bright image of the luminous point formed at the center of the field, and one or two less bright on each side, so close that they cannot be distinguished: after this there will be others, but their intensity will diminish so rapidly (being at one of the *maxima* only $\frac{1}{10000}$ of that of the brightest) that they will be invisible; and at a distance there will be another point as bright as the first: and at an equal distance beyond it, another: and so on. Thus there will be a succession of luminous points at equal distances from each other, with no perceptible light between them. The distance of these points is found by making

$$\theta = 0, \text{ or } \pi, \text{ or } 2\pi \text{ \&c.; or } p = 0, \text{ or } \frac{b\lambda}{e+g}, \text{ or } \frac{2b\lambda}{e+g}, \text{ \&c.}$$

This applies to any one kind of homogeneous light. When there is a mixture of differently coloured lights (as in white light), there will be a union of bright points of all the colours where $p = 0$, but at no other place. For, to arrive at the second bright point, we must go to a distance from the first proportional to λ. Consequently the next blue point will be nearer to the center than the next red point, &c. Thus in the center there will be a bright white point, but at the sides there will be spectra similar to those formed by a prism, their blue ends being nearest to the center. And as each bright point is almost perfectly insulated, the spectrum will be *pure;* that is, there will be no sensible mixture of colours in any part of it. This is verified completely by experiment: the spectra are so pure that, when the solar light is used, the fixed lines or interruptions of the spectrum, which are so delicate that only the best prisms will shew them, may be seen in the spectra formed as we have described.

85. We shall now consider the term

$$\left(\frac{\lambda b}{\pi p e} \cdot \sin \frac{\pi p e}{\lambda b}\right)^2.$$

When p is small, or when e is small, this is $= 1$. When p is increased to any multiple of $\frac{\lambda b}{e}$, it vanishes. Consequently, whenever the same value of p is a multiple of $\frac{\lambda b}{e}$ and of $\frac{\lambda b}{e + g}$, one of the spectra will disappear: that is, whenever e and g are commensurate. This is true in experiment. And at all events, the successive spectra are less bright than the central colourless image, this term having its greatest value when $p = 0$.

It is unnecessary to consider the effect of the first term, as it only points out the law of brightness in the direction of the length of the parallelograms.

86. The whole of the experiments which are the subject of Prop. 20 are easily made by limiting the aperture of the object glass of a telescope, or by placing gratings before it. The appearances which we have investigated are those that would be formed on a screen in the focus of the object glass; but it is well known by common Optics that the appearance presented to the eye, when an eye-glass is applied whose focus coincides with the focus of the object glass, is just the same as if the light had been received on a screen placed

there. Thus it is only necessary to limit the aperture and then to view a bright point (as a star), when the phænomena that we have described will be seen in great perfection.

87. The appearance when the aperture is not limited, or is left in a circular form, is found by making

$$y' = -\sqrt{(e^2 - x^2)},\ y'' = +\sqrt{(e^2 - x^2)}:$$

e being the radius of the object glass. On pursuing the investigation the result will be found to depend on the integral

$$\int_e \sqrt{(e^2 - x^2)} . \cos nx.$$

This cannot be exhibited, and thus we cannot find completely the appearance. But we have found that when the aperture is a square or a parallelogram there are patches of light surrounding the image at distances inversely proportional to its breadth, and thus we might expect *à priori* to find with a circular aperture rings surrounding the image, their diameters being inversely as the aperture. This is observed to be true in experiment. For a complete calculation of this case, the reader is referred to the *Cambridge Transactions*, Vol. v. p. 283.

88. The experiment of (83) &c. is particularly remarkable on this account. It shews that there is light diverging in all directions from the front of the grand wave where it meets the lens, which is insensible only because it is destroyed by other light. For if we view a luminous point with a telescope in its usual state, no side images are seen: on putting a grating on the object glass, which *intercepts* a part of the light, the side images are visible. That this depends simply on the obstruction of the light, and not on any reflection or refraction by the grating, is evident from this circumstance, that it is indifferent whether the grating consist of wire, or silk, or lines scratched on the glass with a diamond point, or lines ruled on a film of soap or grease. The same principle may be used to explain the spectra produced by the reflection of light from metallic surfaces on which lines are engraved at very small equal distances. In fig. 23 if light from F falls on a small reflecting surface Ad and is received on a screen GH, and if F and G be

both distant, then a point G may be found such that the paths FAG, FBG, &c. will not sensibly differ in length; and therefore the small waves which are produced by the same great wave, coming from every part of the surface, will meet in the same phase at G. And this will be true whether any part of the surface is removed or not. But at H there will be no illumination, because we may divide the surface into parts A, a, B, b, &c. such that the path FaH (supposing the surface a continuous plane) is less than FAH by $\frac{\lambda}{2}$, and therefore the small wave coming from a will destroy that coming from A; the small wave coming from b will destroy that coming from B: and so on. Now suppose that we remove the parts a, b, c, d, &c. There is now no wave to destroy any one of those coming from A, B, &c.: and they will not destroy each other, because the path FBH being less than FAH by λ, FCH being less than FBH by λ, &c., they are all in the same phase. Consequently there will be brightness at H. For different values of λ it is evident that we must take points at different distances from G: and thus spectra will be formed nearly as in (84).

For calculations applying to various cases of interference, the reader is referred to several volumes of the *Cambridge Transactions*, and of the *Philosophical Magazine*.

APPLICATION OF THE THEORY OF UNDULATIONS TO THE PHÆNOMENA OF POLARISED LIGHT.

89. In the preceding investigations no reference has been made to the direction in which the particles of the luminiferous ether vibrate. They might, like the particles of air in the transmission of sound, vibrate in the direction in which the wave is passing: or they might, like the particles of a musical string, vibrate perpendicularly to the direction of the wave, but all in one plane passing through that direction. To these or any other conceivable vibrations our investigations would apply equally well: all that is required being that they should be subject to the general law

of undulations, and that for a considerable number of vibrations the extent and time of vibration should be the same. The phenomena of polarization however enable us to point out what is the kind of vibration.

90. The properties of Iceland spar (which, it has since been discovered, are possessed by the greater number of transparent crystals) first pointed out the characteristic law of polarization. If a pencil of common light be made to pass through a rhombohedron of this crystal, it is separated into two of equal intensity. This may be shewn either by viewing a small object through it, when two images will be seen; or by placing it behind a lens on which the light of the Sun or that of a lamp is thrown, when two images will be formed at the focus. A line drawn through those two images is in the direction of the shorter diagonal of the rhombic face of the crystal; the rhombohedron being supposed to be bounded by planes of cleavage, and the pencil of light being incident perpendicular to one of them.

91. On examining the paths of these pencils within the crystal, it is found that one of them obeys the ordinary laws of refraction, but the other follows a more complicated law (which we shall hereafter describe). The first is therefore called the Ordinary pencil, and the other the Extraordinary pencil: and they are frequently denoted by the letters O and E.

92. To the eye no difference is discoverable between the two pencils, or between either of them and a pencil of common light whose intensity is the same. Yet the properties of the light in the two pencils are different, and both are different from common light. For if we take one of the pencils (for instance O) and place a second rhombohedron before it; on turning the first rhombohedron it is found that in general the second crystal separates the pencil O into two of *unequal* intensity, one following the ordinary law and the other the extraordinary law (which we may call Oo and Oe), and that in certain relative positions of the crystal one of the pencils wholly disappears. On ex-

amining the positions it is found that when the two rhombohedrons are in similar positions (that is, when all the planes of cleavage of one are parallel to those of the other), or when they are in opposite positions (that is when, keeping the same surfaces in contact, the first is turned 180° from the position just described), *Oe* disappears, and *Oo* only remains; that is, there is only an ordinary pencil produced by the second crystal. On the contrary, when the first rhombohedron is turned 90° either way from the position first described, *Oo* disappears, and *Oe* only remains: that is, there is only an extraordinary pencil produced. In any intermediate position that pencil is strongest which, in the nearest of the four positions that we have mentioned, does not vanish.

93. Now if instead of *O* we take the pencil *E*, the appearances are wholly changed. In general, as before, the second rhombohedron divides this into two pencils of unequal intensity, one following the ordinary and the other the extraordinary law (which we shall call *Eo* and *Ee*). But when the crystals are in similar or in opposite positions *Eo* vanishes, and *Ee* only remains: that is, there is only an extraordinary pencil produced. When one is turned 90° from the similar position, *Ee* vanishes and *Eo* remains: that is, there is only an ordinary pencil produced.

94. It appears then that neither* of these two pencils is similar to common light; for, when either of them is received on a second rhombohedron, it does not always produce two pencils: common light always does produce two. It appears also that they are not similar to each other; for, in certain positions of the second rhomb, *O* will produce only an ordinary ray, while *E* will produce only an extraordinary ray: in certain other positions, *O* will produce only an extraordinary ray, and *E* only an ordinary ray. The rays therefore have some peculiar properties depending on the

---

* The reader will observe that the term Ordinary pencil does not signify that the pencil is similar in its properties to common light, but merely that it is subject to the same laws of refraction as common light.

position of the crystal. But between the properties of the two rays a remarkable relation can be found. When the rhombohedrons are in similar positions, O will produce only an ordinary ray. When the first is turned 90^0, E will produce only an ordinary ray. Consequently, on turning the crystal 90^0, E has the same properties which O had before turning it. Again, when the rhombohedrons are in similar positions, E will produce only an extraordinary ray. On turning the first through 90^0, O will produce only an extraordinary ray. Consequently, on turning the crystal 90^0, O has the same properties which E had before turning it. This shews clearly that the two pencils have properties of the same kind with reference to two planes passing through their direction and moving with the crystal; and that the two planes are at right angles to each other. If a plane passing through the direction of the pencil and the shorter diagonal of the rhombic face be called the principal plane* of the crystal, then we may assert that the properties of the Ordinary ray possess the same relation to the principal plane which the properties of the Extraordinary ray possess to the plane at right angles to the principal plane. This is commonly expressed thus: the Ordinary ray is *polarized in the principal plane*, and the Extraordinary ray is *polarized in a plane perpendicular to the principal plane.*

95. There are some crystals which possess the property of separating common light into an Ordinary and an Extraordinary pencil, and then absorbing one of them. Thus certain specimens of agate, and plates of tourmaline cut parallel to the axis, allow the Ordinary pencil to pass, and nearly suppress the Extraordinary. This however is only true when the plates have a certain thickness: for, when they are very thin, the Ordinary and Extraordinary pencils are seen to have equal intensities. But the method of exhibiting polarized light which is most extensively used in experiments is, to reflect common light from unsilvered glass or any other transparent substance, solid or fluid. It is found that if the tangent of the angle of incidence

* This term will be used hereafter in a more general sense.

is equal to the refractive index, the whole of the reflected light is *polarized** in the same way as the Ordinary ray produced by the first rhombohedron of Iceland spar when its principal plane is parallel to the plane of reflection from the unsilvered glass, &c. For if the second rhombohedron be placed in that position to receive the reflected ray (instead of receiving the ray from the first rhombohedron), an ordinary ray only is produced: if in the position differing 90^0 from this, an extraordinary ray only is produced: which (92) is exactly the same effect as that produced by *O*, the crystal having the position that we have described. This is expressed by saying that the reflected light is *polarized in the plane of reflection.* The angle of incidence which is proper for this is called the *polarizing angle.* The transmitted light, it is found, possesses in part the properties of the Extraordinary ray (the principal plane of the crystal with which we mentally compare it being still conceived to be parallel to the plane of reflection). For, if the second rhomb be placed in that position, the transmitted light produces both an ordinary and an extraordinary ray, but the former is less bright than the latter. This is expressed by saying that the transmitted light is *partially polarized in the plane perpendicular to the plane of reflection.* If a great number of plates of unsilvered glass be placed together, the reflected light appears completely polarized in the plane of reflection, and the transmitted light appears completely polarized in the plane perpendicular to the plane of reflection.

96. We have here considered the test of polarization to be, the formation of only one ray when the light passes through a crystal of Iceland spar. But as the reflection from unsilvered glass at the polarizing angle gives properties exactly similar to those of the Ordinary ray of Iceland spar (the principal plane of the spar being conceived parallel to the plane of reflection), it may be conjectured that light polarized

* This was the discovery of Malus. It was important at the time, as calling the attention of philosophers to the subject: but all the phænomena of coloured rings, &c. may be exhibited perfectly well without using this property of reflection.

in the plane perpendicular to the plane of reflection, as it would not furnish any Ordinary ray with Iceland spar, will not furnish a reflected ray from unsilvered glass, but will be entirely transmitted. This is verified by experiment: and thus we have an easy practical test of the polarization of light. If light incident at the polarizing angle on unsilvered glass is not susceptible of reflection, it is polarized in the plane perpendicular to the plane of reflection. And if, on turning the glass round the incident ray without varying the inclination, the reflected light does not vanish in any position of the glass, the light is not polarized. In the same manner the polarization of a ray may be ascertained by examining the state of the emergent ray, after incidence on a plate of tourmaline; if in any position of the tourmaline the emergent ray disappears, the plane of polarization of the incident ray is perpendicular to the plane then passing through the ray and through the axis of the tourmaline.

97. From this it will easily be inferred that if two such plates of tourmaline are placed with their axes at right angles to each other, no light can pass through them. For the light which is transmitted by the first is polarized in the plane of its axis, that is, in the plane perpendicular to the axis of the second: and therefore is not allowed to pass through the second. If one of the tourmalines be turned, light is immediately seen: it increases till the axes of the tourmalines are parallel. In the same manner if in fig. 24 *A* be a plate or several parallel plates, of unsilvered glass, and *B* an unsilvered glass* (whose posterior surface is blackened, to prevent reflection there), *B* being fixed on a block which turns round a spindle at *C* in the direction of *AB*: and if each of the glass surfaces make with *AB* an angle of about $33^0. 13'$ (the refractive index of plate glass for mean rays being about $1{,}527 = \tan 56^0. 47'$): then on receiving the light of the clouds on *A* in such a direction that the reflected light falls on *B*, and placing the eye to receive the light reflected from *B*, it is seen that when the planes

* Instead of reflection from an unsilvered glass, transmission through a plate of tourmaline may be used.

of reflection coincide, or nearly coincide, a considerable quantity of light is reflected from B: as the planes of reflection are inclined, less light is reflected: and when (as in the figure) they are perpendicular to each other, no light is reflected. This is strictly true only for the light incident from A on B in direction of the line joining their centers: but it is nearly true for light making a small angle with this. It is strictly true also for light of only one colour (since the polarizing angle, which depends on the refractive index, is different for differently coloured rays), but if the mean angle be used it is nearly true for all. We shall frequently allude to this apparatus: we shall then call A the *polarizing plate* and B the *analyzing plate*.

98. Now in the experiment of Prop. 12 or 13, (fig. 14 and 15) if a plate of tourmaline be placed in each of the pencils of light supposed to start from G and H, the plates of tourmaline being cut from the same large plate which has been carefully worked to uniform thickness, it is found that the existence of the fringes of interference depends entirely on the relative position of the axes of the tourmaline plates. If both axes be parallel, whatever be their position, the fringes of interference are seen well, and the dark bars are perfectly black. If they are not parallel, the dark bars are not so black; and if they are at right angles to each other, the fringes disappear entirely. It appears therefore that pencils of light polarized in planes at right angles to each other cannot interfere (that is cannot destroy each other) in circumstances in which pencils of common light, or pencils of light polarized in the same plane, can destroy each other.

99. From the experiments that we have described, the following general conclusions are drawn.

(1) If from common light we produce, by any known contrivance, light that is polarized in one plane, there is always produced at the same time light more or less polarized in the plane perpendicular to the former.

(2) Light polarized in one plane cannot be made to furnish light polarized in the perpendicular plane.

(3) Light polarized in one plane cannot be destroyed by light polarized in the perpendicular plane.

The first leads at once to the presumption that polarization is not a modification or change of common light, but a resolution of it into two parts equally related to planes at right angles to each other; and that the exhibition of a beam of polarized light requires the action of some peculiar forces (either those employed in producing ordinary reflection and refraction or those which produce crystalline double refraction) which will enable the eye to perceive one of these parts without mixture of the other. This presumption is strongly supported by the phænomena of partially-polarized light. If light falls upon a plate of glass inclined to the ray, the transmitted light, as we have seen, is partially polarized. If now a second plate of glass be placed in the path of the transmitted light, inclined at the same angle as the former plate, but with its plane of reflection at right angles to that of the former plate, the light which emerges from it has lost every trace of polarization; whether it be examined only with the analyzing plate B, or by the interposition of a plate of crystal in the manner to be explained hereafter (144). This seems explicable only on the supposition that the effect of the first plate of glass was to diminish that part of the light which has respect to one plane (without totally removing it), and that the effect of the second plate is to diminish in the same proportion that part of the light which has respect to the other plane, and therefore that, after emergence from the second plate, the two portions of light have the same proportion as before. On considering this presumption in conjunction with the second and third conclusions, we easily arrive at this simple hypothesis explaining the whole:

Common light consists of undulations in which the vibrations of each particle are in the plane perpendicular to the direction of the wave's motion. The polarization of light is the resolution of the vibrations of each particle into two, one parallel to a given plane passing through the direction of the wave's motion, and the other perpendicular to that plane; which (from causes that we shall not allude to at present), become in certain cases the origin of waves that travel in different directions. When we are able to separate one of those from the other, we say that the light of each is polarized. When the resolved vibration parallel to the plane is preserved unaltered, and that perpendicular to the plane is diminished in a given ratio (or vice versa), and not separated from it, we say that the light is partially polarized.

The reader who has possessed himself fully of this hypothesis, will see at once the connection between all the experiments given above.

100. For the general explanation of these experiments, and for the accurate investigation of most of the phænomena to be hereafter described, it is indifferent whether we suppose the vibrations constituting polarized light to take place parallel to the plane of polarization, or perpendicular to it. There are reasons however, connected with the most profound investigations* into the nature of crystalline separation and into the nature of reflection from glass, &c., and confirming each other in a remarkable degree, that incline us to choose the latter: and thus,

When we say that light is polarized in a particular plane, we mean that the vibration of every particle is perpendicular to that plane.

Thus, in the undulation constituting the Ordinary ray of Iceland spar, the vibration of every particle is perpendicular to the principal plane of the crystal: in that constituting the Extraordinary ray, the vibration of every particle is parallel to the principal plane. When light falls upon unsilvered glass at the polarizing angle, the reflected wave is formed entirely by vibrations perpendicular to the plane of incidence: the transmitted wave is formed by some vibrations perpendicular to the plane of incidence, with an excess of vibrations parallel to the plane of incidence.

* M. Cauchy's theoretical investigations have led him to the conclusion that the vibrations constituting polarized light are parallel to the plane of polarization. These investigations however do not connect the different classes of phænomena so completely as Fresnel's, which are adopted in the text.

101. The reader will perceive that it is absolutely necessary to suppose, either that there are no vibrations in the direction of the wave's motion, or that they make no impression on the eye. For if there were such, there ought in the experiment of (98) to be visible fringes of interferences: of such however there is not the smallest trace.

102. As we now suppose light generally to consist of two sets of vibrations which cannot interfere with each other, it becomes important to establish some measure of the intensity of the compound light. It seems that this cannot be any other than the sum of the intensities corresponding to the two sets of vibrations. So that if the displacement from one vibration be represented by $a \sin(vt-x+A)$, and that from the other by $b . \sin(vt-x+B)$, the intensity of the mixed light will be a^2+b^2. This then is the expression which we ought in strictness to have used in our former investigations. But as in all these (except those relating to reflection from plane glasses and lenses) the quantities a and b have in every part of the operation the same proportion, it is evident that the results, considered as giving the proportion of intensities of light, are in every instance correct.

Prop. 21. To explain on mechanical principles the transmission of a wave in which the vibrations are transverse to the direction of its motion.

103. In fig. 25 let the faint dots represent the original situations of the particles of a medium, arranged regularly in square order, each line being at the distance h from the next. Suppose all the particles in each vertical line disturbed vertically by the same quantity; the disturbances of different vertical lines being different. Let x be the horizontal abscissa of the second row; $x-h$ that of the first, and $x+h$ that of the third: let u, $u_{\prime}$, and u' be the corresponding disturbances. The motions will depend upon the extent to which we suppose the forces are sensible. Suppose the only particles whose forces on A are sensible, to be

$$B, \; C, \; D, \; E, \; F, \; G,$$

(omitting those in the same line, as their attractions are equal and in opposite directions): and suppose them to be attractive, and as the inverse square of the distance: and the absolute force of each $= m$. The whole force tending to pull A downwards is

$$\frac{m(h+u-u_{,})}{\{h^2+(h+u-u_{,})^2\}^{\frac{3}{2}}}+\frac{m(u-u_{,})}{\{h^2+(u-u_{,})^2\}^{\frac{3}{2}}}-\frac{m(h-u+u_{,})}{\{h^2+(h-u+u_{,})^2\}^{\frac{3}{2}}}$$

$$+\frac{m(h+u-u')}{\{h^2+(h+u-u')^2\}^{\frac{3}{2}}}+\frac{m(u-u')}{\{h^2+(u-u')^2\}^{\frac{3}{2}}}-\frac{m(h-u+u')}{\{h^2+(h-u+u')^2\}^{\frac{3}{2}}}.$$

Expanding these fractions, and neglecting powers of $u-u_{,}$ and $u-u'$ above the first, the force tending to diminish u is

$$\left(1-\frac{1}{2^{\frac{3}{2}}}\right)\cdot\frac{m}{h^3}(2u-u_{,}-u').$$

Putting for $u_{,}$,

$$u-\frac{du}{dx}h+\frac{d^2u}{dx^2}\cdot\frac{h^2}{2},$$

and for u',

$$u+\frac{du}{dx}h+\frac{d^2u}{dx^2}\cdot\frac{h^2}{2},$$

we find

$$\frac{d^2u}{dt^2}=\left(1-\frac{1}{2^{\frac{3}{2}}}\right)\frac{m}{h}\cdot\frac{d^2u}{dx^2}*$$

an equation of exactly the same form as that for the transmission of sound (10). The solution therefore has the same form: and therefore the transversal motion of particles supposed here follows the same law as the direct motion of the particles of air: that is, it follows the law of undulation.

* If h is so large with regard to the length of a wave that the terms after h^2 cannot be safely neglected, we may, by assuming a form for the function expressing u, integrate the equation

$$\frac{d^2u}{dt^2}=-\left(1-\frac{1}{2^{\frac{3}{2}}}\right)\frac{m}{h^3}(2u-u_{,}-u').$$

If,

104. It seems probable that if we had supposed any other regular arrangement, or taken any other law of force, the same conclusion would have been obtained. And if we suppose the arrangement symmetrical with respect to certain fixed lines, but different in distance of particles, &c. in different directions (as for instance if we suppose every eight adjacent particles to be at the angles of a parallelopiped as in fig. 26), then for vibrations of the particles in different directions the multiplier of $\frac{d^2u}{dx^2}$ will be different, and consequently the velocity of transmission of the wave (which is the square root of the multiplier) will be different. And the velocities of two waves may be different even when they are

---

If, as we usually suppose,

$$u = A \sin \frac{2\pi}{\lambda}(vt - x),$$

$$\text{then } \frac{d^2u}{dt^2} = -\frac{4\pi^2 v^2}{\lambda^2} A \sin \frac{2\pi}{\lambda}(vt - x),$$

$$u_{,} + u' = A \sin \frac{2\pi}{\lambda}(vt - x + h)$$

$$+ A \sin \frac{2\pi}{\lambda}(vt - x - h)$$

$$= 2A \,.\, \cos \frac{2\pi h}{\lambda} \,.\, \sin \frac{2\pi}{\lambda}(vt - x),$$

$$\text{and } 2u - u_{,} - u' = 4A \,.\, \sin^2 \frac{\pi h}{\lambda} \,.\, \sin \frac{2\pi}{\lambda}(vt - x),$$

and by substitution, the equation becomes

$$-\frac{4\pi^2 v^2}{\lambda^2} = -4\left(1 - \frac{1}{2^{\frac{3}{2}}}\right) \cdot \frac{m}{h^3} \,.\, \sin^2 \frac{\pi h}{\lambda};$$

$$\text{whence } v^2 = \left(1 - \frac{1}{2^{\frac{3}{2}}}\right) \cdot \frac{m}{h} \cdot \frac{\lambda^2}{\pi^2 h^2} \sin^2 \frac{\pi h}{\lambda},$$

$$\text{and } v = \sqrt{\left(1 - \frac{1}{2^{\frac{3}{2}}}\right) \cdot \frac{m}{h}} \cdot \frac{\sin \frac{\pi h}{\lambda}}{\frac{\pi h}{\lambda}}.$$

This expression increases as λ increases, that is, undulations consisting of long waves travel with greater velocity than those consisting of short waves. Thus the different refrangibility of differently coloured rays is accounted for. See Article 38. For other modifications of this theory, and their comparison with the observed indices of refraction of different rays in different media, the reader is referred to Professor Powell's treatise *On the Undulatory Theory as applied to the explanation of unequal refrangibility.*

going in the same direction, provided that one of these waves consist of vibrations in one direction, and the other of vibrations in another direction, as if for instance in fig. 26 the directions of both waves were perpendicular to the paper, but if one set of vibrations were in the direction up and down, and the other in the direction right and left. For the force with which the particles act on each other depends on the distance of the particles in the direction of the waves' motion, and on their distance in the direction of the particles' vibration: and in the case supposed, the latter element is different for the two waves, though the former is the same.

105. If the displacement of a particle, considered as in any direction, be resolved into three displacements in the directions of x, y, z, the variations of force in those directions produced by the alteration of a single particle (and consequently the force produced by the whole system) are the same as if the displacements in those directions had been made independently. From this it easily follows that the sum of any number of displacements causes forces equal to the sum of the forces corresponding to the separate displacements: and then, by the reasoning in (10) and (11), any number of undulations, produced by vibrations in different directions, may co-exist without disturbing each other.

PROP. 22. To explain the separation of common light into two pencils by doubly refracting crystals: and to account for the polarization of the two rays in planes at right angles to each other.

106. We shall assume for the state of the particles of ether within a crystal, an arrangement similar to that described in (104), or at least possessing this property, that there are three directions* at right angles to each other, in which if a particle be disturbed, the resultant of the forces

* M. Fresnel has demonstrated that this must be the case when the small displacement of a particle in the direction of any one of the co-ordinates produces forces in the direction of all, represented by multiples of that displacement. This is apparently the most general supposition that can be made. *Memoires de l' Institut*, 1824.

acting on it will tend to move it back in the same line in which the displacement is produced. These lines we suppose to be parallel to some lines determined by the form of the crystal.

107. Now in general the displacement* of a particle or a series of particles will not produce a force whose direction coincides with the line of displacement. For suppose the disturbance in the direction of x to be X; that in the direction of y to be Y: and suppose the corresponding forces to be $a^2 X$ and $b^2 Y$. The tangent of the angle made by the resultant force with the axis of x is $\frac{b^2 Y}{a^2 X}$: but the tangent of the angle made by the direction of displacement with the axis of x is $\frac{Y}{X}$: and these are different if a^2 and b^2 are different. In the same manner if we supposed a displacement Z in the direction of z, and if it produced a force $c^2 Z$, the tangents of the angles, made by the projection of the resultant's direction on the planes of xz and yz with the axis of z, would be $\frac{a^2 X}{c^2 Z}$ and $\frac{b^2 Y}{c^2 Z}$: while those made by the projection of the line of displacement would be $\frac{X}{Z}$ and $\frac{Y}{Z}$.

108. Now suppose that, in fig. 26, MN is the front of a wave: or by the definition of (20) and the assumptions

---

* We have spoken here of *displacements* as if the forces concerned in the transmission of a wave were thus put in play by *absolute* displacements. It is however plain from (103) that the forces on A really put in play are produced by *relative* displacements: but it is evident that these forces are the same as those that would be put in play by the *absolute* displacement

$$\tfrac{1}{2}(2u - u_{,} - u') \text{ or } \frac{d^2 u}{dx^2} \cdot \frac{h^2}{2}.$$

In like manner, when the direction of displacement is any whatever, the quantity $\frac{1}{2}(2u - u_{,} - u')$ in its proper direction may be resolved into the direction of the co-ordinates, and the forces really acting on A will be the forces corresponding to these spaces considered as *absolute* displacements.

of (99) and (101), all the particles in the plane of which *MN* is the projection are moving with equal motions in that plane. In general the force which acts on these particles in consequence of their displacement, is not in the direction of the displacement, and is not even in the plane *MN*. Resolve it into two, one perpendicular to the plane and one parallel to it. The former of these may be neglected, because it can only produce a motion which by (101), is not sensible to the eye. The latter, though in the plane, will not generally be in the direction of the displacement. It is impossible then to find what motions will be caused by this displacement without resolving it. If we can resolve it into two, such that the force produced by each displacement is in the direction of that displacement, then we can find for each of these separately the motions that will result from it. It is clear now that we have fallen on a case exactly similar to that of (104), and the conclusion is the same, namely that there will be two series of waves travelling with different velocities.

109. Now in (34) we have found that the refraction in a transparent medium depends on the velocity of the wave within that medium. Consequently the refraction of the two series of waves will be different; and thus is explained the bifurcation of the ray, when common light is incident on a surface of the crystal. And each of these consists of vibrations parallel to one line, that is, by (99), of polarized light: and, as will appear by subsequent investigations, the lines of vibration are perpendicular to each other, and therefore the planes of polarization (which are perpendicular to the lines of vibration) are perpendicular to each other. This explanation may be considered as the most important step in Physical Science since the establishment of the law of gravitation by Newton.

Prop. 23. To investigate the law of double refraction in uniaxal crystals.

110. By *uniaxal crystals* we mean those in which $b^2 = a^2$, while c^2 is not equal to a^2. The investigation, it is seen from

(108) and (109) reduces itself to these two things; the discovery of those directions of displacement in the plane of a wave in which the resolved part of the force parallel to the plane is in the same direction as the displacement, and the investigation of the velocity of transmission for waves whose vibrations are in those directions. Now the force, produced by a displacement in any direction parallel to the plane of xy, is in the same direction as the displacement; and therefore it is indifferent what line in the plane of xy we take for x. Let x then be perpendicular to the intersection of the front of the wave with the plane xy. In fig. 27 let MN be the projection on the paper of the front of the wave (supposed perpendicular to the paper), AM the axis of x, AN the axis of z, which we shall call the axis* of the crystal: θ the angle made by the front of the wave with the plane of xy. Then it is plain, from the symmetry of the forces with respect to z, that a displacement parallel to the line MN will cause a force whose resolved part parallel to the plane MN is in the line MN; and that a displacement in the plane MN perpendicular to the line MN will cause a force also perpendicular to MN. The vibration then of the wave incident on the crystal must be resolved into two vibrations parallel to these, and these vibrations, as in (108) will produce two rays that will travel with different velocities.

111. Now the force put in play by a displacement perpendicular to the paper is represented by a^2. displacement. Consequently the wave depending on these vibrations moves with the velocity a whatever be the position of the front of the wave. This is the same law as that assumed in (34) for common refracting media, and the resulting law of refraction is therefore the same. Let any plane passing through the axis of z or the axis of the crystal be called a *principal plane* of the crystal; then this conclusion may be stated thus: the waves consisting of vibrations perpen-

---

* This always coincides with the mineralogical axis of the crystal. Thus in Iceland spar it is in the solid angle included by three obtuse angles of the planes of cleavage, and makes equal angles with them; in quartz, tourmaline, beryl, &c. it is the axis of the prism.

dicular to a principal plane of the crystal are refracted according to the ordinary law of refraction. This accounts for the refraction and polarization of the ordinary ray.

112. For the displacement in the plane of the paper: putting D for that displacement, it may be resolved into $D\cos\theta$ parallel to x, and $D\sin\theta$ parallel to z. The resulting forces will be represented by $a^2.D\cos\theta$ parallel to x and $c^2D\sin\theta$ parallel to z; and the sum of the resolved parts of these parallel to MN is represented by

$$D(a^2\cos^2\theta + c^2\sin^2\theta).$$

The velocity of transmission of the wave perpendicular to its own front is therefore

$$\sqrt{(a^2\cos^2\theta + c^2\sin^2\theta)}.$$

This is not the same in all directions; and hence the waves consisting of vibrations parallel to a principal plane of the crystal are not refracted according to the ordinary law.

113. If now the front of a wave produced by such vibrations have at any time the form PQR fig. 28, the form of the front at a succeeding time will be determined by taking Pp perpendicular to the surface at P and proportional to the value of $\sqrt{(a^2\cos^2\theta + c^2\sin^2\theta)}$ there; Qq perpendicular to the surface at Q and having the same proportion to the value of $\sqrt{(a^2\cos^2\theta + c^2\sin^2\theta)}$ there; &c. If the wave were produced originally by an agitation at C, all the successive fronts must be similar; and if we take points of all where their tangents are parallel, that is, points along a radius CQ, the perpendicular distance of each front from the next is proportional to

$$\sqrt{(a^2\cos^2\theta + c^2\sin^2\theta)},$$

and therefore the sum of all, which is the same as the perpendicular on the tangent at Q, must be proportional to

$$\sqrt{(a^2\cos^2\theta + c^2\sin^2\theta)}.$$

Therefore, to find the form of an extraordinary wave diverging from a point, we must solve this problem: To find the curve where the perpendicular on the tangent is proportional to $\sqrt{(a^2\cos^2\theta + c^2\sin^2\theta)}$, θ being the angle made by the tangent with the axis of x. It is well known that this is an ellipse, whose axes in the directions of z and x are in the proportion of $a : c$. Consequently, to discover the path of the extraordinary ray, we must suppose the waves produced by vibrations parallel to a principal plane to diverge in the form of a spheroid of revolution round a line parallel to the axis of z, and must suppose the semi-axes of the spheroid parallel and perpendicular to z to be represented by a and c: and must then proceed as for common light. The radius of the sphere into which the ordinary wave has diverged must at the same time be represented by a.

114. It is easily seen that the motion of an extraordinary wave in the crystal is not generally perpendicular to its front. For let *AB*, fig. 29, be an aperture through which a small part of an extraordinary wave passes: *CD* a line parallel to the axis of the crystal. Consider *A*, *a*, *b*, *c*, &c. as the origins of equal spheroidal waves, the axes of the waves being parallel to *CD*. It is plain that the part between *E* and *F* is the only place in which the waves strengthen each other, as at all points on both sides of this they precede or follow each other by different quantities, and therefore mutually destroy each other, while between *E* and *F* the neighbouring waves meet in nearly the same phase. The wave therefore will seem to travel from *AB* to *EF*. The general rule therefore is this; describe a spheroid whose axis is parallel to the axis of the crystal, and find the point of its surface where the tangent plane is parallel to the front of the wave; then the motion of the wave is parallel to the radius of that point.

115. The general construction for determining the path of both rays is this. In fig. 30, let the plane of the paper be the plane of incidence, *BA′* the projection of the surface of the crystal, *AB* the front of a wave moving in the direction *AA′*. Let *CD* be the axis of the crystal, not necessarily in the plane of the paper. While a part of

the wave moves in vacuum from A to A', suppose that the ordinary wave diverging from B will spread into the sphere Fo, and the extraordinary wave into the spheroid Fe (whose axis of revolution = diameter of sphere). Through the line, of which A' is the projection, draw a plane touching the sphere in o; this plane is the front of the ordinary wave, and Bo represents the direction and velocity of its motion. Through the same line draw a plane touching the spheroid in e; this plane is the front of the extraordinary wave, and Be represents the direction and velocity of its motion. If the axis of the spheroid does not lie in the plane of the paper, the point e will not be in the plane of the paper: and thus the direction of the ordinary ray will not lie in the plane of incidence. The demonstration of this construction is exactly similar to that of (33).

The course of an extraordinary ray after internal reflection is to be found in a manner analogous to that of (32). Thus in fig. 30, suppose that the extraordinary wave whose front is $A'e$ moves in the direction $A'G$ and is wholly or partially reflected at the surface GH. When the part A' has arrived at G, suppose the part e to be at I on its way to H. Then when I reaches H, the small wave caused by the disturbance at G will have extended into a spheroid similar and equal to that which must be described from the center I to pass through H. Let KH be this spheroid (the axis being always parallel to CD) and LM the spheroid equal to it whose center is G: let HL be the tangent plane passing through the line projected in H. Then, as in (32), HL is the front of the reflected wave, and, as above, GL is the direction of the reflected wave. Here the angle of reflection is not generally equal to the angle of incidence, and the angles of incidence and reflection are not generally in the same plane.

116. If we consider the extent of refraction to be determined by the change in the position of the front of the wave (which is sometimes the most convenient way) and if the spheroid be oblate, as it is for Iceland spar, beryl, tourmaline, &c., the extraordinary ray is always less refracted than the ordinary ray, since in fig. 30 the spheroid

includes the sphere. If the spheroid be prolate, as in quartz, uniaxal apophyllite, &c., the extraordinary ray is always more refracted than the ordinary ray. The normal to the front of the wave is always in the plane of incidence.

117. A compound prism which produces great angular separation of the two rays is thus constructed. Let a prism A be cut from Iceland spar with its edge parallel to the axis, and another prism B of equal angle with its edge perpendicular to the axis, and let them be placed as in fig. 31. The vibrations parallel to the plane of the paper will furnish the ordinary ray of A and the extraordinary ray of B; that is, this wave will be most refracted by A towards C, and least by B towards D, and it will therefore on the whole pass towards C. In a similar manner, the wave produced by vibrations perpendicular to the paper will be least refracted by A towards C, and most refracted by B towards D, and it will therefore on the whole pass towards D. If the prisms be cut from quartz, the separation is in the opposite direction; it is smaller also, as the prolate spheroid of quartz differs less from a sphere than the oblate spheroid of Iceland spar.

PROP. 24. To investigate the law of double refraction in biaxal crystals.

118. By *biaxal crystals* are meant those in which a^2, b^2, c^2, are all different. Our limits will not allow us to go through the whole of this investigation, and we shall merely give the principal steps, referring for details to the *Memoires de l' Institut*, 1824; the *Annales de Chimie*, 1828; and the *Cambridge Transactions*, Vol. VI. p. 85.

119. The first thing to be done, as in (108), is to find two directions in the front of a plane wave in which a displacement produces a force in the same direction, neglecting that force which is perpendicular to the front. As we shall only have to calculate the forces in the directions possessing this property, we shall at once resolve the whole

force of displacement into two, one parallel to the direction of displacement, the other perpendicular to it (not necessarily perpendicular to the front of the wave), and shall neglect the latter. If the direction of displacement make angles X, Y, Z, with the axes of x, y, z, this resolved force, as in (111), is

$$\text{displacement} \times (a^2 \cos^2 X + b^2 \cos^2 Y + c^2 \cos^2 Z).$$

Construct a surface of which the latter factor is the radius, which we shall call the surface of elasticity; it is easily seen that the radius is the squared reciprocal of the radius in the ellipsoid whose axes are $\frac{1}{a^2}$, $\frac{1}{b^2}$, $\frac{1}{c^2}$.

120. Make a section by the plane front of the wave through the center of this surface; the radius vector of the section will be the square of the reciprocal of the radius vector in the corresponding section of the ellipsoid, that is in an ellipse; and this section of the surface of elasticity will therefore be a curve symmetrical with respect to its greatest and least diameters, which are at right angles.

121. The radius vector of this section in any direction represents the resolved part, in that direction, of the force produced by displacement in that direction, the neglected part being perpendicular to that direction and not necessarily perpendicular to the front of the wave. If now we examine the direction of displacement in which the neglected part is perpendicular to the front of the wave, it is found that the greatest and least diameters above alluded to are the only ones which satisfy this condition. Consequently the vibrations must be resolved into two, parallel respectively to these diameters; and these will produce the two rays. Their velocities will be represented by the square roots of the values of those semi-diameters.

122. In two positions of the front of the wave, and no more, the section becomes a circle. Whatever then is

the direction of vibration in that front, the velocity of transmission is the same, and there is no separation into two rays. The two lines perpendicular to these circles are called the *optic axes*.

123. The difference between the reciprocals of the squares of the velocities of the two rays is proportional to the product of the sines of the two angles made by the front of the wave with the two circular sections, or to the product of the sines of the angles made by the normal to the front with the two optic axes.

124. The plane of polarization of one ray bisects the angle made by the two planes which pass through the normal and the two optic axes: and the plane of polarization of the other is perpendicular to it. This is easily shewn thus: where the front of the wave cuts the two circular sections, the radii in the section by the front must be equal to the radii of the circles, and therefore must be equal to each other, and therefore must make equal angles on both sides of the longest or shortest diameter: and therefore if the planes be projected on a sphere concentric with the surface of elasticity, the point which is the projection of the longest or shortest diameter will bisect one side of the spherical triangle. Construct the spherical triangle whose angles are the poles of those sides: then the circle drawn from the bisection to the pole of that side (which is the projection of one plane of vibration) will bisect the angle made by the sides joining that pole, or the pole of the front, with the optic axes, inasmuch as those sides are equally inclined to the quadrants joining that pole with the two angles of the first triangle where the front of the wave meets the circular sections.

125. The form into which the wave must be supposed to diverge is determined as in (113), by finding the forms of the surfaces where the perpendicular to the tangent plane is proportional to one of the velocities found in (121). After a very troublesome algebraical process it is found that the

equation expressing the two surfaces (which are in fact one continuous surface) is

$$(x^2+y^2+z^2)(a^2x^2+b^2y^2+c^2z^2)-a^2(b^2+c^2)x^2$$
$$-b^2(a^2+c^2)y^2-c^2(a^2+b^2)z^2+a^2b^2c^2=0.$$

This cannot be resolved into factors, and therefore cannot express a sphere and any other surface, as in (111) and (113). Consequently neither of the rays is subject to the law of ordinary refraction. This conclusion might also have been drawn from the observation that neither of the velocities found in (121) is constant. The direction &c. of the two rays when light is incident on a surface of the crystal are found exactly as in (114), using the surface above mentioned instead of the sphere and spheroid, and finding the two positions of the tangent plane passing through the line projected in A'.

Before leaving this investigation we must remark that this theory is imperfect in the same degree as the explanation of refraction. In every uniaxal crystal, we believe, the axis is the same for all the colours, but the ratio of a to c is not the same for different colours. In biaxal crystals generally the direction of the three axes is the same for different colours, but the ratio of a, b, c, is not the same, and consequently the position of the optic axes (122) is not the same for different colours, though the optic axes for all colours are in the same plane. And it has been discovered by Sir John Herschel that the direction of the three axes is in some instances different for different colours, and then the optic axes for different colours are not all in the same plane.

PROP. 25. Light polarized in the plane of incidence falls on a refracting surface of glass &c.: to find the intensity of the reflected and the refracted ray.

126. The three next investigations which we offer to the reader cannot be considered as wholly satisfactory. The extreme difficulty of mathematical investigation into the state of particles at the confines of two media prevents us from

making them more complete. It is however consolatory to know that they are fully supported by experiment, and that they have given a law to phænomena, of which some appeared inexplicable, and others would never have been reduced to laws by observation alone.

127. Suppose that the particles of ether, retaining the same attractive force*, are in the inside of glass &c. loaded with some matter which increases their inertia in the ratio of 1 : n, without increasing their attraction. The equation of (103) would be changed to this:

$$\frac{d^2u}{dt^2} = \frac{1}{n}\left(1 - \frac{1}{2^{\frac{1}{2}}}\right) \cdot \frac{m}{h}\,\frac{d^2u}{dx^2}.$$

If the solution before were $u = \phi(vt - x)$, the solution would now be

$$u = \phi(vt - x\sqrt{n}).$$

The velocity of transmission is diminished therefore in the ratio of $\sqrt{(n)} : 1$. But we have supposed that the velocity is diminished in the ratio of $\mu : 1$. Consequently $n = \mu^2$.

128. Now suppose that we have a series of equal quantities of the ether in a line, and that a transverse motion is given to the first, which, from the constitution described in (103), it has the power of transmitting to the second, &c. When we arrive at the surface of the glass, we must take volumes of the denser ether, whose dimensions are determined in the direction of the transmission of the wave by lengths proportional to the velocity of transmission, and in the other directions by their correspondence with the quantity of ether which puts them in motion. Thus in fig. 32, if $DF = \dfrac{BD}{\mu}$, the ether in $ABDC$ may be considered as putting $CDFE$ in motion. Put i for the angle of incidence, i' for that of refraction. The proportion of the lengths in the direction of the ray is $\mu : 1$, or $\sin i : \sin i'$. The proportion of the breadths is $\cos i : \cos i'$. The proportion of

---

* Perhaps this supposition is hardly reconcileable with that made in the last propositions.

densities is $1 : \mu^2$, or $\sin^2 i' : \sin^2 i$. Combining these proportions, the proportion of the masses is

$$\sin i' . \cos i : \sin i . \cos i'.$$

Now if an elastic body impinges on an equal elastic body, it loses its own velocity and communicates to the other a velocity equal to its own: this is similar to the action of one mass of the ether in vacuum on the next. Supposing the similarity of action to apply to the different states of ether at the confines of the medium, we must compare this with the motion of two unequal elastic bodies A and B after the impact of A with the velocity V on B originally at rest. It is known that A retains the velocity $\frac{A-B}{A+B}V$, and that B receives the velocity $\frac{2A}{A+B}V$. Substituting for A, $\sin i' . \cos i$, and for B, $\sin i . \cos i'$, we find for the motion retained by the external ether, $\frac{\sin (i'-i)}{\sin (i'+i)} \times$ its previous motion; and for that communicated to the internal ether, $\frac{2 \sin i' \cos i}{\sin (i'+i)}$ $\times$ previous motion of external ether. Now by a succession of numerous impulses of this kind, following a given law, a series of waves with any law of displacement may be produced: and every impulse produces parts in the two media having the proportions given above. If then the original displacement be represented by

$$a . \sin \frac{2\pi}{\lambda}(vt - x),$$

that retained by the external ether, and which produces the reflected ray, must be

$$a . \frac{\sin (i'-i)}{\sin (i'+i)} . \sin \frac{2\pi}{\lambda}(vt - x),$$

and that transmitted to the internal ether, and which produces the refracted ray, must be

$$a \frac{2 \sin i' . \cos i}{\sin (i'+i)} . \sin \frac{2\pi}{\lambda}(vt - \mu x).$$

These formulæ apply equally to refraction from air into glass, and from glass into air, giving i and i' their proper values. The intensities of the rays will be represented by the squares of the coefficients.

PROP. 26. Light polarized perpendicular to the plane of incidence falls on a refracting surface: to find the intensity of the reflected and the refracted ray.

129. We cannot here use the same kind of reasoning as in (128), because the motion of displacement (being in the plane of incidence and perpendicular to the path of the ray) is not in the same direction for any two of the three rays. To overcome this difficulty, M. Fresnel has adopted the following hypotheses. First he supposes that the law of *vis viva* holds: that is, that the sum of the products of each mass by the square of its velocity is constant. (This is certainly true if as in (128) masses are supposed to act nearly as elastic bodies. And in all cases of mechanical action it is equal to the sum of all the integrals of *force* × *space through which it has acted*, which is constant in all the cases of undulation that we can strictly examine, and is probably constant in this). Next he supposes that the resolved parts of the motion perpendicular to the refracting surface will preserve after leaving the surface the same relation which they have there, and which, if they follow the same laws as those of the impact of elastic bodies, would be thus connected: the relative motions before and after impact will be equal in magnitude but opposite in sign. (This is confessed by M. Fresnel to be purely empirical). Adopting these hypotheses, and considering the masses to be as

$$\sin i' . \cos i \ : \ \sin i . \cos i',$$

and representing the displacements in the incident, refracted, and reflected ray, (estimated positive in that direction perpendicular to their respective rays which is nearest to that of a body falling perpendicularly from vacuum on the refracting surface,) by a, b, c, we have the following equations:

$$\sin i' . \cos i . a^2 = \sin i . \cos i' . b^2 + \sin i' . \cos i . c^2$$

$$a \cos i = b \cos i' + c . \cos i.$$

Eliminating b,

$$(\sin 2i' + \sin 2i)\, c^2 - 2\sin 2i\,.\, ac - (\sin 2i' - \sin 2i)\, a^2 = 0,$$

$$\text{or}\quad (c-a)\left\{(\sin 2i' + \sin 2i)\, c + (\sin 2i' - \sin 2i)\, a\right\} = 0.$$

This equation is satisfied by $c = a$: but that would give $b = 0$, and therefore expresses only total reflection, which would require exactly the same mathematical conditions as those that we have used, but would not correspond to the physical circumstances of the problem now before us. The other is the only solution which we want: it gives

$$c = -a\frac{\tan(i'-i)}{\tan(i'+i)},$$

$$\text{and}\quad b = a\,.\frac{\cos i}{\cos i'}\left\{1 + \frac{\tan(i'-i)}{\tan(i'+i)}\right\}.$$

Hence if the displacement produced by the incident wave is

$$a\sin\frac{2\pi}{\lambda}(vt - x),$$

that produced by the reflected wave is

$$-a\,.\frac{\tan(i'-i)}{\tan(i'+i)}\sin\frac{2\pi}{\lambda}(vt-x),$$

and that by the refracted wave is

$$a\frac{\cos i}{\cos i'}\left\{1 + \frac{\tan(i'-i)}{\tan(i'+i)}\right\}\sin\frac{2\pi}{\lambda}(vt - \mu x).$$

130. One of the most remarkable inferences from this expression is obtained by making $i' + i = 90^0$. The displacement produced by the reflected wave is then $= 0$. Suppose now light consisting of transversal vibrations in all directions to be incident at this angle on a surface of glass. Resolve the vibrations into two sets, one parallel to the plane of incidence and the other perpendicular to it. The former (as we have just seen) will furnish no reflected ray: the latter, by (128), will produce a reflected ray. Consequently the

reflected light will consist solely of vibrations perpendicular to the plane of reflection. The condition $i' + i = 90$ gives

$$\sin i' = \cos i, \text{ or } \frac{\sin i}{\mu} = \cos i, \text{ whence } \tan i = \mu:$$

which (95) defines the polarizing angle. Thus the angle of incidence at which, according to theory, the vibrations of the reflected ray are entirely perpendicular to the plane of incidence, is the same as the angle at which, in experiment, the reflected ray is entirely polarized in the plane of incidence. And we have found from theory in (111) that the ray of a uniaxal crystal which undergoes the ordinary refraction, and which (94) is said to be polarized in the principal plane, is produced by vibrations perpendicular to the principal plane. These are the two reasons which induce us to say, as in (100), that light polarized in a particular plane consists of vibrations perpendicular to that plane.

131. Another remarkable inference is this. If the two surfaces of a glass plate are parallel, i and i' at the second surface are the same as i' and i at the first. Consequently, if the light reflected from the first surface is polarized, or if $i + i'$ at the first surface $= 90^0$, $i + i'$ at the second surface also $= 90^0$, and therefore the light reflected internally from the second surface is also polarized. This is true in experiment.

Many investigations applying to these problems are to be found in the *Cambridge Transactions* and other *Transactions*, the *Philosophical Magazine*, and the *Comptes Rendus* of the French Academy.

PROP. 27. Light polarized in a plane inclined by the angle a to the plane of incidence falls on the surface of a refracting medium: to find the position of the plane of polarization of the reflected light.

132. The displacement of a particle of ether before incidence may be represented by

$$a \sin \frac{2\pi}{\lambda}(vt - x)$$

in the direction making with the plane of incidence an angle $(90^0 - \alpha)$:

and this may be resolved into

$$a \cos \alpha \, . \frac{2\pi}{\lambda}(vt - x)$$

perpendicular to the plane of incidence, and

$$a \, . \sin \alpha \, . \sin \frac{2\pi}{\lambda}(vt - x)$$

parallel to the plane of incidence. And these expressions will apply to the reflected ray, giving x the same alteration in both, and altering the coefficients in the ratios determined in (128) and (129). Hence we shall have after reflection,

Displacement perpendicular to the plane of incidence

$$a \cos \alpha \, . \frac{\sin (i' - i)}{\sin (i' + i)} \, . \sin \frac{2\pi}{\lambda}(vt - x).$$

Displacement parallel to the plane of incidence

$$- a \sin \alpha \, . \frac{\tan (i' - i)}{\tan (i' + i)} \, . \sin \frac{2\pi}{\lambda}(vt - x).$$

Since these are in the same ratio whatever be the value of x, it follows that the displacement compounded of these is entirely in one plane, and therefore the reflected light is polarized. And if β is the angle at which the new plane of polarization is inclined to the plane of incidence, or $90^0 - \beta$ the angle at which the new direction of vibration is inclined to the plane of incidence, we have

$$\cot \beta = \frac{a \cos \alpha \, . \dfrac{\sin (i' - i)}{\sin (i' + i)}}{- a \sin \alpha \, . \dfrac{\tan (i' - i)}{\tan (i' + i)}} = - \cot \alpha \, . \frac{\cos (i' - i)}{\cos (i' + i)},$$

$$\text{or } \tan \beta = - \tan \alpha \, . \frac{\cos (i' + i)}{\cos (i' - i)}.$$

When i and i' are both small, β and α have different signs: this shews that the planes of polarization before and after reflection are inclined* on opposite sides of the plane of incidence. If $i + i' = 90^0$, that is if the angle of incidence is the polarizing angle, the plane of polarization of the reflected ray coincides with the plane of incidence: and if i be further increased, β and a have the same signs. These results have been verified by numerous observations and careful measures of M. Arago, and Sir David Brewster.

PROP. 28. Light is incident on the internal surface of glass at an angle equal to or greater than that of total reflection; to find the intensity and nature of the reflected ray.

133. The expressions in (128) and (129) become impossible. Yet there is a reflected ray, whatever be the nature of the vibrations in the incident light. And on the principle of *vis viva* the intensity of the reflected ray ought to be equal to that of the incident ray, since there is no refracted ray to consume a part of the *vis viva*. And indeed in the last state of the expressions of (128) and (129) before becoming impossible, that is when $i' = 90^0$, each of them becomes $= 1$. After this the expression for the coefficient of vibrations perpendicular to the plane of incidence {putting $\mu \sin i$ for $\sin i'$, and $\sqrt{(-1)} \cdot \sqrt{(\mu^2 \sin^2 i - 1)}$ for $\cos i'$} becomes

$$\frac{\mu \sin i \cos i - \sin i \sqrt{(-1)} \sqrt{(\mu^2 \sin^2 i - 1)}}{\mu \sin i \cos i + \sin i \sqrt{(-1)} \sqrt{(\mu^2 \sin^2 i - 1)}},$$

$$\text{or} \quad \cos 2\theta - \sqrt{(-1)} \cdot \sin 2\theta,$$

$$\text{where} \quad \tan \theta = \frac{\sqrt{(\mu^2 \sin^2 i - 1)}}{\mu \cos i},$$

and that for the coefficient of vibrations parallel to the plane of incidence becomes

* The inclinations are considered to be on the same side when (supposing for facility of conception the angle of incidence to be considerable) the upper parts of both planes are on the same side of the plane of incidence.

$$\frac{\sin i \,.\, \cos i - \mu \sin i \sqrt{(-1)}\sqrt{(\mu^2 \sin^2 i - 1)}}{\sin i \,.\, \cos i + \mu \sin i \sqrt{(-1)}\sqrt{(\mu^2 \sin^2 i - 1)}},$$

$$\text{or } \cos 2\phi - \sqrt{(-1)} \,.\, \sin 2\phi,$$

$$\text{where } \tan\phi = \frac{\mu\sqrt{(\mu^2 \sin^2 i - 1)}}{\cos i}.$$

It is improbable that these formulæ are entirely without meaning: what can their meaning be?

134. M. Fresnel seems to have considered that as the direction of the reflected ray and the nature and intensity of the vibration were already established, there remained but one element which could be affected, namely, the phase of vibration. And it seems not improbable that this may be affected, inasmuch as the incident vibration, though it cannot cause a refracted ray, must necessarily cause an agitation among the particles of the ether outside the glass. It would seem to us most likely that the ray would be retarded (though the phænomena to be hereafter described compel us to admit that it is accelerated): and in all probability differently according to the direction in which the vibrations take place. Nothing then seems more likely than that 2θ and 2ϕ should express these accelerations*: and as they are

---

* M. Fresnel's reasoning is of this kind. In several geometrical cases, the occurrence of an impossible quantity indicates a change of 90° in the position of the line whose length is multiplied by $\sqrt{-1}$. It is probable then that here the multiplication by $\sqrt{(-1)}$ denotes that the phase of the vibration which it affects is to be altered (suppose increased) by 90°. Thus the expression

$$\{\cos 2\theta + \sqrt{(-1)} \,.\, \sin 2\theta\} \,.\, \sin\frac{2\pi}{\lambda}(vt - x)$$

is to be interpreted as signifying

$$\cos 2\theta \,.\, \sin\frac{2\pi}{\lambda}(vt - x) + \sin 2\theta \,.\, \sin\frac{2\pi}{\lambda}(vt - x + 90^\circ),$$

$$\text{or } \cos 2\theta \,.\, \sin\frac{2\pi}{\lambda}(vt - x) + \sin 2\theta \,.\, \cos\frac{2\pi}{\lambda}(vt - x),$$

$$\text{or } \sin\left\{\frac{2\pi}{\lambda}(vt - x) + 2\theta\right\}.$$

And similarly for the other.

angles, they must be combined with the angles in the expression for the vibration. Thus for instance, if $a \sin \frac{2\pi}{\lambda}(vt - x)$ were the expression for the vibrations perpendicular to the plane of incidence on the supposition that they were not accelerated, $a \sin \left\{\frac{2\pi}{\lambda}(vt - x) + 2\theta\right\}$ would be the expression on the supposition that they were accelerated.

135. The only thing which concerns us experimentally is the difference $2\phi - 2\theta$ (which we shall call δ) of the accelerations, for vibrations perpendicular to and parallel to the plane of incidence. Now

$$\tan(\phi - \theta) = \frac{\cos i \sqrt{(\mu^2 \sin^2 i - 1)}}{\mu \sin^2 i},$$

$$\text{whence } \cos\delta = \frac{1 - \tan^2(\phi - \theta)}{1 + \tan^2(\phi - \theta)} = \frac{2\mu^2 \sin^4 i - (1 + \mu^2)\sin^2 i + 1}{(1 + \mu^2)\sin^2 i - 1}.$$

It appears from this expression that $\delta = 0$ when $\sin i = \frac{1}{\mu}$, or when $\sin i = 1$: and that δ is greatest when

$$\sin^2 i = \frac{2}{1 + \mu^2}:$$

the value of $\cos\delta$ being then

$$\frac{8\mu^2}{(1 + \mu^2)^2} - 1.$$

If we assume $\delta = 45^0$, we have this equation:

$$\frac{2\mu^2}{(1 + \mu^2)\operatorname{cosec}^2 i - \operatorname{cosec}^4 i} = 1 + \sqrt{\tfrac{1}{2}}:$$

the solution of which, supposing $\mu = 1{,}51$, gives

$$i = 48^0.37'.30'', \text{ or } 54^0.37'.20''.$$

If then light be incident internally on the surface of crown glass at either of these angles, the phase of the vibrations in the plane of incidence is accelerated more than that of the vibrations perpendicular to the plane of incidence by 45^0. If the light be twice reflected in the same circumstances and with the same plane of reflection, the phase of vibrations in the plane of incidence is more accelerated than that of the other vibrations by 90^0.

136. If then we construct a rhomb of glass, fig. 33, two of whose sides are parallel to the plane of the paper, and the others perpendicular to the paper and projected in the lines AB, BC, CD, DA; and if the angles at A and C are $54^0.37'$: then light incident perpendicular to the end at F will be internally reflected at G and H, making at those points angles of incidence $54^0.37'$, and will emerge at I in the direction parallel to that in which it entered at F. The immersion at F and the emersion at I will produce no alteration in the light, but the effect of the two reflections at G and H will be to accelerate the phases of vibration in the plane of the paper more than those perpendicular to that plane by 90^0. A rhomb thus constructed we shall call *Fresnel's rhomb.*

Prop. 29. Polarized light is internally reflected in a refracting medium at an angle of incidence greater than that necessary for total reflection: to find the nature of the reflected ray.

137. Let the plane of polarization make with the plane of incidence the angle α. Then the vibration, represented by

$$a \,.\, \sin \frac{2\pi}{\lambda}(vt - x),$$

is performed in a direction making the angle $90^0 - \alpha$ with the plane of incidence. Consequently the resolved vibrations are

$$a \,.\, \cos \alpha \,.\, \sin \frac{2\pi}{\lambda}(vt - x)$$

perpendicular to the plane of incidence, and

$$a \,.\, \sin \alpha \,.\, \sin \frac{2\pi}{\lambda}(vt - x)$$

parallel to the plane of incidence. The latter of these, by (135), is more accelerated than the former by δ. If then after reflection we use

$$a \,.\, \cos \alpha \,.\, \sin \frac{2\pi}{\lambda}(vt - x)$$

to express the vibration perpendicular to the plane of incidence, we must take

$$a \,.\, \sin \alpha \,.\, \sin \left\{\frac{2\pi}{\lambda}(vt - x) + \delta\right\}$$

for the vibration parallel to the plane of incidence. The same expression holds if the light is internally reflected any number of times, (the planes of incidence being always the same,) if we take care to give the proper value to δ.

138. Now let us examine the motions of a particle of ether in the reflected pencil. We will take y for the ordinate in the plane of reflection, and z for that perpendicular to it, both measured from the place of rest of the particle, in the plane transverse to the direction of the reflected ray.

(1) Let $\alpha = 45^0$, and $\delta = 90^0$. (This represents the case of Fresnel's rhomb when the plane of polarization is inclined 45^0 to that of reflection). Here

$$y = a\sqrt{\tfrac{1}{2}} \,.\, \cos \frac{2\pi}{\lambda}(vt - x), \quad z = a\sqrt{\tfrac{1}{2}} \,.\, \sin \frac{2\pi}{\lambda}(vt - x),$$

$$\text{and } y^2 + z^2 = \frac{a^2}{2}.$$

That is, every particle describes a circle whose radius is $\dfrac{a}{\sqrt{(2)}}$.

(2) Let α have any value, δ being $= 90^0$. (This is the general case of Fresnel's rhomb). Here

$$y = a \sin \alpha \,.\, \cos \frac{2\pi}{\lambda}(vt - x), \quad z = a \cos \alpha \,.\, \sin \frac{2\pi}{\lambda}(vt - x)$$

$$\text{and } \frac{y^2}{a^2 \sin^2 \alpha} + \frac{z^2}{a^2 \cos^2 \alpha} = 1.$$

That is, every particle describes an ellipse, whose semi-axes are $a \sin \alpha$ parallel to the plane of reflection, and $a \cos \alpha$ perpendicular to that plane.

(3) In the general case, α and δ having any values,

$$y = a \sin \alpha \left\{ \sin \frac{2\pi}{\lambda}(vt - x) \,.\, \cos \delta + \cos \frac{2\pi}{\lambda}(vt - x) \,.\, \sin \delta \right\},$$

$$\text{and } z = a \cos \alpha \,.\, \cos \frac{2\pi}{\lambda}(vt - x).$$

$$\text{Hence } \cos \frac{2\pi}{\lambda}(vt - x) = \frac{z}{a \cos \alpha}:$$

and

$$(y - \tan \alpha \,.\, \sin \delta \,.\, z)^2 = a^2 \sin^2 \alpha \,.\, \cos^2 \delta \,.\, \left\{ 1 - \cos^2 \frac{2\pi}{\lambda}(vt - x) \right\}$$

$$= a^2 \sin^2 \alpha \,.\, \cos^2 \delta \,.\, - \tan^2 \alpha \,.\, \cos^2 \delta \,.\, z^2:$$

the equation to an ellipse whose axes are inclined to the plane of reflection.

(4) If we compare the expressions for y and z in the first case with the equations to a circular helix, (t being considered constant) we find that they exactly coincide. That is, a series of particles which were originally in a straight line, will be at any subsequent time in the form of a circular helix. In the other cases, the position of the particles will be what may by analogy be called an elliptic helix.

(5) For all values of δ, if $\alpha = 0$, or if $\alpha = 90^0$, the reflected light has the same polarization as the incident light.

139. The nature of the light in the reflected ray may then be generally expressed by saying that it is *elliptically polarized:* and in the first case by saying that it is *circularly polarized.* Wherever after this we speak of common polarized light we shall for the sake of distinction call it *plane polarized* light. From the investigation of the second case it appears that Fresnel's rhomb, by proper adjustment of position with respect to the plane of polarization, is capable of producing elliptically polarized light of every degree of ellipticity. We will therefore suppose that the circularly or elliptically polarized light is produced by Fresnel's rhomb*. For use, it is convenient to have it mounted in a frame which, without stopping the light, admits of its turning round the axis *HI* fig. 33: this frame may be placed on the board in fig. 24: then the light plane-polarized by *A* is by the rhomb converted into circularly or elliptically polarized light and emerges from the end *DC* opposite to the analyzing plate *B* in fig. 24. If the mounting be graduated so as to determine the angle made by the plane of polarization with the plane of reflection, then when this angle is 0, 90^0, 180^0, 270^0, the plane polarized light is not altered: when it is 45^0, 135^0, 225^0, 315^0, the emergent light is circularly polarized: when it has any other value, the light is elliptically polarized.

140. Now it is evident that circularly polarized light may be resolved into two vibrations parallel and perpendicular to any arbitrary plane, and that the magnitudes of these vibrations are always the same. Consequently this light, when examined only by the analyzing plate *B*, shews no sign of polarization (96). This is experimentally true. But if elliptically-polarized light is resolved in the same way, neither of the resolved parts ever vanishes, though their magnitudes vary: and therefore when examined with the analyzing plate it will appear to be partially polarized. This is also true.

---

* We shall hereafter mention another contrivance which produces nearly but not exactly the same effect, and which has been used more extensively than Fresnel's rhomb.

141. Between two kinds of circularly-polarized light there is an important distinction which we have not yet pointed out. We have seen that if $\alpha = 45^0$ in Fresnel's rhomb, the light is circularly polarized: it is also circularly polarized if $\alpha = -45^0$. For in the latter case

$$y = -a\sqrt{\tfrac{1}{2}}\cos\frac{2\pi}{\lambda}(vt - x),$$

$$z = a\sqrt{\tfrac{1}{2}}\sin\frac{2\pi}{\lambda}(vt - x),$$

$$\text{and therefore } x^2 + y^2 = \frac{a^2}{2}.$$

The difference consists in the difference of direction of each particle's motion. In the former case,

$$\frac{z}{y} = \tan\frac{2\pi}{\lambda}(vt - x);$$

in the latter

$$\frac{z}{y} = -\tan\frac{2\pi}{\lambda}(vt - x).$$

If we suppose the plane of reflection vertical, y measured upwards, and z to the right hand (looking in the direction of the wave's motion) then in the former case the revolution will be in the same direction as that of the hands of a watch, and in the latter case in the opposite direction. In the former case the expressions shew that the particles which were originally in a straight line will be at any time arranged as in a left-handed spiral; in the latter case, as in a right-handed spiral. A similar distinction exists between two kinds of elliptical polarization.

142. One of the most remarkable proofs of the correctness of the theory is this; if a second Fresnel's rhomb be placed to receive the light coming from the first, and if its position be similar, the emergent light is plane-polarized, but the new plane of polarization is inclined 2α to the former plane of polarization. The theoretical explanation is this; the vibrations in the plane of incidence are

accelerated 90^0 by the first rhomb and 90^0 again by the second rhomb, more than those perpendicular to the plane of incidence. Consequently {taking up the investigation of (137)}, the vibration perpendicular to the plane of incidence being

$$a \cos \alpha \cdot \sin \frac{2\pi}{\lambda}(vt - x),$$

that parallel to the plane will be

$$a \sin \alpha \cdot \sin \left\{ \frac{2\pi}{\lambda}(vt - x) + 180^0 \right\},$$

$$\text{or} \quad - a \sin \alpha \cdot \sin \frac{2\pi}{\lambda}(vt - x).$$

As these are always in the same proportion, the vibration is entirely in one plane, or the light is plane-polarized. But as the tangent of the angle made with the plane of reflection is $-\tan \alpha$, instead of $\tan \alpha$, which was its value before incidence, the plane of polarization is inclined on the side of the plane of reflection opposite to that on which it was before, and by the same angle; the change of position therefore is 2α.

If the second rhomb is placed in a position 90^0 different from that of the first, the emergent light is similar to the incident light. For, the vibrations which were most accelerated by the first rhomb are least accelerated by the second rhomb, and *vice versâ*, so that the relation of their phases is not altered.

143. It is remarkable that in the only other case of reflection unaccompanied by refraction whose laws are well known to us, namely reflection at the surfaces of metals, the reflected ray appears to possess properties similar to those of light totally reflected within glass: if the incident light is plane-polarized, the reflected light is in fact elliptically polarized, and the difference of the phases varies with the angle of incidence. It is not however certain that we can refer the physical explanation to the same principles: all that we can seem able to conclude from it is that the

reflection from metallic surfaces is not strictly analogous to the reflection of sound from a wall, but has a closer relation to reflection from a surface terminating a dense medium*. Whether any explanation could be founded on the supposition that the ether is absolutely terminated at the reflecting surface, or that the ether within the metal is in a rarer state than that external to it, is a point that has not been examined.

ON THE COLOURED RINGS PRODUCED BY INTERPOSING A CRYSTALLINE BODY BETWEEN A POLARIZING PLATE AND AN ANALYZING PLATE.

144. In (97) we have mentioned as one of the fundamental facts of polarization that if the planes of reflection of A and B in fig. 24 are at right angles to each other, the angles of incidence at both being the polarizing angles, the

* It appears from Sir David Brewster's experiments that, in reflection from metals, the proportion of the vibrations parallel to the plane of reflection in the reflected ray to those in the incident ray, is less than the proportion of the vibrations perpendicular to the plane of reflection in the reflected ray to those in the incident ray. Consequently, after a great number of reflections from metallic surfaces, the reflections being all performed in the same plane, the vibrations parallel to that plane are diminished in a rapidly decreasing geometrical series, and are soon insensible, and therefore the light appears to be polarized in the plane of reflection. This happens with a much smaller number of reflections from steel than from silver. It appears also that the alteration of phases of the two sets of vibrations is somewhat different for different metals, and different at different angles of incidence; beginning to be sensible at the incidence 40° nearly, and amounting at its maximum (which it has at an incidence of nearly 70°) to about 90°: perhaps to more, but Sir D. Brewster's statements leave it doubtful. It is also doubtful whether the phase of the vibrations parallel to the plane of incidence is accelerated or retarded. See Brewster on Elliptic Polarization, Phil. Trans. 1830. The nature of elliptically polarized light had been sufficiently indicated in a few words by Fresnel, and the merit of this valuable paper consists entirely in shewing that the reflection of polarized light at a metallic surface produces light of that kind. The result of these laws, on the principles of the text, may be thus stated. If the incident vibration perpendicular to the plane of incidence is $a \,.\, \sin(vt - x + A)$; the multiplier of the coefficient in the reflected ray p: the corresponding quantities for the other vibration $b \,.\, \sin(vt - x + B)$ and q: and the acceleration of the latter δ: then after n reflections the vibration perpendicular to the plane of incidence is

$$a \,.\, p^n \,.\, \sin(vt - x + A),$$

and that parallel to the plane of incidence

$$b \,.\, q^n \,.\, \sin(vt - x + B + n\delta).$$

light reflected from A is incapable of being again reflected from B. If the eye be placed near B so as to observe the image of A, a very dark spot is seen at its center, and the whole image, though not quite so black as the central spot, is very obscure.

145. Now if we interpose between A and B a plate which possesses double refraction, the image of A is generally seen bright, but sometimes crossed by one or more dark brushes, and sometimes by rings of circular or more complicated figure, richly coloured. On inclining the plane of the interposed plate, the rings generally shift their places and are succeeded by others; shewing that the peculiar arrangement of colours and brushes depends on the relation of the direction of the rays to some fixed lines in the interposed plate. There are few substances which when interposed present exactly the same phænomena, but nearly all exhibit appearances of the same general character: gorgeous colours, arrayed in symmetrical forms, generally shifting with every change in the position of the interposed plate, and always altering as B is turned round its spindle. This class of phænomena is far the most splendid in Optics.

146. The interposition of a piece of common glass produces no effect. And even a doubly refracting substance produces no effect if it be placed to receive the light either before it is polarized at A, or after it is analyzed at B. It seems therefore that a doubly refracting substance has generally the power of altering polarized light, in such a manner that the light, either from losing the character of polarization, or from a change in the plane, acquires according to certain complicated laws the capability of reflection. It appears however that it exerts no influence on common light which makes it incapable of polarization as usual, and that it does not alter polarized light so as to produce any alteration in the impression made on the eye unless it is subsequently analyzed.

Prop. 30. To explain generally the origin of the coloured rings.

147. The general explanation may be given thus. From experience*, as well from the theory of Prop. 22, &c., it appears that, whatever be the nature of light incident on a doubly refracting crystal, the two rays which it produces are polarized, one in one plane, and the other in the plane perpendicular to the former. That is, the vibrations of the incident ray are resolved into two sets, one in one direction and the other in the direction perpendicular to that, which produce waves that describe different paths; and one of these forms the Ordinary ray and the other the Extraordinary ray. And from the theory of (108) it appears that these two sets of waves will pass through the crystal with different velocities, and therefore on coming out of the crystal will be in different phases. Their union therefore will produce a kind of light not necessarily plane-polarized, or not necessarily polarized in the same plane as before passing through the crystal: and therefore their capability of reflection at the analyzing plate is, generally, restored. But, as the positions of the two planes of polarization, as well as the difference of velocity of the two rays, will depend upon the direction of the paths through the crystal, the nature of the light produced by the union of the two emergent streams will vary as the directions vary; and consequently the intensity of the light coming to the eye after analyzation will vary with the direction of the ray. Thus bright patches or curves of different intensity will be seen. The difference of phases, it may be easily conceived, is generally a function of λ, and thus the form or size of the curves may be different for differently coloured light. From the mixture of these differently sized and differently coloured curves, curves will be produced in which the mixture of colours is different at almost every point, as in the fringes of interference and in Newton's rings.

148. We have supposed here that neither plane of polarization of the rays in the crystal coincides with the plane of polarization of the light reflected from *A*. But

* Quartz is the only well established exception to this rule. It appears that neither the ordinary nor the extraordinary ray of quartz is strictly plane-polarized.

conceive that in one direction of the ray, the plane of polarization of the ordinary ray coincides with the plane of polarization of light reflected from *A*. In that case the light reflected from *A* will produce in the crystal only the ordinary ray (92), and consequently the crystalline separation of the rays is of no consequence, because only one of the rays exists. The ordinary ray emerges therefore from the crystal just as it entered, unmixed with any other ray, and therefore falls upon *B* in the same state as if it had not passed through the crystalline plate, and therefore, is not reflected. The same would be true, *mutatis mutandis*, if for another direction of the ray the plane of polarization of the extraordinary ray in the crystal coincided with the plane of polarization of light reflected from *A*. Thus if we determine all the directions of rays in which the plane of polarization of either the ordinary or the extraordinary ray coincides with the plane of reflection from *A*, the rays passing in those directions will not be capable of reflection from *B*, and the appearance presented to the eye by the rays passing in all these directions will be that of one or more black lines not necessarily straight, cutting the coloured curves before mentioned.

149. If *B* be turned round its spindle till its plane of reflection coincides with that of *A*, the positions determined by the conditions of (148) will define the directions in which the light is most highly susceptible of reflection from *B*, and therefore one or more bright lines will be seen cutting the curves. If *B* be turned to any intermediate position, it will be found in the same way that the directions of rays, which make the plane of either the ordinary or the extraordinary ray to coincide with the plane of reflection either at *A* or at *B*, determine the form of lines which cut all the rings, and in which the intensity of light is uniformly the same as if the crystal were not interposed.

These particular cases are pointed out merely as matters of interest in the general explanation. The determination of the form of the uncoloured curves will be included in the general investigation of the intensity of light reflected in all directions from *B*.

PROP. 31. A plate of Iceland spar (or other uniaxal crystal, except quartz) is bounded by planes perpendicular to the axis of the crystal: light is incident nearly in the direction of the axis; to find the position of the front, and the velocity perpendicular to the front, of the ordinary and extraordinary waves: and the retardation of each produced by passing through the plate.

150. First, for the extraordinary ray. In fig. 34 let AB be the normal to the front of the incident wave, or the direction of the incident ray: BC the normal to the front of the extraordinary wave, which is not generally the same as the direction of the extraordinary ray: CD the direction of emergence parallel to AB: i the angle of incidence made by AB, i' the angle of refraction made by BC: v the velocity before incidence, v' the velocity of the extraordinary wave perpendicular to its front: T the thickness of the plate. The time of describing BC is $\frac{T}{v' \cos i'}$; the space which the wave would in the same time have described in air is $\frac{Tv}{v' \cos i'}$. But since the front of the wave at incidence was perpendicular to AB at B, and at emergence perpendicular to CD at C, the whole space which the wave really has advanced is

$$BE = \frac{T \cos (i - i')}{\cos i'},$$

and therefore it has been retarded by a space in air equal to

$$\frac{T}{\cos i'} \left(\frac{v}{v'} - \cos i \, . \cos i' - \sin i \, . \sin i' \right).$$

151. Now $\sin i' = \frac{v'}{v} \sin i$. For if GH be a position of the front before incidence and BK after entrance, GB and HK must have been described in the same time, and therefore GB : HK (or $\sin i$: $\sin i'$) :: velocity of incident wave : velocity of extraordinary wave perpendicular to its

surface :: v : v'. And as the perpendicular to the refracting surface coincides with the axis, we have by (112)

$$v' = \sqrt{(a^2 \cos^2 i' + c^2 \sin^2 i')}.$$

From these equations we find

$$\sin i' = \frac{a \sin i}{\sqrt{\{v^2 - (c^2 - a^2) \sin^2 i\}}},$$

$$\cos i' = \frac{\sqrt{(v^2 - c^2 \sin^2 i)}}{\sqrt{\{v^2 - (c^2 - a^2) \sin^2 i\}}},$$

$$v' = \frac{av}{\sqrt{\{v^2 - (c^2 - a^2) \sin^2 i\}}}.$$

Substituting, the retardation

$$= T\left\{\frac{\sqrt{(v^2 - c^2 \sin^2 i)}}{a} - \cos i\right\}.$$

152. Next for the ordinary ray. This may be deduced from the last by putting a for c. For the expression in (112) is changed to that in (111) by this alteration. Consequently the retardation for the ordinary ray is

$$T\left\{\frac{\sqrt{(v^2 - a^2 \sin^2 i)}}{a} - \cos i\right\}.$$

153. The only quantity that concerns us is the excess* of the latter above the former. Its value is

$$\frac{T}{a}\left\{\sqrt{(v^2 - a^2 \sin^2 i)} - \sqrt{(v^2 - c^2 \sin^2 i)}\right\}.$$

When i is small, this is nearly

$$= T \cdot \frac{c^2 - a^2}{2av} \cdot \sin^2 i.$$

This we shall call I.

---

* c is greater than a for Iceland spar, beryl, and all the crystals termed by some writers *negative*; and less than a for some varieties of apophyllite, and all crystals of their *positive* class.

154. In estimating then the displacement in the ether produced by these two separate pencils after emergence from the plate of crystal, if we represent that which is produced by the ordinary ray by a multiple of

$$\sin \frac{2\pi}{\lambda}(vt - x),$$

we must represent that produced by the extraordinary ray by a multiple of

$$\sin \frac{2\pi}{\lambda}\{vt - (x - I)\} \text{ or } \sin . \left\{\frac{2\pi}{\lambda}(vt - x) + \frac{2\pi I}{\lambda}\right\}.$$

For the extraordinary ray is less retarded than the ordinary ray by the space I in air, and therefore the displacement really caused by the extraordinary ray will correspond to that which would have been produced at a space less advanced by I* if they had been equally retarded.

Prop. 32. A plate of Iceland spar &c. bounded by planes perpendicular to the axis of the crystal (as in Prop. 31) is placed between the polarizing and analyzing plates fig. 24: to investigate the intensity of the light in various parts of the image seen after reflection at B.

155. In fig. 35 conceive the direction of any ray to be perpendicular to the paper: let the plane passing through this ray and through the axis of the crystal {which in (111) we have termed the *principal plane* for that ray} make with the plane of first polarization the angle ϕ: and let the plane of polarization at the analyzing plate (which we shall call the *plane of analyzation*) make with the plane of first polarization the angle α. Let the vibration in the rays as first polarized be represented by

---

* The same expression applies when two plates cut in the same way from crystals either of the same or of different kinds are applied together, I being now the space by which in the combination the extraordinary ray is less retarded than the ordinary ray. If in both plates the extraordinary ray is less retarded than the ordinary ray, or in both more, the effect of the combination is that of a thick plate: if in one it is less and in the other more retarded, the effect is that of a thin plate.

$$a \,.\, \sin \frac{2\pi}{\lambda}(vt - x),$$

perpendicular to the plane of first polarization. On entering the crystal this is resolved into

$$a \,.\, \cos \phi \,.\, \sin \frac{2\pi}{\lambda}(vt - x)$$

perpendicular to the principal plane (which produces the Ordinary ray), and

$$a \,.\, \sin \phi \,.\, \sin \frac{2\pi}{\lambda}(vt - x)$$

parallel to the principal plane (which produces the Extraordinary ray). The former of these expressions may be assumed to be true after the Ordinary ray has emerged from the crystal, provided that we make the proper alteration in the value of x or t: but then for the Extraordinary ray we must, by (154), take the expression

$$a \,.\, \sin \phi \,.\, \sin \left\{ \frac{2\pi}{\lambda}(vt - x) + \frac{2\pi I}{\lambda} \right\}.$$

If the rays entered the eye in this state, there would be no variation of intensity in the light coming in different directions through the crystal. For the intensity of the ordinary wave $= a^2 \cos^2 \phi$, and that of the extraordinary wave $= a^2 \sin^2 \phi$, the sum of which, or a^2, represents the intensity of the united waves (102) and this is constant. Now the analyzing plate being applied, those resolved parts only of the vibrations are preserved which are perpendicular to the plane of analyzation. That* furnished by the ordinary ray is

$$a \,.\, \cos \phi \,.\, \cos (\phi + \alpha) \,.\, \sin \frac{2\pi}{\lambda}(vt - x):$$

and that furnished by the extraordinary ray is

---

* As the analyzing plate does not transmit to the eye the whole of the vibrations perpendicular to its plane of polarization, we ought in strictness to multiply these expressions, in this and similar investigations, by a constant. The omission is of no consequence in comparing the intensities of different parts of the image.

$$a \, . \sin\phi \, . \sin(\phi + a) \, . \sin\left\{\frac{2\pi}{\lambda}(vt - x) + \frac{2\pi I}{\lambda}\right\}.$$

The sum of these represents the displacement produced by the wave that enters the eye. Adding them, and expanding $\sin\left\{\frac{2\pi}{\lambda}(vt - x) + \frac{2\pi I}{\lambda}\right\}$, we find for the coefficient of $\sin\frac{2\pi}{\lambda}(vt - x)$,

$$a \, . \cos\phi \, . \cos(\phi + a) + a\sin\phi \, . \sin(\phi + a) \, . \cos\frac{2\pi I}{\lambda},$$

and for the coefficient of $\cos\frac{2\pi}{\lambda}(vt - x)$,

$$a\sin\phi \, . \sin(\phi + a) \, . \sin\frac{2\pi I}{\lambda}.$$

The sum of the squares of these coefficients is to be taken for the measure of the intensity {as in (17) and (23)}. This sum is

$$a^2 . \cos^2 . \phi \, . \cos^2(\phi + a) + a^2\sin^2\phi \, . \sin^2(\phi + a)$$

$$+ 2a^2\sin\phi \, . \cos\phi \, . \sin(\phi + a) \, . \cos(\phi + a) \, . \cos\frac{2\pi I}{\lambda},$$

$$\text{or } \frac{a^2}{2}\left\{1 + \cos 2\phi \, . \cos(2\phi + 2a)\right.$$

$$\left. + \sin 2\phi \, . \sin(2\phi + 2a) \, . \cos\frac{2\pi I}{\lambda}\right\},$$

$$\text{or } a^2\left\{\cos^2 a - \sin 2\phi \, . \sin(2\phi + 2a) \, . \sin^2\frac{\pi I}{\lambda}\right\}.$$

156. This gives the intensity of the light that enters the eye in a given direction, or the brightness of one point of the visible image. To determine what point of the image it is, we have only to remark that this ray makes the angle i with the ray that passes in the direction of the axis, in a plane that is inclined $\phi + a$ to the plane of analyzation (supposing that we look in the direction of the ray's motion),

measuring from the top to the right. By the reflection at the analyzing plate, this course of the rays is inverted with regard to *up* and *down*, while it is not altered with regard to *right* and *left*: but then, as the eye is placed to receive the light in the direction opposite to that in which we look in studying figure 35, there is another inversion with regard to *right* and *left*, but none with regard to *up* and *down*. On the whole therefore, this ray comes from a point whose apparent angular distance from a certain point through which the rays pass parallel to the axis is i, which distance is measured in a direction that makes the angle $\phi + a$ or ψ with the plane of analyzation, measuring from the upper part of the plane to the right. In the image presented to the eye, i may be considered as a radius vector, and ψ the angle that it makes with the upper part of the line that represents the plane of analyzation. The brightness, putting ψ for $\phi + a$, is

$$a^2 \left\{\cos^2 a - \sin(2\psi - 2a) \, . \, \sin 2\psi \, . \, \sin^2 \frac{\pi I}{\lambda}\right\}.$$

157. Let $a = 90^0$, or let the analyzing plane be in the position in which no light is reflected without the interposition of the crystal. The expression becomes

$$a^2 \, . \, \sin^2 2\psi \, . \, \sin^2 \frac{\pi I}{\lambda}.$$

This is 0, whatever be the value of λ and of I, when $\sin^2 2\psi = 0$: that is, when

$$\psi = 0, \quad \text{or} \ = 90^0, \quad \text{or} \ = 180^0, \quad \text{or} \ = 270^0.$$

This shews that, whatever be the kind of light, there is a black cross, passing through the point of the image formed by the light that is parallel to the axis. For all intermediate values of ϕ it vanishes only when

$$\frac{\pi I}{\lambda} = 0, \ \pi, \ 2\pi, \ \&c., \quad \text{or} \quad I = 0, \ \lambda, \ 2\lambda, \ \&c.,$$

$$\text{or} \ \sin i = 0, \ \sqrt{\frac{2av\lambda}{(c^2 - a^2)T}}, \ \sqrt{\frac{4av\lambda}{(c^2 - a^2)T}}, \ \sqrt{\frac{6av\lambda}{(c^2 - a^2)T}}, \ \&c.$$

and the light is brightest and $= a^2 \sin^2 2\psi$ when

$$I = \frac{\lambda}{2}, \quad \frac{3\lambda}{2}, \quad \&c.,$$

$$\text{or} \quad \sin i = \sqrt{\frac{av\lambda}{(c^2 - a^2)T}}, \quad \sqrt{\frac{3av\lambda}{(c^2 - a^2)T}}, \quad \&c.$$

The four spaces between the arms of the cross are therefore occupied by bright and dark rings, the radii of the bright rings being as $\sqrt{1}$, $\sqrt{3}$, $\sqrt{5}$, &c. and those of the dark rings as $\sqrt{2}$, $\sqrt{4}$, $\sqrt{6}$, &c. The radii are inversely as $\sqrt{T}$, and the rings are therefore smaller with a thick plate than with a thin one. The radii are also inversely as $\sqrt{\frac{c^2 - a^2}{av}}$;* and as this expression may conveniently be taken as a measure of the doubly refracting power of the crystal, the rings are less with a powerful doubly refracting crystal than with one which has that property in a feeble degree. The radii vary also as $\sqrt{\lambda}$ (considering $\frac{av}{c^2 - a^2}$ as independent of λ), and thus are larger for red light than for blue light. This produces exactly the same effect which we have noticed in speaking of interference fringes and Newton's rings, (51) and (66): the rings which at first are black and white have very soon a mixture of colours, different at every successive ring, and finally disappear from the mixture of all in almost equal proportions. If $\frac{av}{c^2 - a^2}$ were constant, the proportion of the radii for different colours, and consequently the mixture of colours, would be nearly the same as in Newton's rings. But $\frac{av}{c^2 - a^2}$ is generally a function of λ: the radii of rings of different colours vary therefore as $\sqrt{\frac{av\lambda}{c^2 - a^2}}$, and the colours are

* The reader must not infer from this expression that there are no rings when c^2 is less than a^2. On going through the whole of the investigation it will be seen that the very same expressions will apply, putting only $a^2 - c^2$ instead of $c^2 - a^2$.

not the same as those of Newton's scale. To such an extent, and so differently in different crystals, does $\frac{av}{c^2 - a^2}$ vary with λ, that in one variety of the uniaxal apophyllite Sir John Herschel found that $\frac{av\lambda}{c^2 - a^2}$ was almost exactly constant, so that more than 35 rings were visible: while in another variety $c^2 - a^2$ was positive for the rays from one end of the spectrum and negative for those from the other end, and $= 0$ for the intermediate rays, and only one or two rings were visible.

158. Let $\alpha = 0^0$, or the plane of reflection at the analyzing plate coincide with that at the polarizing plate. The expression for the intensity is

$$a^2 \left(1 - \sin^2 2\psi \,.\, \sin^2 \frac{\pi I}{\lambda}\right).$$

This expression, added to that discussed in (157), produces a sum a^2. Consequently the intensity at any point of the image in this case is complementary to that in the case of (157). Thus, instead of a black cross interrupting the rings, there is a bright cross interrupting the rings: instead of the dark rings having radii

$$\sqrt{\frac{2av\lambda}{(c^2 - a^2)T}}, \quad \sqrt{\frac{4av\lambda}{(c^2 - a^2)T}}, \quad \&c.$$

and the bright rings having the radii

$$\sqrt{\frac{av\lambda}{(c^2 - a^2)T}}, \quad \sqrt{\frac{3av\lambda}{(c^2 - a^2)T}}, \quad \&c.$$

the bright rings have the former and the dark ones the latter.

159. In the general case, there is no variation of the intensity with different values of I or i, (that is, there is a brush of some sort with light of uniform intensity throughout, interrupting the rings), when

$$\sin(2\psi - 2\alpha) \,.\, \sin 2\psi = 0.$$

For, the succession of rings depends on the alteration of values of $\sin^2\frac{\pi I}{\lambda}$: and this is removed by the evanescence of its multiplier. This equation gives

$$\psi = 0, \text{ or } = 90^0, \text{ or } = 180^0, \text{ or } = 270^0, \text{ or } = a, \text{ or } = 90^0 + a,$$
$$\text{or } = 180^0 + a, \text{ or } = 270^0 + a.$$

Consequently there are two rectangular crosses, inclined a to each other, which interrupt the rings: and the intensity of the light in these crosses is $a^2 \cos^2 a$. For the parts between $\psi = 0$, $\psi = a$, or between $\psi = 90^0$, $\psi = 90^0 + a$, or between $\psi = 180^0$, $\psi = 180^0 + a$, or between $\psi = 270^0$, $\psi = 270^0 + a$, the multiplier of $\sin^2\frac{\pi I}{\lambda}$ is positive, and the light is therefore greatest when $I = \frac{\lambda}{2}$, $= \frac{3\lambda}{2}$, &c., and least when $I = \lambda$, $= 2\lambda$, &c.: these four sectors are therefore occupied by portions of rings nearly similar to those in (157), the intensity for the portions of the bright rings being

$$\frac{a^2}{2}\{1 + \cos(4\psi - 2a)\} \text{ or } a^2\cos^2(2\psi - a),$$

and that for the portions of the darker rings $a^2 \cos^2 a$.

But for the parts between $\psi = a$, $\psi = 90^0$, &c., the multiplier of $\sin^2\frac{\pi I}{\lambda}$ is negative: the light is least when $I = \frac{\lambda}{2}$, $\frac{3\lambda}{2}$, &c., and greatest when $I = \lambda$, 2λ, &c.: these sectors therefore are occupied by portions of rings nearly similar to those in (158), the intensity of the portions of the bright rings being $a^2 \cos^2 a$, and that of the fainter rings

$$a^2\cos^2(2\psi - a).$$

The brighter rings in the last mentioned sectors have the same radii and the same brightness as the darker rings in those first mentioned: and this brightness is the same as the brightness in the eight rays of the crosses.

PROP. 33. In the last experiment, Fresnel's rhomb is placed between the polarizing plate and the plate of crystal, with its plane of reflection inclined 45^0 to the plane of polarization, so that the light incident on the crystal is circularly polarized: to find the intensity of the light after reflection from B, and the form of the coloured rings.

160. Resolve the vibration

$$a \,.\sin \frac{2\pi}{\lambda}(vt - x)$$

which is perpendicular to the plane of first polarization,

$$\text{into } \frac{a}{\sqrt{2}} \sin \frac{2\pi}{\lambda}(vt - x)$$

perpendicular to the plane of reflection in the rhomb

$$\text{and } \frac{a}{\sqrt{2}} \sin \frac{2\pi}{\lambda}(vt - x)$$

parallel to that plane. The latter of these, by (135), has its phase increased by 90^0, and therefore on coming out of the rhomb the vibrations may be represented by

$$\frac{a}{\sqrt{2}} \sin \frac{2\pi}{\lambda}(vt - x)$$

perpendicular to the plane of reflection

$$\text{and } \frac{a}{\sqrt{2}} \cos \frac{2\pi}{\lambda}(vt - x)$$

parallel to that plane. Resolving these in directions perpendicular and parallel to the principal plane of the crystal, we find; Vibration which produces Ordinary ray

$$= \frac{a}{\sqrt{2}} \,.\cos(45^0 - \phi) \sin \frac{2\pi}{\lambda}(vt - x)$$

$$+ \frac{a}{\sqrt{2}} \sin(45^0 - \phi) \,.\cos \frac{2\pi}{\lambda}(vt - x)$$

$$= \frac{a}{\sqrt{2}} \sin \left\{ \frac{2\pi}{\lambda}(vt - x) + 45^0 - \phi \right\}.$$

Vibration which produces Extraordinary ray

$$= - \frac{a}{\sqrt{2}} . \sin (45^0 - \phi) . \sin \frac{2\pi}{\lambda}(vt - x)$$

$$+ \frac{a}{\sqrt{2}} . \cos (45^0 - \phi) . \cos \frac{2\pi}{\lambda}(vt - x)$$

$$= \frac{a}{\sqrt{2}} . \cos \left\{ \frac{2\pi}{\lambda}(vt - x) + 45^0 - \phi \right\}.$$

On emerging from the crystal, the Ordinary vibration being represented by the same expression, the Extraordinary vibration must be represented by

$$\frac{a}{\sqrt{2}} \cos \left\{ \frac{2\pi}{\lambda}(vt - x) + 45^0 - \phi + \frac{2\pi I}{\lambda} \right\}.$$

The resolved parts of these perpendicular to the plane of analyzation (which are the only parts that reach the eye) are

$$\frac{a}{\sqrt{2}} \cos (\alpha + \phi) . \sin \left\{ \frac{2\pi}{\lambda}(vt - x) + 45^0 - \phi \right\}$$

$$+ \frac{a}{\sqrt{2}} \sin (\alpha + \phi) . \cos \left\{ \frac{2\pi}{\lambda}(vt - x) + 45^0 - \phi + \frac{2\pi I}{\lambda} \right\}.$$

Expanding the last term, the coefficients of

$$\sin \left\{ \frac{2\pi}{\lambda}(vt - x) + 45^0 - \phi \right\} \text{ and } \cos \left\{ \frac{2\pi}{\lambda}(vt - x) + 45^0 - \phi \right\}$$

are

$$\frac{a}{\sqrt{2}} \cos (\alpha + \phi) - \frac{a}{\sqrt{2}} . \sin (\alpha + \phi) . \sin \frac{2\pi I}{\lambda}$$

$$\text{and } \frac{a}{\sqrt{2}} . \sin (\alpha + \phi) . \cos \frac{2\pi I}{\lambda}$$

and the intensity of the light, or the sum of the squares, is

$$\frac{a^2}{2}\left\{1 - \sin(2\alpha + 2\phi) \cdot \sin\frac{2\pi I}{\lambda}\right\},$$

$$\text{or} \quad \frac{a^2}{2}\left\{1 - \sin 2\psi \cdot \sin\frac{2\pi I}{\lambda}\right\}.$$

161. Since α does not enter into this expression, the appearance will not be altered on turning B round its spindle. When $\sin 2\psi = 0$, that is when $\psi = 0$, or 90^0, or 180^0, or 270^0, the intensity is $\frac{a^2}{2}$: this shews that there is a cross with light of mean intensity interrupting the rings. When ψ is $> 0 < 90^0$, or $> 180^0 < 270^0$, the expression is maximum when

$$\frac{2\pi I}{\lambda} = \frac{3\pi}{2}, \quad \frac{7\pi}{2}, \quad \&c.$$

and minimum when

$$\frac{2\pi I}{\lambda} = \frac{\pi}{2}, \quad \frac{5\pi}{2}, \quad \&c.$$

or maximum when

$$I = \frac{3\lambda}{4}, \quad \frac{7\lambda}{4}, \quad \&c.$$

and minimum when

$$I = \frac{\lambda}{4}, \quad \frac{5\lambda}{4}, \quad \&c.$$

When ψ is $> 90^0 < 180^0$ or $> 270^0 < 360^0$, the expression is maximum when

$$I = \frac{\lambda}{4}, \quad \frac{5\lambda}{4}, \quad \&c.$$

and minimum when

$$I = \frac{3\lambda}{4}, \quad \frac{7\lambda}{4}, \quad \&c.$$

Thus it appears that of the four quadrants into which the cross divides the image, each opposite pair is similar, but each adjacent pair is dissimilar: the bright rings in one quadrant having the same radii as the dark rings in the next quadrant. And on comparing these expressions with those in (157), it will be seen that the effect of placing Fresnel's rhomb has been to push the rings outward by $\frac{1}{4}$ of an order in two opposite quadrants, and to pull them in by $\frac{1}{4}$ of an order in the other two opposite quadrants. At the same time the cross which was perfectly black has now some light. The most important difference of character however which the use of Fresnel's rhomb produces is the unchangeability of appearances as B is turned round.

If we compared the rings produced with the same position of Fresnel's rhomb by two crystals, in one of which c^2 was $> a^2$ and in the other of which c^2 was $< a^2$, then for a given order of rings, that is for those in which the magnitude of $\frac{I}{\lambda}$, without respect to its sign, is the same, $\sin\frac{2\pi I}{\lambda}$ would be positive for the first and negative for the second, or *vice versâ*. Consequently the bright rings of one crystal would correspond to the dark ones of the other. But we have seen that the bright rings of one quadrant correspond to the dark rings of the neighbouring quadrant. Consequently the rings presented by one of the crystals would be the same as those presented by the other, supposing the latter rings turned round 90^0. This affords a convenient method of determining whether the double refraction of a given uniaxal crystal is of the same kind as that of a standard crystal (for instance Iceland spar) or of the opposite kind.

PROP. 34. A plate of a biaxal crystal whose optic axes make a small angle with each other (as nitre or arragonite) is bounded by planes perpendicular to the plane passing through the axes and nearly perpendicular to each axis; light is incident at a small angle of incidence: to find the difference of retardation of the two rays.

162. The accurate solution of this problem leads to some rather complicated expressions: and we shall therefore content ourselves with a very approximate solution analogous to that found in Prop. 31. We have found there that the difference of retardation was nearly

$$= T.\frac{c^2 - a^2}{2av}.\sin^2 i, \text{ or nearly } = T\frac{v(c^2 - a^2)}{2a^3}\sin^2 i'$$

where the difference of the squares of velocities of the two rays was $(c^2 - a^2)\sin^2 i'$: or that the difference of retardations was nearly $= \frac{Tv}{2a^3} \times$ difference of squares of velocities of two rays. As the difference of retardation arises solely from the difference of velocities, we shall suppose the same proportion to be true here. Now by (125) neither of the rays undergoes Ordinary refraction, or has a constant velocity. Still, even in extreme cases, the velocity of one is so nearly a constant $= a$, that in a calculation depending almost wholly on the difference there will be no sensible error in considering one as constant and $= a$. And by (123), putting v' for the velocity of the other,

$$\frac{1}{v'^2} - \frac{1}{a^2} = C.\sin m'.\sin n',$$

where m' and n' are the angles made by the normal to its front with the two optic axes of the crystal (C being always small), whence

$$v'^2 = a^2 - C.a^4.\sin m'.\sin n', \text{ or } a^2 - v'^2 = C.a^4.\sin m'.\sin n'.$$

Hence the difference of retardations

$$= \frac{T.C.av}{2}\sin m'.\sin n'.$$

Now let us consider the system of rays in air which on entering the crystal will pass in the directions that we have described. Let m and n be the angles made by the same ray in air with the rays which on entering the crystal will pass in the directions of the optic axes. As all the refracted rays (represented by the normals to the fronts of the waves) are in the same planes perpendicular to the refracting surface as the incident rays, and as all the angles of refraction are very nearly in the same proportion to the angles of incidence, it follows that all the other small angles depending on them, and their sines, are nearly in the same ratio.

Hence

$$\sin m' = \frac{a}{v} \sin m \text{ nearly}, \quad \text{and } \sin n' = \frac{a}{v} \sin n \text{ nearly}:$$

and therefore the difference of retardations is

$$\frac{T \cdot C \cdot a^3}{2v} \sin m \cdot \sin n.$$

This as before we shall call I.

PROP. 35. A plate cut from a biaxal crystal (as in Prop. 34) is placed between the polarizing and analyzing plates: to investigate the intensity of the light in different points of the image seen after reflection from B.

163. Let ϕ be taken now to represent the angle made by the plane of polarization of either ray with the plane of first polarization, and the expression of (155), which is founded on no supposition except that the planes of polarization of the two rays are perpendicular to each other, will apply to this case. The intensity of light is therefore

$$a^2 \left\{ \cos^2 \alpha - \sin 2\phi \cdot \sin (2\phi + 2\alpha) \cdot \sin^2 \frac{\pi I}{\lambda} \right\}.$$

Conceive fig. 36 to be the projection of the directions of the rays and planes on a sphere (or rather on the tangent plane to a sphere) of which the eye is the center and whose radius is r. Let A, B, thus represent the optic axes, P any ray under consideration, DE the plane of first polarization. Put β for the angle made by the plane passing through the optic axes of the crystal with the plane of first polarization. Let PQ bisect the angle APB: then, by (124), PQ represents the plane of polarization of one ray, and therefore

$$PQA = \phi + \beta.$$

$$\text{Also } \sin m = \frac{AP}{r} \text{ nearly}, \quad \sin n = \frac{BP}{r} \text{ nearly},$$

and therefore

$$I = \frac{T \cdot C \cdot a^3}{2vr^2} \cdot AP \cdot BP,$$

164. Now the form of the brushes interrupting the rings will be discovered by making the multiplier of $\sin^2 \frac{\pi I}{\lambda} = 0$. This gives

$$\sin 2\phi = 0, \quad \text{or} \quad \sin (2\phi + 2a) = 0.$$

Consequently

$$\tan (2\phi + 2\beta) = \tan 2\beta, \quad \text{or} \quad = \tan (2\beta - 2a).$$

Now refer P to the point C bisecting AB, by rectangular co-ordinates, x being measured in the direction CA and y perpendicular to it; let $CA = b$. Then $\tan PAF = \frac{y}{x - b}$: $\tan PBF = \frac{y}{x + b}$: whence $\tan (2\phi + 2\beta) = \tan 2\,PQA$ $= \tan (PBF + PAF)$ because PQ bisects the angle at P

$$= \frac{2xy}{x^2 - b^2 - y^2}.$$

Hence the brushes are determined by these equations

$$\frac{2xy}{x^2 - b^2 - y^2} = \tan 2\beta, \quad \text{or} \quad (x^2 - b^2 - y^2) \tan 2\beta - 2xy = 0;$$

$$\frac{2xy}{x^2 - b^2 - y^2} = \tan (2\beta - 2a),$$

$$\text{or} \quad (x^2 - b^2 - y^2) \tan (2\beta - 2a) - 2xy = 0.$$

These are evidently equations to hyperbolas, of which C is the center. As in both of them $y = 0$ when $x = \pm b$, the hyperbolas defined by both equations pass through A and B. The position of the asymptotes will be determined by supposing x and y very great compared with b: this gives in the first equation

$$\frac{y^2}{x^2} + 2 \cot 2\beta \frac{y}{x} - 1 = 0, \quad \text{or} \quad \frac{y}{x} = + \tan \beta \quad \text{or} \quad - \cot \beta,$$

and similarly in the second

$$\frac{y}{x} = + \tan (\beta - a) \quad \text{or} \quad - \cot (\beta - a).$$

This shews that both hyperbolas are rectangular, and that the asymptotes of one are parallel and perpendicular to the plane of first polarization, and those of the other inclined to them by α. The intensity of light in the brushes is

$$a^2 \cos^2 \alpha.$$

165. When $\beta = 0$, or $= 90^0$, $\tan 2\beta = 0$, and the first hyperbolas are changed into two straight lines, one in the direction of FG, and the other perpendicular to it, passing through C. Similarly when $\beta = \alpha$, or $= 90^0 + \alpha$, the second hyperbolas are changed into a similar cross. Whatever be the value of β, if $\alpha = 0$ or $= 90^0$, the two pairs of hyperbolas coincide: but the value $\alpha = 0$ gives for the intensity a^2, or the brush is bright: and the value $\alpha = 90^0$ gives for the intensity 0, or the brush is black.

166. The nature of the rings is determined by the variations of value of the last term

$$-\sin 2\phi \,.\, \sin(2\phi + 2\alpha) \,.\, \sin^2 \frac{\pi I}{\lambda}.$$

$$\text{When } \phi > 0 < 90^0 - \alpha, \quad \text{or } > 90^0 < 180^0 - \alpha,$$

$$\text{or } > 180^0 < 270^0 - \alpha, \quad \text{or } > 270^0 < 360^0 - \alpha,$$

(the limits of which are determined by the hyperbolas already described) the brightness is greatest when

$$I = 0, \quad \text{or } = \lambda, \quad \text{or } = 2\lambda, \text{ \&c.},$$

and least when

$$I = \frac{\lambda}{2}, \quad \frac{3\lambda}{2}, \quad \text{\&c.}$$

When $\phi > 90^0 - \alpha < 90^0$, the brightness is greatest when

$I = \frac{\lambda}{2}$, $\frac{3\lambda}{2}$, &c. and least when $I = 0$, λ, 2λ, &c.

The cases when $\alpha = 0$ or $\alpha = 90^0$ will be very easily investigated by the reader.

167. It is plain then that the general appearance will be rings interrupted by the brushes in such a way that the bright rings on one side of the brushes correspond to the dark rings on the other side (except $\alpha = 0$ or $\alpha = 90^0$, when the number of brushes is diminished, and the rings on opposite sides correspond); and that the form of these rings will be determined by the equation $I = \lambda \times \text{constant}$; or, by (163),

$$\frac{T \,.\, C \,.\, a^3}{2vr^2} AP.\,BP = \lambda \times \text{constant},$$

$$\text{or } AP.\,BP = \frac{2vr^2\lambda}{TCa^3} \times \text{constant},$$

where the constant determines the order of the rings. The curves determined by this equation are of the kind called *lemniscates*. If the constant is small, they will be nearly circles of which A and B are centers. As the constant is increased, the circles become elongated towards C: at last they become a single curve like the figure 8 crossing at C: then a single curve like a ring nipped so as almost to meet in the middle: and afterwards a ring slightly flattened. Combining this determination with those of (164) and (166) it is seen that, supposing values of β and α similar to those in figs. 35 and 36, the general appearance is that of fig. 37, where the curves represent the dark rings.

168. Since $AP \,.\, BP \propto$ constant that determines the order, the radii of the successive rings, when they are small and nearly circular, are as 1, 2, 3, &c. nearly. In this respect they differ from those of a uniaxal crystal, which are as $\sqrt{1}$, $\sqrt{2}$, $\sqrt{3}$, &c. (157). And since $AP.\,BP \propto \frac{1}{T}$, the rings are smaller with a thick plate of a given crystal than with a thin plate. And since $AP.\,BP \propto \frac{1}{C}$, the rings are smaller with a crystal that produces a great difference in the velocity of the two rays than with one whose energy is feeble. And since $AP \,.\, BP \propto \lambda$ or rather $\propto \frac{v\lambda}{Ca^3}$, the

curves are *cæteris paribus* larger for red than for blue rays.

169. There is however one difference between the curves for the different colours which in its nature is unlike any thing else that we have yet seen. It is that the optic axes for different colours do not coincide. In every instance however the alteration of place is symmetrical with regard to the two axes. Thus the red axes may make with each other a smaller angle than the blue axes, or *vice versâ*, but the angle between one red and one blue axis is the same as that between the other red and the other blue. In one or two instances this amounts to nearly 10^0. The consequence is that the colours are not the same in different parts of the rings of the same order. Suppose for instance (as in nitre) the red axes are less inclined than the blue. As the red rings are larger than the blue, we shall on taking points exterior to A and B find positions where all the colours are mixed or all are absent, and therefore the rings are nearly white and black. If we trace the same rings to the positions between A and B, the red rings will very much over-shoot the blue rings, and therefore the rings have the colour peculiar perhaps to a high order in Newton's scale.

170. It was till very lately supposed that the axes of the different colours were all in the same plane. Sir J. Herschel has discovered that in some instances (in borax for example) this is not true: the planes, however, as far as yet observed, all pass through the line bisecting the angle formed by the two axes. The reader will have little difficulty in conjecturing the nature of the alteration which this irregularity produces in the colour of the curves.

Prop. 36. In the experiment of Prop. 35, Fresnel's rhomb is interposed between the polarizing plate and the crystal: to find the form, &c. of the coloured curves.

171. As in (160), the intensity of light at any point is

$$\frac{a^2}{2}\left\{1 - \sin(2\phi + 2a) \, . \sin\frac{2\pi I}{\lambda}\right\}.$$

There is a brush interrupting the rings where

$$\sin(2\phi + 2a) = 0:$$

this is the same equation as that which determines the second hyperbolas in (164), and which when $\beta = a$ or $= 90^0 + a$ becomes a cross. When $\sin(2\phi + 2a)$ is positive, the intensity is maximum if $I = \frac{3\lambda}{4}, \frac{7\lambda}{4}$, &c., and minimum if $I = \frac{\lambda}{4}, \frac{5\lambda}{4}$, &c.: and the contrary when $\sin(2\phi + 2a)$ is negative. These spaces are separated by the brush: consequently the bright rings on one side of the brush correspond to the dark rings on the other side. The form of the rings is just the same as in (167).

Prop. 37. A plate of uniaxal or biaxal crystal, cut in any direction different from those of Prop. 31 and 34, is placed between the polarizing and analyzing plates: to find the appearance presented to the eye.

172. The general expression for the brightness in (155),

$$\frac{a^2}{2}\left\{1 + \cos 2\phi \cdot \cos(2\phi + 2a) + \sin 2\phi \cdot \sin(2\phi + 2a) \cdot \cos\frac{2\pi I}{\lambda}\right\}$$

applies to this case. To confine ourselves to the most important instances we will make $a = 90^0$, which reduces the expression to

$$\frac{a^2}{2}\sin^2 2\phi\left(1 - \cos\frac{2\pi I}{\lambda}\right).$$

By I is meant here the space that one ray (which whether in uniaxal or biaxal crystals we shall call the Ordinary ray) is retarded more than the Extraordinary ray, and to which the two expressions in (162) still apply, observing only that in the former i' is the angle made with the axis. The essential difference between this case and that of Propositions 31 and 34 is, that here I is large for all rays which pass nearly perpendicular to the plate.

(1) If the plate be thick, all traces of colours will disappear (as we have seen in several cases of interference where one ray had gained many multiples

of λ on another.) For I being considerable, a very small variation of λ will make $\frac{2\pi I}{\lambda}$ vary by 2π: and thus, for the various rays included in every small portion of the spectrum, $\cos\frac{2\pi I}{\lambda}$ will have all its values, positive and negative, and the sum of all these values will be $= 0$. The intensity of light will therefore be $\frac{1}{2}\sin^2 2\phi \times$ incident light, a proportion which is the same for all the colours. On turning the plate round in its plane, ϕ will vary from 0 to 360, and the light will disappear four times. It will be greatest when $\phi = 45^0$, 135^0, &c.

(2) Colours may however be produced by crossing two plates of very nearly the same thickness cut in the same manner from the same crystal. For let I be the retardation of the Ordinary above the Extraordinary ray in the first, I' that in the second; I' will be very nearly equal to I. And, the plates being at right angles to each other, the Ordinary ray of the first will be the Extraordinary ray of the second. I' therefore will be the acceleration in the second plate of the same vibrations for which I was the retardation in the first: and therefore the whole retardation is $I - I'$, and the brightness is now

$$\frac{a^2}{2}\sin^2 2\phi \left\{1 - \cos\frac{2\pi(I - I')}{\lambda}\right\},$$

$$\text{or } a^2 \sin^2 2\phi \,.\, \sin^2\frac{\pi(I - I')}{\lambda}.$$

This may be so small that $\frac{\pi(I - I')}{\lambda}$ may differ little (not more than a fraction of π or a small multiple of π) for differently coloured rays, and then there will be vivid colours.

(3) Colours may also be produced by applying together, with their axes parallel, two plates cut from uniaxal crystals, one of the positive and one of the negative class (as quartz and beryl). For in one of these the Ordinary ray is most retarded, and in the other the Extraordinary ray is most retarded: and as the Ordinary ray in one forms the Ordinary ray in the other, the ray which is most retarded in the first is least retarded in the second, and thus the difference of retardations may be made as small as we please.

(4) From the bodies which crystallize in laminæ it is frequently possible to detach a plate so thin that it will exhibit colours: for instance sulphate of lime, or mica. Both these are biaxal: in the former the axes are in the plane of the laminæ: in the latter they are in a plane perpendicular to it, but widely separated.

(5) In all these cases, the colours do not form small rings, as in the cases that we have treated at length, but are diffused in broad sheets. This arises merely from the circumstance that the expression for I or $I - I'$ varies very slowly with the variation of incidence. In sulphate of lime, for instance, a ray perpendicular to the laminæ makes

$$m = 90^0, \qquad n = 90^0:$$

a ray inclined to this will produce very little alteration in $\sin m' . \sin n'$ on which I depends. The same is true in mica, where the ray makes equal angles with the two axes. If it be inclined in the plane of the axes, $\sin m' . \sin n'$ is diminished: if perpendicular to that plane, it will be increased.

173. If the thickness of a lamina of sulphate of lime or mica is such that, for a ray perpendicular to the lamina, $I = \frac{\lambda}{4}$ for mean rays, the lamina may be used instead of Fresnel's rhomb. For here the light which is incident is

resolved into two sets of vibrations at right angles to each other, and one of these is retarded in its phases by 90^0 more than the other; which is precisely the effect of Fresnel's rhomb. There is however this difference between them. In Fresnel's rhomb, whatever be the colour of the light, the retardation of the phase is exactly 90^0, or the corresponding retardation in space is exactly $\frac{\lambda}{4}$ whatever* the value of λ may be. In the crystallized plate, on the contrary, the retardation for mean rays is exactly $\frac{\lambda}{4}$, but it is greater than $\frac{\lambda}{4}$ for blue rays, and less than $\frac{\lambda}{4}$ for red rays. This is seen most distinctly on putting several such laminæ together, when the light which is reflected from the analyzing plate is coloured: whereas on putting together several Fresnel's rhombs, there is no such colour. It is plain that in substituting such a lamina for Fresnel's rhomb, the plane of polarization of that ray which is least retarded corresponds to the plane of reflection in the rhomb.

If a thicker plate (for instance one that produces a difference of retardations amounting to 6λ for mean rays) be placed with its planes of polarization inclined 45^0 to that of first polarization, the effect on the rings of a uniaxal crystal is very remarkable. In two opposite quadrants, the ray which is most retarded furnishes the Ordinary ray, and in the other two the same furnishes the Extraordinary ray. In two quadrants therefore the difference of the paths of the rays of the uniaxal crystal is increased, and in the others diminished; and therefore in two the colours are those belonging to distant rings, while in the other two the colours of the fifth or sixth ring are pure white and black, as if they were close to the center.

174. The investigations which we have given will apply to all the crystalline bodies whose laws of double refraction

---

* This is not strictly true, as the same angle of incidence in the rhomb does not produce exactly the same effect for all the rays: but it is much more exact than the other.

are accurately known. Quartz has been mentioned as an exception to the common laws of uniaxal crystals. It appears that the phænomena which it exhibits may be perfectly represented by supposing the Ordinary ray to consist of elliptically-polarized light whose greater axis is perpendicular to the principal plane, and the Extraordinary ray to consist of elliptically-polarized light whose greater axis is in the principal plane: these two rays having also the difference* mentioned in (141): and the ellipses being changed to circles when the direction of the rays coincides with the axis of the crystal. It is also necessary to suppose that the axis of revolution of the spheroid (prolate for quartz) in which the Extraordinary wave is supposed to diverge (115) is less than the radius of the sphere into which the Ordinary wave diverges. For these investigations we must refer the reader to the *Cambridge Transactions*, Vol. IV.

PROP. 38. In every case where the interposed crystal resolves the light into two rays polarized in planes at right angles to each other, on turning the analyzing plate 90^0 the intensity of the light at each point is complementary to what it was before.

175. This is seen from the expression of (155). On putting $90^0 + a$ for a, the expression becomes

$$\frac{a^2}{2}\left\{1 - \cos 2\phi \cdot \cos(2\phi + 2a) - \sin 2\phi \cdot \sin(2\phi + 2a) \cdot \cos\frac{2\pi I}{\lambda}\right\},$$

which added to that in (155) makes a^2. Thus if in one case there is black, in the other there will be white: if in one there is an excess of red at any point and an absence

* The crystal is right-handed or left-handed according as the Ordinary or the Extraordinary ray is of the first of these kinds. Sometimes (as in macled quartz, or amethyst) the two species of quartz are mixed: the optical phænomena which the mixture presents are very remarkable. It is to be observed (as a consequence of what is stated in the text) that in the direction of the axis the two rays, circularly polarized in opposite ways, are transmitted with different velocities: no mechanical theory has yet been formed which will completely account for this. (See however Mr Tovey's papers in the *Philosophical Magazine.*) It is remarkable that several fluids (as turpentine, sugar and water, &c.) possess this property, and even the vapour of turpentine: and apparently in all directions.

of blue, in the other there will be an absence of red at the same point and an excess of blue, &c.

If instead of an analyzing plate we use the doubly-refracting prism described in (117), two images are seen at once in different positions, every part of one of which is complementary to the corresponding part of the other. For, one pencil emerging from the prism consists only of vibrations perpendicular to the plane of refraction of the prism, and therefore presents to the eye the same image as the analyzing plate in a given position: the other consists only of vibrations in the plane of refraction, and therefore presents the same image as the analyzing plate in the position differing 90^0 from the former.

PROP. 39. Glass under pressure possesses double refraction.

176. This was experimentally shewn by M. Fresnel in the following manner. A number of prisms were placed as in fig. 38, and to prevent loss of light a fluid of nearly the same refractive power was dropped between the adjacent surfaces. The ends of A, B, C, D, were then violently pressed by means of screws. On passing a ray of light through the combination it was divided into two, one polarized in the plane of the paper and the other in the perpendicular plane.

177. It is found also that pressure affects the separation of the two rays in crystals which possess the property of double refraction. This leads to the presumption that double refraction is produced generally by a state of mechanical constraint in the particles of bodies.

178. According to our preceding theories, since compressed glass possesses double refraction, it ought, when properly interposed between the polarizing and analyzing plate, to exhibit colours. This may be seen on squeezing by means of a screw a piece of glass and holding it in the apparatus. But it may be best exhibited by taking a thick piece of plate glass which is polished at the edges, and

bending it by a weight or a screw pressing the middle, and in that state placing it edgeways between the polarizing and analyzing plates at an angle of 45^0 to each plane of reflection. A black line is seen along the middle extending the whole length, with stripes more and more coloured on each side: the number of stripes is greatest in the middle of the length (perhaps six dark and as many bright): towards the ends the stripes become broader and fewer, and the ends are wholly black. It is plain here that the central black line is seen in those parts which suffer no strain; and that those which are extended as well as those which are compressed possess double refraction. On putting a plate of mica across it, with the plane of its axes either parallel or perpendicular to the plane of the glass, and comparing its effects on the fringes with its effects on the rings of Iceland spar &c., it is found that the double refraction of the compressed parts is of the same kind as that of a negative crystal, and that of the extended parts of the same kind as that of a positive crystal, the axis being supposed to lie in the direction of the length of the glass plate.

179. It is found also that if glass is heated in one part, or if it is heated generally and cooled in one part, or if it is made nearly red hot and suddenly cooled by placing between cold irons &c., it possesses the property of exhibiting beautiful colours divided by black brushes &c., when placed between the polarizing and analyzing plates. There is no doubt that the glass is here in a state of mechanical constraint. On turning the glass, the black brushes are seen to pass by turns over every part. This determines the plane of polarization of the rays at every part of the glass, since at that point (172) ϕ must $= 0$, 90^0, &c.: that is, the two planes of polarization are then parallel and perpendicular to the plane of first polarization.

180. Between constrained glass and crystals there is however one important difference. The rings &c. exhibited by crystals respect a direction which is independent of the size of the specimen or the part from which it is taken. The smallest fragment of crystal properly shaped may be

made to exhibit the rings as well as the largest. In constrained glass, on the contrary, the rings and brushes cannot be seen, except the specimen is placed so far from the eye that the whole can be seen at once: and then we perceive an effect, not similar to that produced by rays passing in different directions through the same crystal, but to that produced by a number of crystals of different doubly-refractive power arranged in different positions and then united into one system.

181. It may be useful to examine the effects of the three agents which are necessary, and in a particular order, for the exhibition of these appearances: namely the polarizing plate, the analyzing plate, and the interposed crystal. We shall begin with the last.

It is necessary that there should be an interposed body for the purpose of altering the nature of the ray in order to make it reflexible at the analyzing plate. We know of only two ways in which this can be done. One is, by resolving the vibrations into two parts and suppressing one of them. This is done if a plate of tourmaline with its axis inclined 45^0 to both planes of reflexion is interposed; light is then reflected from the analyzing plate. The other is, by resolving the vibrations into two parts and retarding one; the retardation being either constant (measured by its effect on the phases), or varying with the colour of the light and with the direction of the ray. The first of these is done when Fresnel's rhomb is interposed: the second when a crystal or any body possessing double refraction is interposed. In either of these cases, the nature of the light compounded of these two parts as they emerge is different from that of the light which enters. When the planes of reflection at the polarizing and analyzing plate coincide, the same contrivance is necessary in order to make the light less capable or wholly incapable of reflexion at the analyzing plate.

It is necessary for the exhibition of coloured rings that there should be an analyzing plate, because, in all the cases of resolution that we know, the intensity of the light produced by the union of the resolved parts after one of them has been retarded is constant (155), and equal to that of the light before it was resolved. By again resolving this united light according to a certain law into two parts, and preserving one, it is probable that in different directions the preserved part will have different values, inasmuch as the nature of the light emerging from the crystal in different directions is different. And this in fact is the origin of the coloured curves. But there is no necessity that the resolution should be (as we have commonly supposed) into two sets of vibrations at right angles to each other, of which only one is preserved. For instance, if Fresnel's rhomb is interposed between the crystal and the analyzing plate, coloured rings* of a different kind are seen. This artifice amounts to the same as resolving the light when it emerges from the crystal into two rays, both elliptically-polarized, of the opposite kinds mentioned in (141), and preserving only one. If any other kind of resolution could be conceived, it would serve as well for exhibiting coloured rings, of forms probably different.

With respect to the necessity for a polarizing plate, there is a little more difficulty. We see from theory as well as from observation that light consisting of vibrations parallel to a certain plane will, after passing through the

---

* There is no difficulty in investigating their form. Each of the rays emerging from the crystal is to be resolved into two sets of vibrations, one parallel and one perpendicular to the plane of reflection in the rhomb: the phase of the former is to be accelerated 90^0; and the light emerging from the rhomb is to be resolved as usual at the analyzing plate. If the incident light is circularly polarized, and a uniaxal crystal is interposed, the rings are circular without any brush or cross: the center is bright or black according as the two Fresnel's rhombs have their planes of reflection coincident or at right angles to each other. If a biaxal crystal is interposed, the rings are uninterrupted, as there is no brush of any kind. This experiment is worthy of notice, as being the only one (so far as we know) in which the rings, and especially the lemniscates, are seen in their whole extent without interruption. Instead of placing a Fresnel's rhomb between the crystal and the analyzing plate, it is more convenient to use a plate of mica which retards one ray more than the other by $\frac{\lambda}{4}$: and to place it with the plane of its axes inclined 45^0 to the plane of polarization.

crystal and undergoing analyzation, exhibit coloured rings. But we see also that other kinds of light will do as well. For instance, circularly or elliptically-polarized light (of which one has lost all trace of polarization according to the usual tests, and the other has but imperfect traces) will exhibit rings. Common light however will not exhibit rings.

182. It becomes then a matter of interest to inquire what is the difference between common light, and the class comprehending plane-polarized, circularly-polarized, and elliptically-polarized light. Now for a given colour of light, (that is, where the length of waves is invariable), and for a given intensity of light (that is, where the coefficient of vibration is invariable), the most general kind of light that we can conceive is elliptically-polarized light; inasmuch as the union of any number of vibrations in any directions and following each other at any intervals will produce elliptically-polarized light. Common light therefore must be elliptically-polarized (including in this term plane and circularly-polarized). The phænomena of interference, which are exhibited in every respect as well with common light as with polarized light, require us to allow that many successive vibrations are exactly similar to each other. For instance, on examining Newton's rings, when the incident light is that of a spirit lamp, or that from any one point of the solar spectrum (which are nearly homogeneous), fifty or sixty rings may be seen very well, and perhaps more in favourable circumstances. Since the sixtieth of these rings is produced by the interference of a wave with the sixtieth following wave, we must conclude not only that sixty successive waves are exactly similar, but that a large multiple of sixty successive waves are exactly similar. The state of the investigation then at present is this. The phænomena of interference, combined with our most general ideas on the nature of light, compel us to suppose that common light consists of elliptic vibrations many of which in succession are exactly similar. The difference between the phænomena of polarization (with a crystal and an analyzing plate) exhibited by common light and elliptically-polarized light,

shews that common light* does not consist of an indefinite succession of similar elliptic vibrations.

183. The only supposition that seems able to reconcile these conclusions is this. *Common light consists of successive series of elliptical vibrations (including in this term plane and circular vibrations), all the vibrations of each series being similar to each other, but the vibrations of one series having no relation to those of another. The number of vibrations in each series must amount to at least several hundreds; but the series must be so short that several hundred series enter the eye in every second of time.*

It must be observed that a gradual change in the nature of the vibrations is inadmissible. If, for instance, we supposed the vibrations elliptical and supposed the ellipse to revolve uniformly about its center, it would be found that the vibrations in each plane could be resolved into two whose lengths of wave were different; and, compounding the corresponding vibrations in perpendicular planes, we should have two rays of elliptically-polarized light of different colours.

As a simple instance of our general supposition, suppose 1000 similar vibrations in one plane to be followed by 1000 vibrations, of magnitudes equal to the former, in the plane at right angles to the former plane; then 1000 in the same plane as at first, &c. The succession of similar waves would be sufficient to give all the phænomena of interferences in perfection. At the same time, no colours would be exhibited with a crystal and an analyzing plate. For the first series alone would give rings and colours, but the second would give rings &c. with intensities exactly complementary† to

* We have not mentioned here the law discovered by the French philosophers, that if two streams of common light from the same source were polarized in planes perpendicular to each other, and afterwards brought to the same plane of polarization, they would not interfere; but if two streams of polarized light from the same source were treated in the same way, they would interfere. The fact is, that the observing of rings &c. in crystals is far the best way of making the experiment: the crystal which has double refraction exhibits the two rays polarized in perpendicular planes, and the analyzing plate brings them to the same plane of polarization.

† This is seen in our expressions (155) by putting $\phi+90^\circ$ instead of ϕ, and $\alpha-90^\circ$ instead of α.

the former: and as these would enter the eye in such rapid succession that we could not distinguish them, we should only perceive the combined effect, which would be a uniform white.

184. We have always spoken of the *colour* of a ray as if it alone were sufficient to identify the nature of the ray. Perhaps, however, it would have been better to consider a ray as defined by its *refrangibility*. The remarkable experiment of Fraunhofer (84), from which it appeared that the interruptions in the spectrum formed by interference correspond exactly to those in the spectrum formed by refraction, seems decisive as to this point, that rays of the same refrangibility are produced by waves of the same length. It has, however, long been the opinion of some philosophers that there are rays of different colours which have the same degree of refrangibility, and that there are rays of the same colour with different degrees of refrangibility. Taking this as established, the conclusion seems to be, that colour does not depend on the length of a wave, but probably on some other circumstance, as perhaps the nature of the vibration. The law of vibration may be, not that of a cycloidal pendulum (as we have all along supposed), but something slightly different. It may be that the effect of an absorbing medium is to suppress all that part of the vibration which follows that law, and to allow only the other to pass. These, however, are very vague conjectures, which can scarcely be examined till our knowledge of the subject in question is much more extensive than it is at present.

TABLE of the NUMERICAL VALUES of the INTEGRALS

$$\int_s \cos \frac{\pi}{2} s^2 \quad \text{and} \quad \int_s \sin \frac{\pi}{2} s^2$$

[referred to in the note to p. 300].

Limits of integration.	$\int_s \cos \frac{\pi}{2} s^2$	$\int_s \sin \frac{\pi}{2} s^2$	Limits of integration.	$\int_s \cos \frac{\pi}{2} s^2$	$\int_s \sin \frac{\pi}{2} s^2$
From $s = 0$			From $s = 0$		
to $s = 0,1$	0,0999	0,0006	to $s = 2,9$	0,5627	0,4098
0,2	0,1999	0,0042	3,0	0,6061	0,4959
0,3	0,2993	0,0140	3,1	0,5621	0,5815
0,4	0,3974	0,0332	3,2	0,4668	0,5931
0,5	0,4923	0,0644	3,3	0,4061	0,5191
0,6	0,5811	0,1101	3,4	0,4388	0,4294
0,7	0,6597	0,1716	3,5	0,5328	0,4149
0,8	0,7230	0,2487	3,6	0,5883	0,4919
0,9	0,7651	0,3391	3,7	0,5424	0,5746
1,0	0,7803	0,4376	3,8	0,4485	0,5654
1,1	0,7643	0,5359	3,9	0,4226	0,4750
1,2	0,7161	0,6229	4,0	0,4986	0,4202
1,3	0,6393	0,6859	4,1	0,5739	0,4754
1,4	0,5439	0,7132	4,2	0,5420	0,5628
1,5	0,4461	0,6973	4,3	0,4497	0,5537
1,6	0,3662	0,6388	4,4	0,4385	0,4620
1,7	0,3245	0,5492	4,5	0,5261	0,4339
1,8	0,3342	0,4509	4,6	0,5674	0,5158
1,9	0,3949	0,3732	4,7	0,4917	0,5668
2,0	0,4886	0,3432	4,8	0,4340	0,4965
2,1	0,5819	0,3739	4,9	0,5003	0,4347
2,2	0,6367	0,4553	5,0	0,5638	0,4987
2,3	0,6271	0,5528	5,1	0,5000	0,5620
2,4	0,5556	0,6194	5,2	0,4390	0,4966
2,5	0,4581	0,6190	5,3	0,5078	0,4401
2,6	0,3895	0,5499	5,4	0,5573	0,5136
2,7	0,3929	0,4528	5,5	0,4785	0,5533
2,8	0,4678	0,3913	to $s = \infty$	0,5	0,5

AIRY'S

LUNAR THEORY.

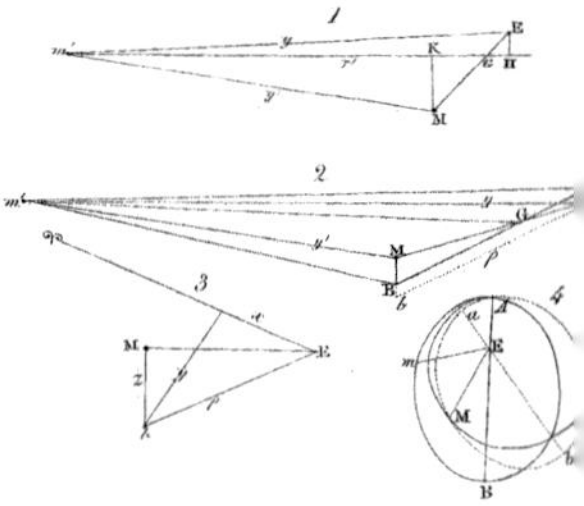

FIGURE OF THE EARTH.

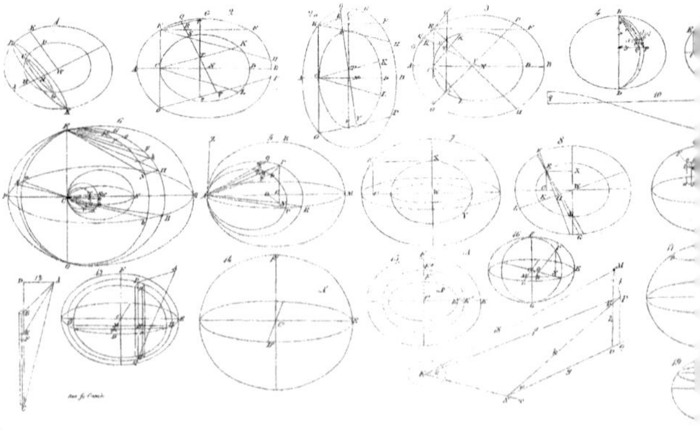

PRECESSION AND NUTATION.

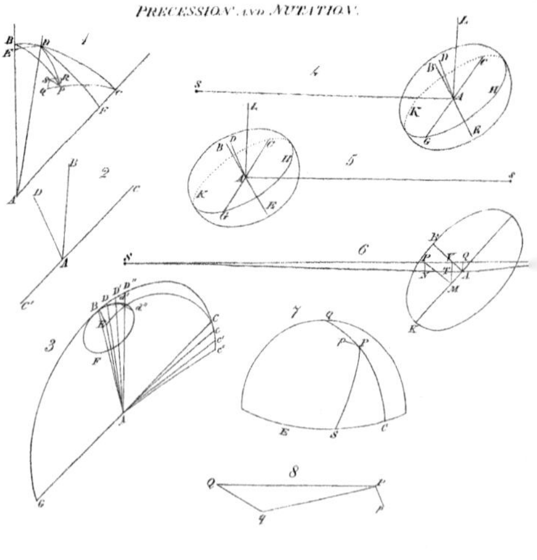

CALCULUS OF VARIATIONS.

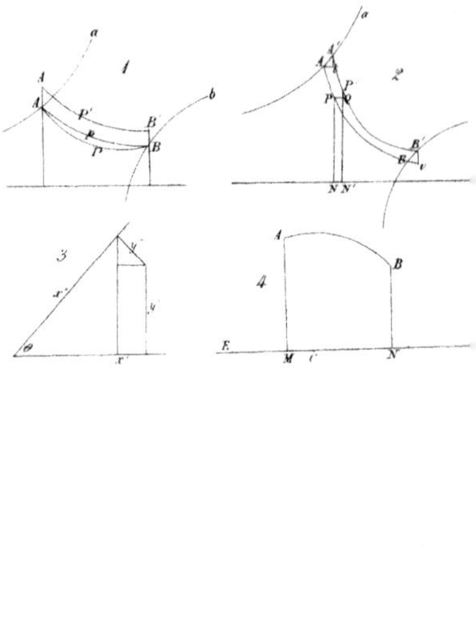

UNDULATORY THEORY.

AIRY'S TRACTS

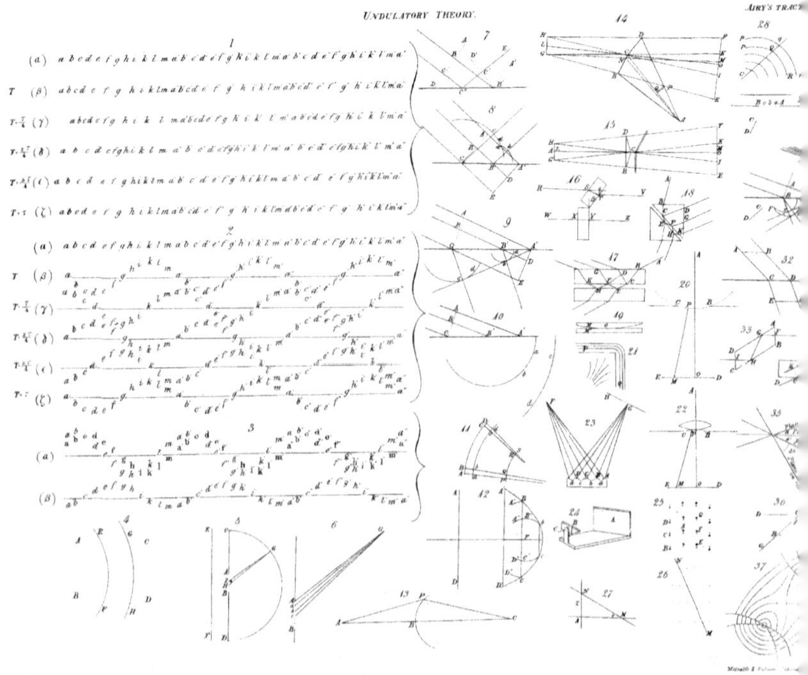

Printed in Dunstable, United Kingdom

91220114R00238